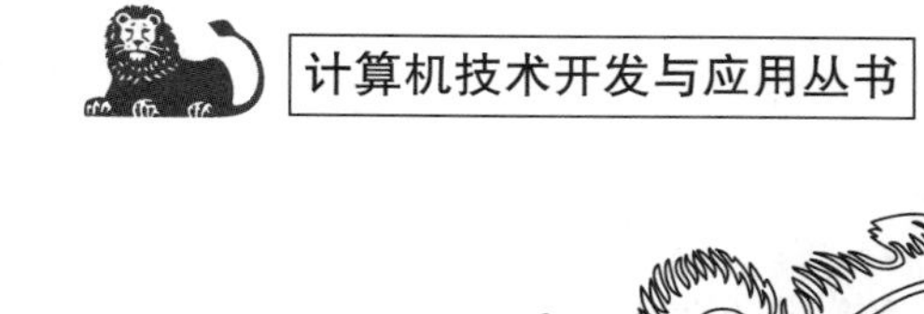

云原生开发实践

高尚衡◎编著

清華大學出版社
北京

内 容 简 介

本书以一个示例应用的云原生化实践过程为主线，系统讲述云原生的理念和主流的云原生相关技术。全书共11章。第1章介绍贯穿全书的示例应用，包括功能和使用的编程语言。第2章介绍传统软件生产流程在应对高频发布、高频部署及规模化等场景中凸显出的问题和不足。第3章介绍以Docker为代表的容器技术在软件开发、测试和交付方面的颠覆性优势。第4章介绍对示例应用进行容器化的详细步骤，包括镜像的定义、构建、发布、使用及通过CI/CD自动化与镜像相关的操作。第5章介绍Docker生态中的容器编排技术，包括Docker Compose与Docker Swarm，并讲解如何利用这两种编排技术分别对示例应用进行容器编排的详细过程。第6章介绍云计算和云原生的理念及两者的密切关系，分析云原生的优势与劣势。第7章介绍作为云原生基础设施的Kubernetes的基础知识，包括常用的资源对象和搭建Kubernetes集群的方式。通过具体的示例体现出Kubernetes强大而灵活的编排能力和优秀的设计理念。第8章介绍将示例应用部署到Kubernetes的详细步骤和常见问题的解决方案，包括提升观测性的日志解决方案和提升安全性的TLS解决方案。第9章介绍Kubernetes应用的打包工具Helm的基础知识，以及为示例应用创建完整Helm Chart的详细步骤。第10章介绍云原生技术中的服务网格的理念和服务网格众多实现中颇具代表性的Linkerd技术，以及在Kubernetes集群中使用Linkerd提升示例应用的可用性、可观测性的具体步骤。第11章介绍云原生现状和展望。

本书适合计算机科学、软件工程等领域的软件工程师和大学生，探索使用容器和Kubernetes相关技术的开发者和技术管理者及希望理解云原生技术基本原理的技术管理者和开发者阅读。

图书在版编目(CIP)数据

云原生开发实践/高尚衡编著. —北京：清华大学出版社，2022.7
(计算机技术开发与应用丛书)
ISBN 978-7-302-60027-5

Ⅰ.①云… Ⅱ.①高… Ⅲ.①云计算 Ⅳ.①TP393.027

中国版本图书馆CIP数据核字(2022)第021626号

责任编辑：赵佳霓
封面设计：吴 刚
责任校对：时翠兰
责任印制：丛怀宇

出版发行：清华大学出版社
网　　址：http://www.tup.com.cn，http://www.wqbook.com
地　　址：北京清华大学学研大厦A座　　**邮　　编**：100084
社 总 机：010-83470000　　**邮　　购**：010-62786544
投稿与读者服务：010-62776969，c-service@tup.tsinghua.edu.cn
质量反馈：010-62772015，zhiliang@tup.tsinghua.edu.cn
课件下载：http://www.tup.com.cn，010-83470236
印 装 者：三河市金元印装有限公司
经　　销：全国新华书店
开　　本：186mm×240mm　　**印　　张**：28　　**字　　数**：627千字
版　　次：2022年8月第1版　　**印　　次**：2022年8月第1次印刷
印　　数：1～2000
定　　价：109.00元

产品编号：092676-01

前言
PREFACE

现代软件的复杂度和部署规模呈指数级膨胀，同时迭代周期越来越短，甚至在一天内会发布多个新版本。传统软件生产方式的缺点日益凸显，尤其在大规模集群的生产场景下软件开发和维护越来越困难。云原生技术在提高软件生产和部署效率及节约IT成本方面有很大优势，所以云原生迅速被广大开发者和软件企业接受，产生了许多大规模落地的案例。

无论是开发工程师还是运维工程师及技术决策者，了解云原生的技术理念并进行开发实践都会对以后的工作产生积极的帮助。

书中以一个用户认证应用程序的实战项目为例，详细讲述应用开发从传统方式一步步向云原生化过渡的过程，其中包括容器化，分别使用Compose、Swarm、Kubernetes部署应用及使用Helm打包等。结合项目实践讨论大规模集群环境下传统软件生产方式的缺陷和云原生方式的优势及其带来的巨大价值。

为贴近生产实践，书中有关示例应用的操作区分了开发环境和生产环境。另外，书中的大部分知识点都附带了通过运行检验的示例，百闻不如一见，相信直观的例子可以使晦涩的知识变得易于理解，可以最大程度地降低学习的难度，提高学习效率和乐趣。

扫描下方二维码，可获取书中示例源代码。

本书源代码

最后感谢家人和朋友的全力支持，让笔者可以全身心投入创作。

由于作者水平有限，书中难免存在疏漏，敬请读者批评指正，以便及时改正和更新。

高尚衡

2022年5月

目录

CONTENTS

第 1 章　用户认证项目 …… 1

1.1　项目介绍 …… 1

1.2　开发环境 …… 1

1.2.1　前端 …… 1

1.2.2　后端 …… 2

第 2 章　传统软件生产流程 …… 4

2.1　开发 …… 4

2.2　测试 …… 4

2.3　计算资源评估 …… 5

2.4　部署 …… 5

2.5　更新和升级 …… 6

第 3 章　容器技术的流行 …… 7

3.1　容器的优势 …… 7

3.2　Docker 简介 …… 7

3.3　Docker 安装 …… 10

3.4　Docker 在开发领域的价值 …… 13

3.5　Docker 在测试领域的价值 …… 14

第 4 章　容器化 …… 15

4.1　容器化简介 …… 15

4.2　保持简单和专注 …… 15

4.3　容器镜像 …… 16

4.4　镜像定义 …… 17

4.4.1　Dockerfile 基础知识 …… 17

4.4.2　定义后端镜像 …… 22

4.4.3 定义前端镜像 …… 27
4.5 构建镜像 …… 29
4.5.1 docker build 命令 …… 30
4.5.2 镜像缓存 …… 30
4.5.3 构建用于开发环境的后端镜像 …… 31
4.5.4 构建用于生产环境的后端镜像 …… 32
4.5.5 构建数据迁移镜像 …… 33
4.5.6 构建用于开发环境的前端镜像 …… 33
4.5.7 构建用于生产环境的前端镜像 …… 34
4.6 启动镜像 …… 34
4.6.1 启动 MySQL 镜像 …… 34
4.6.2 数据迁移 …… 36
4.6.3 启动后端镜像 …… 37
4.6.4 bridge 网络 …… 38
4.6.5 自定义网络 …… 40
4.6.6 启动前端镜像 …… 43
4.7 发布镜像 …… 43
4.7.1 Docker Hub …… 43
4.7.2 私有镜像仓库 …… 44
4.8 CI/CD …… 45
4.8.1 后端引入 CI/CD …… 46
4.8.2 前端引入 CI/CD …… 51

第 5 章 容器编排 …… 53

5.1 容器编排简介 …… 53
5.2 Docker Compose …… 53
5.2.1 Compose 文件 …… 54
5.2.2 Compose 环境变量 …… 57
5.2.3 Compose 运行应用 …… 59
5.2.4 Compose 更新应用 …… 63
5.3 Docker Swarm …… 63
5.3.1 创建 Swarm 集群 …… 64
5.3.2 将样例服务部署到 Swarm 集群 …… 68
5.3.3 伸缩样例服务 …… 70
5.3.4 更新样例服务 …… 71
5.3.5 维护 Swarm 节点 …… 74

5.3.6 Swarm 路由网格 …… 77
5.3.7 开发环境 Swarm 部署 …… 80
5.3.8 生产环境 Swarm 部署 …… 83
5.3.9 约束服务调度 …… 91
5.3.10 日志收集 …… 93

第 6 章 云原生软件生产流程 …… 104

6.1 云原生简介 …… 104
6.2 云计算的能力 …… 105
6.3 云原生的优势 …… 106
6.4 云原生的劣势 …… 107

第 7 章 云原生基础设施 …… 108

7.1 Kubernetes 是什么 …… 108
7.2 客户端工具 kubectl …… 109
7.2.1 kubectl 简介 …… 109
7.2.2 使用 HomeBrew 安装 kubectl …… 109
7.2.3 使用 apt 安装 kubectl …… 110
7.2.4 使用 curl 安装 kubectl …… 111
7.2.5 设置 kubectl 命令自动补全 …… 112
7.3 本地启动 Kubernetes …… 112
7.4 使用 kubeadm 创建 Kubernetes 集群 …… 114
7.4.1 环境要求 …… 114
7.4.2 安装容器运行时 …… 115
7.4.3 安装 kubeadm、kubelet、kubectl …… 115
7.4.4 初始化 …… 116
7.4.5 设置 kubeconfig …… 119
7.4.6 安装网络插件 …… 121
7.4.7 部署样例程序 …… 122
7.4.8 将 Node 添加到集群 …… 123
7.5 创建托管的 Kubernetes 集群 …… 125
7.6 Kubernetes 对象 …… 128
7.6.1 Kubernetes 对象简介 …… 128
7.6.2 如何描述 Kubernetes 对象 …… 128
7.6.3 如何管理 Kubernetes 对象 …… 129
7.7 Node …… 134

7.7.1 Node 简介 …… 134
7.7.2 管理 Node …… 135
7.7.3 Node 状态 …… 135
7.7.4 Node 控制器 …… 137
7.7.5 Node 容量 …… 137
7.8 Pod …… 138
7.8.1 Pod 简介 …… 138
7.8.2 Pod 使用模式 …… 138
7.8.3 Pod 示例 …… 138
7.8.4 Pod 模板 …… 139
7.8.5 Pod 生命周期 …… 140
7.8.6 Pod 中的容器状态 …… 141
7.8.7 Probe …… 141
7.8.8 Init 容器 …… 142
7.9 ReplicaSet …… 144
7.9.1 ReplicaSet 简介 …… 144
7.9.2 ReplicaSet 示例 …… 145
7.9.3 获取模板以外的 Pod …… 148
7.9.4 缩放 ReplicaSet …… 149
7.10 Deployment …… 149
7.10.1 Deployment 简介 …… 149
7.10.2 Deployment 示例 …… 149
7.10.3 更新 Deployment …… 153
7.10.4 回滚 Deployment …… 158
7.10.5 缩放 Deployment …… 162
7.10.6 暂停和恢复 Deployment …… 162
7.11 StatefulSet …… 168
7.11.1 StatefulSet 简介 …… 168
7.11.2 StatefulSet 示例 …… 169
7.11.3 稳定的网络标识 …… 172
7.11.4 稳定的存储 …… 173
7.12 DaemonSet …… 175
7.12.1 DaemonSet 简介 …… 175
7.12.2 DaemonSet 示例 …… 175
7.12.3 DaemonSet 扩缩容 …… 176
7.13 Job …… 177

7.13.1 Job 简介 …… 177
7.13.2 Job 示例 …… 177
7.13.3 Job 清理 …… 179
7.14 ConfigMap …… 180
7.14.1 ConfigMap 简介 …… 180
7.14.2 ConfigMap 示例 …… 181
7.14.3 Pod 使用 ConfigMap …… 182
7.15 Secret …… 186
7.15.1 Secret 简介 …… 186
7.15.2 创建 Secret …… 187
7.15.3 查看 Secret 数据 …… 188
7.16 Kubernetes 存储 …… 188
7.16.1 Volume …… 188
7.16.2 PersistentVolume …… 189
7.16.3 PersistentVolumeClaim …… 191
7.16.4 Pod 使用 PersistentVolumeClaim …… 193
7.16.5 StorageClass …… 195
7.16.6 动态卷供应 …… 197
7.16.7 AWS EBS 使用示例 …… 200
7.17 Kubernetes Service …… 202
7.17.1 Service 简介 …… 202
7.17.2 Service 示例 …… 203
7.17.3 代理模式 …… 209
7.17.4 服务发现 …… 210
7.17.5 Service 类型 …… 211
7.17.6 ClusterIP 类型 …… 211
7.17.7 NodePort 类型 …… 212
7.17.8 LoadBalancer 类型 …… 213
7.17.9 ExternalName 类型 …… 216
7.17.10 headless Service …… 216
7.18 Kubernetes DNS …… 220
7.18.1 DNS 服务 …… 220
7.18.2 Service DNS …… 221
7.18.3 PodDNS …… 221
7.19 Kubernetes Ingress …… 223
7.19.1 Ingress 简介 …… 223

7.19.2 Ingress 示例 …… 224
7.19.3 Ingress 规则 …… 225
7.19.4 Ingress 控制器 …… 225
7.19.5 默认后端 …… 226
7.19.6 资源后端 …… 226
7.19.7 fanout 示例 …… 227
7.19.8 虚拟主机示例 …… 228
7.19.9 TLS 示例 …… 228
7.20 Kubernetes 身份认证 …… 231
7.20.1 Kubernetes 用户 …… 231
7.20.2 认证策略 …… 231
7.20.3 证书认证方式 …… 232
7.21 Kubernetes 授权 …… 238
7.21.1 授权模式 …… 239
7.21.2 RBAC …… 239
7.21.3 常用命令 …… 241
7.21.4 Service Account …… 242
7.22 Kubernetes 调度 …… 245
7.22.1 调度简介 …… 245
7.22.2 约束 Node 选取 …… 246
7.22.3 亲和性和反亲和性 …… 247
7.22.4 nodeName …… 249
7.22.5 污点和容忍 …… 250
7.22.6 Pod 优先级 …… 252
7.22.7 Pod 抢占 …… 254
7.22.8 Pod 拓扑分布 …… 255
7.23 Kubernetes 日志 …… 257
7.23.1 Kubernetes 基础日志功能 …… 258
7.23.2 节点级日志 …… 259
7.23.3 集群级日志 …… 259
7.24 Kustomize …… 264
7.24.1 Kustomize 简介 …… 264
7.24.2 生成 ConfigMap …… 264
7.24.3 生成 Secret …… 268
7.24.4 生成器选项 …… 271
7.24.5 设置横切字段 …… 272

7.24.6　组合 …… 274
7.24.7　定制 …… 276
7.24.8　变量注入 …… 279
7.24.9　基准和覆盖 …… 281
7.24.10　应用、查询和删除对象 …… 284

第8章　Kubernetes 部署应用 …… 286

8.1　环境 …… 286
8.1.1　开发环境 …… 286
8.1.2　生产环境 …… 286
8.2　MySQL 服务 …… 287
8.2.1　开发环境 …… 287
8.2.2　生产环境 …… 292
8.3　数据迁移 …… 294
8.4　后端服务 …… 298
8.5　前端服务 …… 300
8.6　Ingress …… 302
8.7　DNS …… 304
8.7.1　开发环境 …… 304
8.7.2　生产环境 …… 305
8.8　TLS …… 307
8.8.1　证书管理软件 …… 307
8.8.2　ACME …… 308
8.8.3　Ingress TLS …… 310
8.9　日志 …… 314
8.9.1　方案简介 …… 314
8.9.2　ElasticSearch …… 315
8.9.3　Fluentd …… 316
8.9.4　Kibana …… 320
8.10　Kustomize …… 325

第9章　Helm …… 327

9.1　安装 Helm …… 327
9.2　Helm Chart …… 329
9.2.1　Chart 简介 …… 329
9.2.2　安装 Chart …… 330

9.2.3 定制 Chart …… 333
9.2.4 Release …… 336
9.2.5 升级和回滚 …… 336
9.2.6 卸载 Release …… 337
9.2.7 搜索 Chart …… 338
9.3 Chart 模板 …… 339
9.3.1 模板示例 …… 339
9.3.2 模板调用 …… 340
9.3.3 内置对象 …… 342
9.3.4 值文件 …… 343
9.3.5 模板函数和管道 …… 347
9.3.6 流程控制 …… 351
9.3.7 变量 …… 358
9.3.8 命名模板 …… 361
9.3.9 访问文件 …… 366
9.3.10 NOTES.txt …… 370
9.3.11 helmignore 文件 …… 371
9.3.12 Debug …… 372
9.3.13 最佳实践 …… 373
9.4 Chart 依赖 …… 376
9.4.1 简介 …… 376
9.4.2 值覆盖 …… 377
9.4.3 全局值 …… 378
9.5 Chart Hook …… 379
9.5.1 简介 …… 379
9.5.2 Hook 示例 …… 380
9.5.3 Hook 权重 …… 381
9.5.4 Hook 删除策略 …… 381
9.6 Chart 测试 …… 382
9.6.1 测试简介 …… 382
9.6.2 测试示例 …… 382
9.6.3 运行示例测试 …… 383
9.7 库 Chart …… 384
9.7.1 简介 …… 384
9.7.2 示例 …… 384
9.7.3 使用库 Chart …… 386

9.8 创建自己的 Chart …… 389
9.8.1 后端服务 …… 390
9.8.2 MySQL 服务 …… 395
9.8.3 前端服务 …… 399
9.8.4 数据迁移任务 …… 400
9.8.5 Ingress …… 402
9.8.6 安装 Chart …… 405

第 10 章 服务网格 …… 408

10.1 服务网格简介 …… 408
10.2 Linkerd …… 408
10.2.1 Linkerd 简介 …… 408
10.2.2 安装 Linkerd …… 409
10.2.3 网格化 …… 416
10.2.4 代理自动注入 …… 417
10.2.5 暴露仪表盘 …… 423

第 11 章 云原生现状和展望 …… 427

11.1 云原生在企业的落地情况 …… 427
11.2 云厂商对云原生的支持 …… 428
11.3 云原生趋势展望 …… 429

第 1 章

用户认证项目

为了直观地演示从传统方式到云原生方式的软件生产流程的变化，书中用到了一个用户认证项目，下面会具体介绍项目的功能和运行方式。这些细节关系到之后对应用进行容器化和容器编排所需的操作。

1.1 项目介绍

为了聚焦云原生相关的开发实践，书中使用了功能简单的用户认证项目作为示例项目。用户认证项目的功能主要分为用户注册和用户登录，采用前后端分离的架构，由前端和后端两部分构成，前端项目提供用户界面，后端项目提供 API。

项目网址：https://github.com/bitmyth/accounts-frontend。

基于 Vue 开发，使用的 Vue 版本是 2.6.11。后文中把这个前端项目简称为 accounts-frontend。

项目编译后产生的文件会写入项目根目录下 dist 文件夹内。

项目根目录下有.env 文件，用于设置项目用到的环境变量。文件中的变量 VUE_APP_BACKEND 表示后端项目的接口地址。在编译项目之前需要按照实际后端项目接口的地址进行修改，默认为 localhost：8080。

1.2 开发环境

1.2.1 前端

2min

首先下载前端项目源代码，命令如下：

```
git clone https://github.com/bitmyth/accounts-frontend
```

下载结束后，使用如下命令进入项目根目录：

```
cd accounts-frontend
```

项目部署前需要编译，使用 Node.js 编译，命令如下：

```
npm run build
```

也可以使用其他工具(例如 YARN)来编译项目，命令如下：

```
yarn build
```

运行前端项目，命令如下：

```
npm run serve
```

运行命令后会输出的类似信息如下：

```
> accounts - frontend@0.1.0 serve
> Vue - CLI - service serve

INFO Starting development server...
98 %  after emitting CopyPlugin

DONE Compiled successfully in 3967ms \
7:39:39 PM

App running at:
 - Local: http://localhost:8081/
 - Network: http://192.168.0.40:8081/
```

访问输出信息中的链接即可访问前端项目页面。

1.2.2 后端

首先在控制台输入命令，以便下载后端项目源码，命令如下：

```
git clone https://github.com/bitmyth/accounts
```

项目使用 MySQL 数据库来存储数据，所以需要在项目启动前运行 MySQL 数据库。数据库相关信息需要写入配置文件 config/plain.yaml 中。与数据库相关的示例配置如下：

```
database:
  host: 127.0.0.1
  port: 3306
  username: root
```

```
password: 123
schema: accounts
```

相关配置项需要按照实际情况填写。

在项目根目录下运行命令 go run src/server/main.go 来启动程序。程序启动的 HTTP 服务器的监听地址默认为 8080 端口，可以通过修改配置文件 src/config/plain.yaml 中的 server.port 字段来改变端口值。

第 2 章

传统软件生产流程

2.1 开发

在传统的软件生产流程中,项目在完成需求分析之后,会进入编码实现阶段,在开发过程中,需要搭建开发环境,然后在开发环境下频繁地调试程序,使程序可以正常运行。

在裸机上搭建开发环境是一个简单却烦琐的过程,通常会有很多步骤,包括安装编程语言,以及安装依赖的库文件等,并且有些步骤可能有严格的顺序要求,但是搭建环境的流程并没有像代码一样加入版本控制系统中,导致搭建流程的步骤无法很好地记录下来。下次需要用到同样的开发环境的时候,只能从头开始做重复的工作。

也许有的开发者会把过程记录到文档中,下次可以参考文档来重现开发环境,但是这个过程依然烦琐。

当然也可以把搭建流程写入脚本文件。这样下次可以运行脚本来省去部分手工操作。用脚本来完成环境搭建仍然无法彻底解决环境的一致性问题,并且脚本的跨平台兼容性较差,提高脚本的兼容性又会使脚本的复杂度变大,从而提高了维护的难度。

当开发团队中人数变多之后,搭建开发环境过程中遇到的问题会更加棘手,不同的人可能使用不同的操作系统,每个人安装的开发语言的版本也可能有差异,这些都会让保持同样的开发环境变得困难,而保持一致的开发环境对协作开发与软件质量控制有非常重要的意义。

无法便捷地重现开发环境对新加入团队的伙伴也带来了很大的困扰。新伙伴在熟悉代码前,必须花精力搭建和团队成员一致的开发环境,这也许会花上一两天的时间,增加了上手成本。

长期来看,在传统的软件生产流程中搭建开发环境的重复劳动降低了组织的整体开发效率。

2.2 测试

软件测试同样需要一个让程序可以正常运行的环境,称为测试环境。在传统的软件生产流程中,测试环境同样无法便捷地搭建和复制,一般情况下只能费时费力地从头搭建。

随着时间的推移，开发环境也许会发生变化，有时安装了新的依赖，有时可能对某些配置参数做了调整，这时需要对测试环境做相应的升级，使应用程序可以在测试环境下继续运行，如果开发环境的变化比较频繁，则这个同步的过程是非常烦琐的。在开发实践中经常会遇到因为测试环境没有及时同步而导致测试失败的情况，测试人员发现程序无法正常运行，于是求助开发人员，而开发人员在自己的环境里程序可以正常运行。排查之后，发现失败的原因是测试环境没有安装某个库文件。这种因为环境的差异而导致的问题浪费了很多团队测试的时间，开发者却可能没有充分意识到这个问题的严重性。毕竟环境的变更不会每天都发生，可事实是，如果担心某件坏事会发生，则它会有极大的概率发生。修复环境不一致的问题所花的时间甚至会影响对项目进度评估的准确性。

测试环境有时会部署大量的机器以支持压力测试等消耗计算资源较大的测试场景，让大量的机器保持相同的测试环境是一项艰巨的任务。在传统的软件生产流程中，这个问题并没有很理想的解决方法。

2.3 计算资源评估

软件项目发布之前，通常会对这个项目需要的计算资源进行评估，计算资源包括中央处理器、内存、存储所需的磁盘等。

评估计算资源，一方面为了保证项目可以正常运行，不会因为计算资源缺乏而使性能下降甚至崩溃，另一方面是出于节省成本的考虑，如果申请的计算资源远超软件项目所需，则为空闲的计算资源支付的费用就是巨大的浪费。

考虑到用户请求的时间分布不均，需要合理规划计算资源，为软件项目应对请求峰值做出预估后留出合理余量。

常用的评估方法是根据现有用户数和以往的用户活跃时间计算出每秒请求数，并根据测试环境下软件处理相应量级的每秒请求数所需要的计算资源做预估。

2.4 部署

测试通过之后，软件项目一般就可以进入发布流程了，发布也就是把软件项目部署到生产环境的服务器上。

生产环境同样面临和开发环境同步的问题，如果开发环境的变更没有与生产环境同步，则在发布新版本时就有巨大的事故隐患，所以不得不在每次发布新版本前，除了做代码的审查，还要谨慎地检查生产环境需要完成的更新。

随着用户规模的增长，生产环境的服务器数量也会快速增加，部署软件的过程就会越来越复杂，无论是手工还是使用脚本去完成服务器部署都会是一个冗长并且容易出错的过程。

2.5 更新和升级

大部分软件项目需要根据用户的需求定期地更新和升级。在一个软件项目的生命周期中，第一次开发所占的时间往往较短，大部分时间都会用来更新和维护。一行代码写一次可能要被读上万次，这也是为什么软件工程师需要敬畏代码。

在传统的软件生产方式下，如果软件更新涉及环境的变更，则会产生较大的风险。如果软件运行在集群中的多台服务器，则需要同时完成环境的更新和代码的更新。如果环境没有更新成功而代码更新了，程序就可能会崩溃。如果服务器数量很多，则完成这样的环境升级是个巨大的挑战，并且由于高度依赖手动操作，出现误操作的概率也非常高，这会大大增加生产事故的发生概率。

第3章 容器技术的流行

在货物运输历史上，安全快捷地交付货物一直是个难题。集装箱是个伟大的发明，如果没有集装箱，当把不同类型的货物堆在一起进行运输的时候，质地软的货物就容易被压坏。集装箱的出现还使货物可以更快地完成装卸。如同集装箱对货物运输的变革，容器技术的出现带来了软件交付的巨大变革。

3.1 容器的优势

在软件开发实践中，通常希望软件可以运行在和宿主机隔离的环境中，像虚拟机一样，容器技术可以使进程运行在一个隔离的环境中，但是虚拟机的隔离机制是通过在宿主机上启动另一个完整的操作系统实现的，而容器技术只是在宿主机的操作系统之上借助内核接口做了资源视图的隔离，并没有启动另一套操作系统，所以更轻量。这个优势体现在启动和销毁的速度上，容器的启动时间一般只有数秒，而虚拟机的启动时间通常为数分钟。人类对速度的追求从未止步，从蜗行牛步的畜力车到日行千里的动车再到超声速的飞机，从鸿雁飞书到电报、电话、短信再到现在各种实时通信软件，都在印证人类对速度提升的追求。快速启动也是让容器技术广为流行的一个重要原因。

限制进程使用计算资源是容器技术的另一个重要功能。在资源受限的情况下，如果某个进程使用了过多的内存或者占用了过多的 CPU 时间，则会严重影响在同一台机器上的其他进程的正常工作，系统可能因此而崩溃。产生这种情况的原因可能是进程运行的代码有严重的缺陷，也可能是遭受了某种恶意攻击。进程使用的计算资源超过预期会引起系统不稳定。这时就可以利用容器技术来解决这个问题。容器技术提供了细粒度的资源约束能力，通常包括约束进程占用的 CPU 时间，以及使用的内存大小等。

3.2 Docker 简介

Docker 是容器技术的代表，在 2013 年开源，随着 Docker 的迅速普及，它成为容器技术的标准。

2020年知名技术网站StackOverflow举办的问卷调查的结果表明，Docker在“最喜欢的平台”这一项中的排名仅次于Linux。在“未来最想使用的平台”这一项的排名位列第一。这次问卷调查有接近65000名开发者参与，取得这样高的排名可见Docker的受欢迎程度之高。

Docker深刻改变了软件的交付方式，在主流的开源平台上比较受欢迎的软件项目几乎提供了Docker运行的方式。在使用这类提供Docker运行方式的项目时，不需要准备软件所需的环境，只需通过运行开发者提供的镜像就可以把程序运行起来，十分方便。这样的交付方式显著降低了用户使用软件前的学习成本，节省了用户准备软件运行环境的时间成本，Docker帮助开发者实现了一次构建，随处运行的高质量交付。Docker革命性地解决了开发者交付和分享软件的难题。采用Docker运行软件的用户从此不会再遇到环境问题引起的无法运行程序的尴尬。

在各大开源代码托管平台可以看到越来越多的开发者在发布他们的软件项目时，会同时提供运行软件的Docker容器镜像，意味着开发者不仅交付了代码，并且交付了最佳的运行代码的环境。把镜像定义文件加入软件版本控制系统也逐渐成为一种最佳实践。

Docker容器具有高度的可移植性，为容器技术创造了工业标准，Docker屏蔽了基础设施的差异，让应用不管运行在Windows还是Linux平台都能表现出一致的行为。

现在Docker拥有世界上最大的容器镜像仓库Docker Hub，作为Docker默认的镜像仓库，目前它保存的容器镜像已经基本包含了所有常见的应用程序对应的镜像。可以在Docker Hub中查看各个镜像的使用文档。在注册账号以后，就可以把自己构建的镜像上传到Docker Hub来和世界各地的用户分享。

大量的软件公司已经把Docker融入日常工作流程中，用Docker容器作为软件项目的开发环境，用Docker容器运行自动化测试和持续集成。

另外，得益于Docker容器对计算资源的细粒度约束能力，Docker可以帮助软件公司更加集约地使用计算资源，从而有效降低成本，产生良好的经济效益。

Docker提供了非常完善的CLI工具，可以让用户轻松地和Docker交互，这个工具让Docker的易用性大大增加，并且Docker还提供了RESTful API，让用户可以远程管理Docker。Docker友好的交互接口大幅降低了使用容器的技术门槛。

成功没有偶然，Docker有着丰富详尽的文档，简洁易用的交互工具，有活跃的项目更新升级，稳定且优异的性能，Docker镜像有着极佳的移植性和便携性。Docker提供了诸多容器交付的方式，可以上传到镜像仓库，也可以打包成压缩文件。这些都是Docker获得成功的重要因素。

有些工程师接触Docker技术时，对容器技术并没有充分理解，只是把它当作虚拟机来用。虚拟机配置烦琐，启动时间相对漫长，而Docker往往只需一条简单的命令就可以运行需要的程序。例如只用下面的命令就可以启动一个MySQL服务，代码如下：

```
docker run -e MYSQL_ROOT_PASSWORD=123 mysql
```

Docker的设计使它很容易上手，上面这条命令使Docker自动下载最新的MySQL镜像，然后启动这个镜像，用户用一条命令就可以把MySQL数据库运行起来。这甚至比在宿主机上通过下载并安装MySQL安装包，然后运行还要方便。当有更复杂的应用场景时，这种优势会更加明显。举个例子，如果需要同时启动5.6、5.7、8.0这3个版本的MySQL数据库，在宿主机上则需要先找到这3个版本的MySQL并安装，然后指定不同的端口参数来分别启动，不然会因端冲突而无法同时启动，并且同时安装3个版本的MySQL可能因为安装位置冲突而引起文件覆盖等诸多问题，而使用Docker只需下面的三条命令：

```
docker run -e MYSQL_ROOT_PASSWORD=123 mysql:5.6
docker run -e MYSQL_ROOT_PASSWORD=123 mysql:5.7
docker run -e MYSQL_ROOT_PASSWORD=123 mysql:8.0
```

只要指定对应的版本号作为镜像的标签，Docker就会启动对应版本的MySQL容器。

相信Docker的绝大部分初学者在接触之后都会对它的简洁和易用性大加赞赏。作为容器时代的操作系统，Docker正在并将继续为软件的高效交付赋能。

Docker是由Go语言编写的，通过docker version命令可以查看Docker的版本信息，包括开发时使用的Go语言版本等。

查看命令如下：

```
docker version
```

会看到类似如下的输出信息：

```
Client: Docker Engine - Community
 Cloud integration: 1.0.7
 Version:           20.10.2
 API version:       1.41
 Go version:        go1.13.15
 Git commit:        2291f61
 Built:             Mon Dec 28 16:12:42 2020
 OS/Arch:           darwin/amd64
 Context:           default
 Experimental:      true

Server: Docker Engine - Community
 Engine:
  Version:          20.10.2
  API version:      1.41 (minimum version 1.12)
  Go version:       go1.13.15
  Git commit:       8891c58
  Built:            Mon Dec 28 16:15:28 2020
  OS/Arch:          Linux/amd64
```

```
  Experimental:     false
 container:
  Version:          1.4.3
  GitCommit:        269548fa27e0089a8b8278fc4fc781d7f65a939b
 runc:
  Version:          1.0.0-rc92
  GitCommit:        ff819c7e9184c13b7c2607fe6c30ae19403a7aff
 docker-init:
  Version:          0.19.0
  GitCommit:        de40ad0
```

可以看到输出信息包括两大部分,分别是 Client 和 Server,也就是客户端和服务器端。这是因为 Docker 采用了常见的客户端/服务器端的软件架构。版本信息值得引起重视,如果 Docker 客户端版本高于服务器端版本,则客户端的有些命令就无法运行成功。这是因为服务器端的版本较低,所以对某些 API 还未支持。

服务器端指 Docker Engine,表现为一个叫 dockerd 的进程。它提供了容器运行和管理的核心功能,还提供了 API,使用户可以通过 API 来和 Docker 交互,称为 Docker Engine API。

客户端指 Docker CLI 命令行工具,常见的 docker run、docker exec 等命令就是客户端的工具。

3.3 Docker 安装

在个人计算机上通过安装 Docker Desktop 应用程序来使用 Docker。不同的平台需要安装不同的版本。Windows 平台安装的是 Docker Desktop for Windows 版本,Mac 平台安装的是 Docker Desktop for Mac 版本。Docker Desktop 安装过程比较简单,只需下载相应的安装包,然后运行安装包并按提示完成操作。

Docker Desktop 内置了 Kubernetes 等许多组件,并且提供了友好的 UI 界面,包括简洁的仪表盘等,方便可视化地对容器进行管理。

在 Linux 系统安装 Docker,建议使用流行的包管理工具来安装,也可以直接安装 Docker 二进制文件,二进制文件的下载网址为 https://download.docker.com。

下面介绍在 Ubuntu 上使用包管理工具 apt-get 安装 Docker 的步骤。

老版本的 Docker 在 apt 仓库中的名字是 docker、docker.io 或者 docker-engine,如果已经安装了这些包,则在安装最新版本的 Docker 前,需要先删除它们,命令如下:

```
sudo apt-get remove docker docker-engine docker.io containerd runc
```

首先更新 apt 仓库,命令如下:

```
sudo apt-get update
```

接着安装常用的依赖包，命令如下：

```
sudo apt-get install \
    apt-transport-https \
    ca-certificates \
    curl \
    gnupg \
    lsb-release
```

现在添加 Docker 官方的 GPG key，命令如下：

```
curl -fsSL https://download.docker.com/linux/ubuntu/gpg | sudo gpg --dearmor -o /usr/
share/keyrings/docker-archive-keyring.gpg
```

完成以上操作后，开始安装 Docker Engine，命令如下：

```
sudo apt-get update \
      && apt-get install docker-ce docker-ce-cli containerd.io
```

到这里已经完成了 Docker 安装，需要进一步配置 Docker，实现开机自动启动。

大部分的 Linux 发行版使用 systemd 来管理系统服务，下面把 Docker 设置为开机自动启动，命令如下：

```
sudo systemctl enable docker.service
sudo systemctl enable containerd.service
```

最后通过运行 docker version 命令来验证 Docker 是否可以运行。

如果希望以非 root 用户的身份运行 Docker 命令，可以把当前用户加入 docker 组中，命令如下：

```
sudo usermod -aG docker $USER
```

需要注意的是加入 docker 组中的用户具有较高的权限等级，可以用来运行获取宿主机 root 权限的容器，存在一定的安全风险。

熟悉 Ansible 的用户可以参考下面的 ubuntu.yaml 文件，通过命令 ansible-playbook 来自动完成在 Ubuntu 系统上 Docker 的安装。文件内容如下：

```
---
- name: install docker
  hosts: dockers
```

```
  remote_user: Ubuntu

  tasks:
  - name: Run "apt-get update"
    apt:
      update_cache: yes
    become: yes

  - name: install tools
    apt:
      pkg:
        - apt-transport-https
        - ca-certificates
        - curl
        - gnupg
    become: yes

  - name: Add Docker's official GPG key
    ansible.builtin.shell: curl -fsSL https://download.docker.com/linux/ubuntu/gpg | sudo
gpg --dearmor -o /usr/share/keyrings/docker-archive-keyring.gpg
    args:
      executable: /bin/bash

  - name: Set up repo
    ansible.builtin.shell: echo "deb [arch=amd64 signed-by=/usr/share/keyrings/docker-
archive-keyring.gpg] https://download.docker.com/linux/ubuntu $(lsb_release -cs) stable"
| sudo tee /etc/apt/sources.list.d/docker.list > /dev/null
    args:
      executable: /bin/bash
    become: yes

  - name: Run "apt-get update"
    apt:
      update_cache: yes
    become: yes

  - name: ensure docker is at the latest version
    apt:
      pkg:
        - docker-ce
        - docker-ce-cli
        - containerd.io
    become: yes

  - name: ensure docker daemon can autostart
    ansible.builtin.command: systemctl enable docker
```

```
    become: yes

  - name: ensure docker daemon is running
    ansible.builtin.command: systemctl start docker
    become: yes

  - name: ensure user is in docker group
    ansible.builtin.command: usermod -aG docker {{ ansible_ssh_user }}
    become: yes
```

3.4 Docker 在开发领域的价值

Docker 很好地解决了在第 2 章中提到的传统软件生产方式下开发环境带来的问题。在构建完一个可以让软件正常运行的 Docker 容器镜像之后，软件就可以在任何支持 Docker 运行的环境下正常运行了。

如果很不幸在开发的过程中，计算机坏了，需要换另一台计算机继续开发，这时 Docker 容器镜像的优势就体现出来了，可以轻松地把软件所需的开发环境通过 Docker 容器镜像恢复回来，避免了从头开始搭建开发环境的麻烦。对于有些复杂项目而言，开发环境的搭建可能需要几天时间，所以把开发环境构建到一个容器镜像中是一个明智的选择，容器镜像可以把搭建过程中付出的劳动保存下来。

在开发团队多人协作的场景下，Docker 的价值更加凸显，公司都会有正常的人员流动，当新成员加入团队的时候，团队希望新成员能够尽快熟悉软件项目，尽快开始贡献代码。如果团队已经把开发环境构建成了容器镜像，新成员就可以把搭建环境的时间节省下来，通过运行和调试代码来熟悉项目。这个看似不起眼的技术可以提高新成员在磨合期的幸福感，减小熟悉项目的压力。

当一个软件项目开发团队中的成员对开发环境做出变更后，可以及时地发布新的开发环境的容器镜像，这样团队中的其他成员就可以第一时间无感地获取一个统一的变更后的开发环境。这大大节省了为了同步开发环境所付出的协作成本，提高了整个团队的工作效率，也减少了因为环境问题而导致的错误。

Docker 的使用也节省了大量的软件部署时间成本，当新版本的软件发布以后，可以使用镜像快速完成部署，如果新版本的发布不幸存在 Bug，需要紧急回滚，也可以使用上一版本的镜像快速完成回滚，避免代码回滚而环境停留在最新的版本的情况。

为了更好地协作或者为了更高的安全性，有的企业需要由一台企业内网的服务器提供公共的开发环境。这种情况下，开发人员共用一个开发环境会有互相干扰的潜在问题，例如网络端口冲突，文件被覆盖及资源占用不均等。可以通过为每个开发人员分配一个 Docker 容器作为开发环境来解决。

在容器提供的隔离环境下，每个人在自己的容器中看到的进程空间都与自己相关，别人在开发环境启动的某个进程不会被自己看到，这样降低了操作时的心智负担。在容器中误杀的进程也只会影响到自己的开发环境，其他人不会受牵连。通过限制容器的计算机资源的使用可以避免资源被独占。这些都提高了开发团队的工作效率，带来了更好的开发体验。团队成员离职后，只需中止分配的对应容器就完成了回收开发环境的工作。

在软件开发实践中，通常会引入 CI/CD，也就是采用持续集成和持续交付的方法来加速软件的开发和交付流程。在持续集成和持续交付领域，需要用脚本实现一定程度的自动化。持续集成一般会在每次代码发布后自动触发。因为代码的发布比较频繁，所以提供持续集成的服务需要部署到不间接运行的服务器上，这样就需要在服务器搭建一个和开发环境保持同步的环境。Docker 可以用来保证环境的一致性，从而很好地支持自动化脚本来完成软件集成和交付工作。

3.5 Docker 在测试领域的价值

软件测试往往需要在多个目标平台运行软件项目来发现软件的兼容性问题，使用 Docker 可以方便地模拟目标平台。不必为了运行一个历史版本的软件而卸载及重装，或者重新购置服务器，可以用 Docker 模拟一个目标平台，这将会节省很多时间和 IT 成本。

例如某个软件在开发时使用了旧版本 5.0 的 Redis 数据库，后期想升级数据库，当使用新版本 6.0 的 Redis 数据库时，需要安装 6.0 版本的 Redis 并测试，来检查软件的功能是否和之前表现一致。如果使用传统方式可能需要卸载旧版本的 Redis，然后安装新版本的 Redis，而利用 Docker 容器，只需启动新版本的 Redis 镜像，就可以开始测试了。

使用 Docker 容器作为测试环境之后，测试环境和开发环境可以轻松地实现同步。测试人员只需下载最新的镜像就可完成环境的同步，不用关心环境发生了什么变更或如何进行同步。实践中经常会发生因为环境变更而导致之前运行正常的测试用例运行失败的情况。

对于需要运行在远端服务器的自动化测试场景而言，容器化的测试环境可以方便地使用脚本来管理起来，对于提高测试自动化程度有着重要作用。

第4章 容器化

4.1 容器化简介

容器化的意思是把应用程序和它运行时的依赖都打包到容器镜像中，使应用程序可以在容器中运行。完成容器化需要充分了解应用程序的架构、技术栈、环境依赖等。

在传统软件生产流程中，软件运行环境的问题没有得到足够重视，也没有很好的解决方案。随着容器技术的成熟，环境问题得到了有效解决。应用容器化后也具有了更好的上云条件，所以容器化也是云原生方式生产软件流程的重要步骤。

容器化是应用程序未来接入容器编排平台的第一步，而容器编排可以让软件工程师以更高效更智能并且更低成本的方式来利用云上的计算资源，这也是对应用程序做容器化的动力之一。

4.2 保持简单和专注

Docker 推荐使用单进程模型的容器，也就是让每个容器专注于做好一件事，这样可以有效降低容器管理的复杂性，并且可以提高容器的复用率。这与软件设计中的高内聚低耦合、关注点分离的原则相似，在面向对象语言中对类的设计上也有单一职责的设计模式。所谓大道至简，保持简单和专注是一个广泛适用的优秀设计理念。

水平伸缩能力是一个应用保持健壮的重要因素，简单的容器可以更方便地进行水平伸缩，因为简单的容器往往具备无状态的特性。

在升级容器时，需要关闭当前容器，然后重新运行升级后的镜像。简单的容器可以快速地独立升级。对于复杂的容器，当仅需对容器中的某个程序升级时，容器中运行的其他程序会受到无辜牵连，经历不必要的重启。同样地，在需要对复杂容器中的某个程序进行替换时，也会遇到同样的问题。

容器只会启动一个进程，如果想运行多个进程，则会使进程管理的难度增加。如果某个进程意外退出了，则很难让它自动恢复。这就需要容器中再安装一个进程管理器，从而进一步增加了容器的复杂性。

单个进程的容器可以由Docker方便地管理进程，如果定义了容器失败时自动重启的策略，则当容器中的进程意外退出而导致容器关闭时，Docker可以自动把容器重启，从而恢复容器中的进程。

对于静态网站这样简单的应用程序，只需一个Web服务器进程就可以正常工作，这样的应用可以用一个容器运行起来，这样的应用程序称为单容器应用，而对于技术栈复杂的应用程序，往往需要多个进程协同工作。当对复杂的应用容器化的时候，最好的方案是让各个进程在单独的容器中运行，这样容器化后的应用程序称为多容器应用。

4.3 容器镜像

容器镜像包含运行应用程序所需的一切：代码和它需要的所有依赖的库文件等。可以把镜像比作一个软件包，不管在哪里运行这个软件包，都会得到一个相同的容器。容器镜像保证了容器的不可变性，它是容器的模板和蓝图。

容器化的直接产物就是一个容器镜像，Docker容器的镜像使用一个分层的文件系统，每一层的内容中保存了对上一层的增量修改，也就是说无法修改已有的层级中的内容，换言之，每个层级都只能读取，不能写入。这样所有的层级有序地堆起来就形成了最后完整的镜像。Docker在构建过程中会对镜像文件系统层级进行缓存，在下次构建同样的镜像时就会利用缓存来加快整个构建过程。

在启动一个容器镜像后，新生成的容器会在容器镜像的文件系统中的所有层级的最上方添加一个可以写入的层级，称为容器层。容器中对文件修改和建立新文件或者删除文件都保存在这一层，这个机制使容器的启动时间极大地缩短。如果没有这个机制，每次启动容器就需要把镜像中全部的文件系统层级复制出来，然后才开始运行，以免容器写文件时会影响镜像中的文件。这样复制整个镜像的过程会花费很多时间。

Docker使用copy-on-write策略来提升读写性能。也就是说当容器或者镜像中的某一层访问下层的文件时，会直接访问，当对下层中的文件修改或删除时，会把对应的文件从原来的层级复制到当前层级中，然后进行修改或删除。这种策略有效地提高了文件的访问效率，减少了IO操作，并且减小了后续文件层级的大小，从而也减小了镜像的大小。

不同的容器之间可以共享镜像中相同的文件系统层级，使用同一个镜像启动的两个容器会共享全部的镜像文件系统层级。这也显著减少了容器占用的存储空间。

可以通过docker ps命令查看容器可写层大小的预估值，命令如下：

```
docker ps -s
```

-s参数会使命令输出信息中包含容器占用存储空间的信息，命令输出中包含的SIZE列示例如下：

```
SIZE
108MB (virtual 1.13GB)
```

SIZE一列表示容器的可写层中的数据的大小，virtual表示可写层加上只读层占用空间大小的总和。之所以称为virtual是因为容器的只读层中可能有部分和其他容器共享，所以在计算所有容器占用存储空间之和时，不能简单地把所有容器的SIZE列中的virtual值相加。这样得到的结果会远远超过真实值。

镜像的大小会影响镜像通过网络传输的时间，在把镜像上传到镜像仓库或者从镜像仓库下载镜像时，体积小的镜像会比大的镜像花费更少的时间。除此之外，启动镜像时需要把镜像读取到内存，所以镜像的大小还会影响镜像启动的时间，所以努力把镜像变小是个好主意。

构建Docker容器镜像需要一个镜像定义文件，容器镜像定义文件是一个文本文件，它包括了搭建一个完整的程序运行环境的步骤和程序如何运行的指令。Docker容器镜像文件有一个默认名字叫Dockerfile，没有扩展名。一般会把这个文件放到软件项目的根目录，并且加入git等版本控制系统。为了构建出尽可能小的镜像，需要精心设计镜像的构建过程，谨慎编写构建指令。

4.4　镜像定义

本节开始会详细讲述使用Docker容器化第1章所介绍的软件项目的具体步骤。首先了解一下编写镜像定义文件Dockerfile需要用到的基础知识。

4.4.1　Dockerfile基础知识

任何镜像都必须基于某个基础镜像来构建，所以Dockerfile的第一行指令必须是FROM。唯一的例外就是ARG指令，不过它也只是用来替换FROM指令中的变量。指令如下：

```
ARG VERSION = latest
FROM base: ${VERSION}
```

这样FROM指令就会解析为FROM base：latest

镜像必须指定启动后需执行的程序。如果没有指定需执行的程序，则运行构建出来的镜像将会失败。例如下面的Dockerfile：

```
FROM scratch
COPY ..
```

构建这个镜像，运行命令如下：

```
docker build -t test .
```

然后运行这个镜像，命令如下：

```
docker run test
```

输出信息提示镜像运行失败了，原因是没有指定要运行命令，信息如下：

```
docker: Error response from daemon: No command specified.
See 'docker run --help'.
```

指定运行镜像后容器中执行的命令需要用到 CMD 或者 ENTRYPOINT 指令。

CMD 和 ENTRYPOINT 都有两个格式，一个是 exec form，另一个是 shell form。以 CMD 为例，exec form 如下：

```
CMD ["可执行程序","参数 1","参数 2"]
```

命令中所有的参数都以字符串的形式提供。exec form 格式是推荐使用的格式。这个格式本质上采用了 JSON 数组的格式，在 Docker 解析的时候会按照 JSON 数据来解析，所以 exec form 中的引号必须是双引号而不能使用单引号。

另外 exec form 格式下，命令中的环境变量将不能被解析，因为解析和扩展环境变量的功能是由 shell 提供的。解决这个问题可以指定 sh 作为命令，命令如下：

```
CMD ["sh","-c","echo $PATH"]
```

另一个格式是 shell form，同样以 CMD 指令为例，shell form 如下：

```
CMD 可执行程序 参数 1 参数 2
```

shell form 意味着命令会被 shell 解析执行，这样程序会以/bin/sh -c 的方式运行。

在 Dockerfile 中最多只允许有一个 CMD 指令，如果写了多条 CMD 指令，则只有最后出现的 CMD 会生效。

CMD 用于指定默认执行的命令，但是当 Dockerfile 中同时使用了 CMD 和 ENTRYPOINT 指令时，CMD 用于为 ENTRYPOINT 指令提供默认参数，这种情况下，CMD 和 ENTRYPOINT 指令都必须使用 exec form，示例如下：

```
FROM Ubuntu
ENTRYPOINT ["top"]
CMD ["-b","-H"]
```

这个镜像运行起来后执行的命令为 top -b -H。

在使用 docker run 命令运行只使用了 CMD 指令的镜像时，docker run 命令中指定的参数会覆盖镜像中的 CMD 指令。

例如下面的 Dockerfile：

```
FROM Ubuntu
CMD ["top","-b","-H"]
```

构建镜像,把镜像命名为 test,命令如下:

```
docker build -t test .
```

接着运行这个镜像,命令如下:

```
docker run test ls /
```

会看到输出的信息是由 ls /命令产生的,输出信息如下(部分截取):

```
bin
boot
dev
etc
home
...
```

这就说明 Dockerfile 中的 CMD ["top","-b","-H"]被覆盖了,并没有执行 top 命令。

如果不希望 Dockerfile 中指定的命令被覆盖,则需要使用 ENTRYPOINT 指令。ENTRYPOINT 可以使容器每次都执行相同的命令。避免被 docker run 中的参数覆盖。如果需要覆盖,则可以指定--entrypoint 参数。docker run 的参数会附加到 ENTRYPOINT 中的命令后面作为参数,如果同时使用了 CMD 指令为 ENTRYPOINT 提供默认参数,则 docker run 的参数会覆盖掉所有 CMD 指令提供的参数。以 Dockerfile 为例,代码如下:

```
FROM Ubuntu
ENTRYPOINT ["top", "-b"]
CMD ["-c"]
```

基于这个 Dockerfile 来构建名为 test 的镜像,命令如下:

```
docker build -t test .
```

然后启动 test 镜像,命令如下:

```
docker run --rm test
```

可以看到输出信息中 COMMAND 一列的内容如下:

```
top -b -c
```

重新运行这个容器，加上参数－1，命令如下：

```
docker run -- rm test -1
```

在输出信息中可看到 top 命令把各个 cpu 的数据分行展示了出来，说明-1 参数生效了，并且 COMMAND 一列变成了如下命令：

```
top
```

这说明 Dockerfile 中定义的 CMD ["-c"]被覆盖，所以失效了。可以用下面的命令再次验证：

```
docker top test
```

输出信息中可看到 CMD 一列的内容如下：

```
CMD
top -b -1
```

这就再次验证了容器中此时执行的命令是 top -b -1。-c 参数被覆盖。

ENTRYPOINT 只能在 Dockerfile 中出现一次，如果出现多次，则只有最后一条会生效。

当使用 docker stop 命令中止一个容器时，Docker 会向容器中 pid(进程 ID)为 1 的进程发送 SIGTERM 信号，如果中止容器超时，则会发送 SIGKILL 强行关闭容器。

Dockerfile 中为镜像指定的命令，如果在运行起来后进程 pid 不是 1，则这个容器就无法通过 docker stop 来优雅地中止，因为运行指定的命令的进程无法收到系统信号，所以会在超时后被 SIGKILL 强行关闭。shell form 运行的程序是通过/bin/sh 来启动的，所以命令进程的 pid 不会是 1。exec form 则是直接启动命令，生成 pid 为 1 的进程，不需要 shell 参与，这就是推荐使用 exec form 来写 CMD 或者 ENTRYPOINT 的原因。

下面做实验来加深理解。使用了 shell form 格式的示例 Dockerfile 如下：

```
FROM Ubuntu
ENTRYPOINT top -b -c
```

现在用这个 Dockerfile 构建镜像，命令如下：

```
docker build -t test .
```

运行这个镜像，命令如下：

```
docker run -- rm test
```

看到输出如下信息，表示容器中 top 进程的 pid 并不是 1，信息如下：

```
PID USER      PR  NI    VIRT    RES    SHR S  %CPU  %MEM      TIME+ COMMAND
  1 root      20   0    2612    592    524 S   0.0   0.0   0:00.07 /bin/sh -+
  8 root      20   0    5972   3032   2604 R   0.0   0.1   0:00.11 top -b -c
```

用下面的命令也可以观察容器中的所有进程：

```
docker top test
```

输出信息中会看到两个进程：

```
UID    PID     PPID     C     STIME    TTY   TIME      CMD
root   83548   83520    1     01:11    ?     00:00:00  /bin/sh -c top -b -c
root   83587   83548    0     01:11    ?     00:00:00  top -b -c
```

现在测试一下使用 docker stop 命令中止 test 容器会发生什么情况，输入命令如下：

```
time docker stop test
```

这里使用 time 统计命令运行的时间。运行命令会卡住一段时间，接着容器被关闭了。出现这个现象的原因是 top 进程的 pid 不是 1，无法接收 Docker 发送的信号，导致中止容器的操作超时之后被强行关闭。

命令输出的信息如下：

```
docker stop test  0.13s user 0.08s system 2% cpu 10.398 total
```

可以看到命令运行时间长达 10s 多。

对于 shell form 格式的命令可以用 exec 来使命令成为容器中 pid 为 1 的进程。

在 shell 中运行程序时，shell 会生成一个新的进程来运行这个程序，而 exec 可以直接替换当前的 shell 进程，并且继承当前 shell 进程的 pid。可以做个实验帮助理解。

在命令行中输入 echo $$ 获取当前 shell 进程的 pid，命令如下：

```
echo $$
31150
```

用命令 ps 31150 查看 pid 是 31150 的进程信息来再次确认：

```
ps 31150
PID   TT   STAT      TIME COMMAND
31150 s009  S+     0:00.01 bash
```

现在运行 exec sleep 100 替换当前的 shell 进程：

```
exec sleep 100
```

重新打开一个新的命令行窗口，查看 pid 为 31150 的进程信息，输入命令如下：

```
ps 31150
PID    TT   STAT       TIME COMMAND
31150 s009   S+        0:00.02 sleep 100
```

可以看到 31150 进程由之前的 bash 变成了 sleep 100。说明 exec 把 bash 替换成了 sleep 100，并且保持了 bash 进程的 pid。

利用 exec 命令把上面例子中用到的 Dockerfile 修改如下：

```
FROM Ubuntu
ENTRYPOINT exectop -b -c
```

然后用同样的步骤构建镜像，运行镜像。

运行镜像会看到下面的输出信息：

```
PID USER      PR  NI    VIRT    RES    SHR S  %CPU  %MEM        TIME+ COMMAND
  1 root      20   0    5972   3040   2616 R   0.0   0.1   0:00.09 top -b -c
```

输出信息显示这次容器中只有一个进程在运行，它的 pid 是 1。这样就达到了最初的期望。

现在重新测试一下使用 docker stop 命令中止 test 容器会发生什么情况，输入命令如下：

```
time docker stop test
```

同样使用 time 统计命令运行时间。会发现命令马上就执行结束了。输出信息如下：

```
docker stop test  0.13s user 0.08s system 59% cpu 0.350 total
```

仅用时 0.350s。

Dockerfile 中常见的 RUN 指令用于定义构建镜像时运行的命令，RUN 指令会在镜像文件系统中产生新的层级，所以 RUN 指令出现次数越多，一般会使目标镜像体积越大。

需要注意 CMD 指令的作用和 RUN 完全不同，CMD 指令中的命令在镜像构建完成之后运行起来成为容器实例才会执行。RUN 指令中的命令是在构建过程中运行的。Dockerfile 中允许出现多次 RUN 指令，它们会按照从上到下的顺序依次在构建镜像过程中执行，而 CMD 指令只允许出现一次。如果出现多次，则以最后出现的 CMD 为准。

4.4.2 定义后端镜像

了解了容器镜像的构建原理和 Dockerfile 的基础知识后就可以开始编写镜像，以及定义文件 Dockerfile 了。

一个项目通常需要在多个环境中运行，所以需要用到多个不同版本的镜像。这样需要定义多个 Dockerfile 来构建多个版本的镜像以满足不同环境下对容器的要求。

在生产环境需要格外关注镜像的大小和性能，因为镜像的大小会影响上传/下载镜像用时和启动用时，所以需要努力减小生产环境镜像的体积。

在开发环境，为了方便开发和调试，镜像中往往会包含源代码和一些工具用于调试、编译等，镜像体积的优先级在便捷性之后，所以用于开发环境的镜像体积通常会大一些。

1. 用于生产环境的后端镜像

首先为后端项目定义用于生产环境下的镜像所用到的 Dockerfile，在项目 accounts 根目录下创建文件 Dockerfile，并写入以下内容：

```
FROM scratch
COPY dist/accounts /
EXPOSE 80
VOLUME ["/config"]
ENTRYPOINT ["/accounts"]
```

接下来详细介绍每一行指令的含义。

第 1 行 FROM 指令表明将要基于哪个镜像构建。FROM 指令引用的镜像也叫作基础镜像。Docker 中的镜像可以有继承关系，新构建的镜像会继承基础镜像中的所有文件。

FROM scratch 表示这个镜像基于 scratch 镜像构建。scratch 是一个特殊的镜像，它是 Docker 世界里最小的镜像，一般用它来构建基础镜像或者体积非常小的镜像。引用 scratch 镜像并不会在目标镜像的文件系统中引入新的层级，Dockerfile 中 FROM scratch 的下一行指令是目标镜像的文件系统中的第一层级。有意思的是，英文短语 from scratch 的中文意思是从零开始。

Docker 官方建议 FROM 指令引用的基础镜像最好是官方发布的镜像，alpine 是一个不错的选择，它的体积只有 5MB 左右，却拥有一套完整的 Linux 发行版。

第 2 行指令 COPY dist/accounts / 会把运行 docker build 命令时所在的文件夹下的 dist/accounts 文件复制到容器镜像中的根目录下。这里的 dist/accounts 文件是一个编译后的可执行程序。COPY 指令还支持将多个文件复制到目标文件夹下，例如 COPY file1 file2 /somedir/就可以把文件 file1 和 file2 复制到镜像中的/somedir/文件夹下。如果要复制的文件比较多，并且它们的文件名可以用通配符匹配，就可以用通配符来指定文件名，例如 COPY files * /somedir/ 就可以将具有 files 前缀的所有文件都复制到镜像中的/somedir/文件夹下。

除了 COPY 指令，ADD 指令也可以完成文件的复制。官方推荐使用 COPY 指令完成本地文件的复制，理由是 COPY 指令的语义在这种本地文件复制场景下相比 ADD 指令更清晰，ADD 指令除了支持 COPY 指令的功能外还支持把远程网络上的文件复制到镜像中，以及将本地压缩文件复制到镜像，并且自动解压缩等。

第 3 行指令 EXPOSE 80 表示镜像运行起来以后，容器中的应用程序会监听 80 网络端口，提供 HTTP 服务。网络端口的选择建议遵循常见的网络端口定义惯例，例如容器运行的是 MySQL 数据库，那么网络端口最好定义为 3306，如果容器运行的是 Redis 数据库，则网络端口就定义为 6379。EXPOSE 指令并不会使容器自动暴露端口，需要在运行镜像时指定-p 参数来把宿主机的某个端口映射到容器暴露的某个端口，所以 EXPOSE 指令规定的端口只是提示容器中哪些端口可以被暴露出来以提供服务。对容器使用者起到提示的作用。

第 4 行指令 VOLUME ["/config"]声明了一个容器文件系统中的挂载位置，在启动这个容器镜像的时候，可以把其他文件系统挂载到这个位置中以方便容器中的应用程序在挂载位置写入数据或者读取数据。

第 5 行指令 ENTRYPOINT ["/accounts"] 指定了运行容器镜像生成的容器中默认执行的程序是/accounts。当运行 docker run 命令来启动容器时，会产生和直接在宿主机运行 accounts 程序几乎一样的效果。

这样 Dockerfile 定义就完成了，但是用它来构建的容器会有不一致的问题。注意 Dockerfile 中的第 2 行指令：

```
COPY dist/accounts /
```

所复制的 dist/accounts 文件是在宿主机上编译生成的。这意味着不同的编译结果构建的镜像会有差异。不同的编译工具或者相同编译工具的不同版本都可能导致不同的编译结果。

如何使不同的机器上编译的结果都严格保持一致呢？可以把编译流程迁移到构建镜像的过程中，也就是在容器中完成代码的编译。

2. 用于开发环境的后端镜像

为了解决这个问题，需要借助一个编译代码用的容器镜像。为了防止覆盖已有的 Dockerfile 文件，新建一个文件并命名为 Dockerfile. build，写入以下内容：

```
FROM golang:1.15 - alpine
RUN apk -- update upgrade \
&& apk -- no - cache -- no - progress add git mercurial bash gcc musl - dev curl tar ca -
certificates tzdata \
&& update - ca - certificates \
&& rm - rf /var/cache/apk/ *
WORKDIR /go/src/github. com/bitmyth/accounts
COPY go. mod .
COPY go. sum .
RUN GO111MODULE = on GOPROXY = https://goproxy. cn go mod download

COPY ..
```

下面介绍每一行指令的含义。

第 1 行表示把 golang：1.15-alpine 作为基础镜像：

```
FROM golang: 1.15 - alpine
```

第 2 行运行 apk upgrade 命令来更新和升级 apk，接着运行 apk add 命令安装一些编译过程中用到的工具和系统库：

```
RUN apk -- update upgrade \
&& apk -- no - cache -- no - progress add git mercurial bash gcc musl - dev curl tar ca -
certificates tzdata \
&& update - ca - certificates \
&& rm - rf /var/cache/apk/ *
```

第 3 行把工作目录设置为/go/src/github.com/bitmyth/accounts。WORKDIR 指令会改变后续指令中的相对路径的基准：

```
WORKDIR /go/src/github.com/bitmyth/accounts
```

第 4 行表示把 go.mod 复制到镜像中的当前工作目录下，这里的句号(.)表示 WORKDIR 指令中定义的位置：

```
COPY go.mod .
```

第 5 行把 go.sum 文件复制到镜像中的工作目录下：

```
COPY go.sum .
```

第 6 行执行 go mod download 下载项目的所有依赖包。为了解决网络原因可能引起的下载失败问题，指定了 GOPROXY 环境变量来为 go 设置网络代理：

```
RUN GO111MODULE = on GOPROXY = https://goproxy.cn go mod download
```

最后一行把宿主机当前目录下的所有文件(不包括.dockerignore 中列出的文件和文件夹)复制到镜像中的当前工作目录下：

```
COPY ..
```

注意，这一行指令其实会把前面的 COPY go.mod 和 COPY go.sum 指令所复制的文件覆盖。为什么不把前面的 COPY go.mod 和 COPYgo.sum 指令去掉，然后把 COPY ..移到 COPY go.mod 的位置呢？这是因为 go.mod 和 go.sum 文件的内容发生变动的频率会比代码文件低。这样把 COPY go.mod 和 COPY go.sum 放到前面就可以使它们产生的对应文件系统层级缓存被有效利用。包括它们下面的这一行指令：

```
RUN GO111MODULE = on GOPROXY = https://goproxy.cn go mod download
```

产生的缓存也可以被有效利用起来。从而可以加快构建镜像的速度。

而如果把 COPY ..这个指令移到前面，替换掉 COPY go.mod 和 go.sum。一发生代码的变化，对应的缓存就失效了。这时后面的指令对应的缓存也就失效了。即使代码的变化没有引入新的依赖，也就是说 go.mod 和 go.sum 的内容保持原样，仍然会使位于后面的指令重新执行，而这个 go mod download 指令会从网络下载依赖，相对比较费时，所以为了更好利用构建产生的缓存，把 COPY ..放在了 go mod download 之后。这样当代码的变更不涉及依赖时，go mod download 指令就可以利用之前的缓存了，从而得以快速执行完毕。

在这个镜像启动后生成的容器中可以运行 go build 来编译生成 accounts 文件。这样，团队中的所有开发者就可以通过使用这个镜像实现在一个完全相同的环境中编译 accounts 项目了。

3. 后端多阶段构建镜像

目前构建生产环境的后端项目镜像需要分两步完成，第一步使用开发环境的镜像完成编译，生成可执行的程序。第二步把编译的结果复制到最终的镜像中。可以把构建后端项目镜像的这两个步骤写到一个 Dockerfile 中，利用多阶段构建实现运行一次 docker build 便可构建出最终的 accounts 镜像。新建一个文件 Dockerfile.multi 并写入以下内容：

```
FROM golang:1.15 - alpine as build - stage

RUN apk -- update upgrade \
&& apk -- no - cache -- no - progress add git mercurial bash gcc musl - dev curl tar ca -
certificates tzdata \
&& update - ca - certificates \
&& rm - rf /var/cache/apk/ *

WORKDIR /accounts

COPY go.mod .
COPY go.sum .
RUN GO111MODULE = on GOPROXY = https://goproxy.cn go mod download

COPY ..

RUN script/make.sh

FROM scratch
COPY -- from = build - stage /accounts/dist/accounts /
EXPOSE 80
VOLUME ["/config"]
ENTRYPOINT ["/accounts"]
```

可以看到文件内容大致相当于把前面的两个 Dockerfile 合并起来的结果。每个 FROM 指令标志着一个构建阶段。下面解释和之前的两份 Dockerfile 相比发生变化的指令，指令如下：

```
FROM golang: 1.15 - alpine as build - stage
```

这一行表示把这个构建阶段命名为 build-stage，这样其他构建阶段就可以方便地引用这个阶段构建的镜像中的文件了。接着需要解释的指令如下：

```
COPY -- from = build - stage /accounts/dist/accounts /
```

这一行表示把 build-stage 构建阶段中产生的临时镜像中的文件/accounts/dist/accounts 复制到这个阶段所构建的镜像中的文件夹下。

4. 数据迁移镜像

后端项目 accounts 会把数据保存在 MySQL 数据库中，在应用运行之前，需要提前准备好数据中用到的表。另外在开发过程中会经常对数据库结构进行修改，例如在表中加入新的一列，开发团队的成员之间需要共享具体的修改流程来使其他成员知道数据库发生了哪些变化，这样其他成员就可以通过执行同样的修改流程来使数据库的结构保持一致。这类任务称为数据迁移。

有很多数据迁移工具可用来完成这类任务。后端项目 accounts 中使用的是一个开源的工具 goose，下面创建一个运行迁移工具用的容器镜像。为了防止覆盖已有的 Dockerfile 文件，新建一个文件并命名为 Dockerfile. goose，然后写入以下内容：

```
FROM golang:1.15 - alpine

RUN apk -- update upgrade \
&& apk -- no - cache -- no - progress add git gcc musl - dev\
&& rm - rf /var/cache/apk/ *

RUN go get - u github.com/pressly/goose/cmd/goose
```

下面分析文件中每一行指令的含义。

第一行表示基于 golang：1.15-alpine 镜像构建。第二行表示更新系统包管理器 apk，然后安装一些编译用的工具。最后一行运行 go get 命令来安装 goose 数据迁移工具。

4.4.3 定义前端镜像

1. 用于生产环境的前端镜像

下面的前端项目 accounts-frontend 用于定义生产环境下的镜像 Dockerfile：

```
FROM nginx:alpine
COPY dist /app
COPY docker/default.conf /etc/nginx/conf.d/default.conf
```

接下来介绍每一行指令的含义。

第1行 FROM nginx：alpine 表示基于 nginx：alipine 构建镜像。前端项目 accounts-frontend 需要通过 Web 服务器来提供访问，所以使用 nginx：alpine 镜像作为基础镜像。它包含了提供 Web 服务的 Nginx 程序。

第2行 COPY dist /usr/share/nginx/html 指令表示把位于 dist 文件夹下的编译后的前端资源文件复制到镜像中的/app 文件夹下。这个文件夹会被配置成 Nginx 提供 Web 服务时的文档根目录。

第3行 COPY docker/default. conf/etc/nginx/conf. d/default. conf 指令表示把 docker/default. conf 文件复制到镜像中/etc/nginx/conf. d/default. conf。用来更改 Nginx 的默认配置。

至此 Dockerfile 的定义就完成了，但是用它构建的容器存在不一致的问题。注意 Dockerfile 中的第二行指令：

```
COPY dist /app
```

所复制的 dist 文件夹的内容是在宿主机上编译生成的。这意味着不同的编译结果所构建的镜像会有差异。不同的编译工具或者相同编译工具的不同版本都可能导致不同的编译结果。

如何使不同的机器上编译的结果都严格保持一致呢？可以把编译流程迁移到构建镜像的过程中，也就是在容器中完成代码的编译。下面定义一个可以用于代码编译的镜像。

2. 用于开发环境的前端镜像

下面再定义一个 Dockerfile，用来构建开发环境下的镜像。为了防止覆盖已有的 Dockerfile 文件，新建一个文件并命名为 Dockerfile. build，在文件中写入以下内容：

```
FROM node:15
WORKDIR /app
COPY package*.json .
RUN npm install
COPY ..
RUN npm run build
```

下面介绍每一行指令的含义。

第1行 FROM node：15 表示基于 node：15 镜像构建镜像。

第2行 WORKDIR /app 表示将当前工作目录设置为/app。

第3行 COPY package*.json . 表示把 package. json 复制到镜像中的/app 文件夹下。

这个文件定义了项目的所有依赖包。

第 4 行 RUN npm install 表示运行 npm install 安装依赖包。

第 5 行 COPY ..表示把当前文件夹下的所有内容复制到镜像中的/app 文件夹(因为项目中有.dockerignore 文件,所以复制的内容并不包括这个文件中定义的文件或文件夹)。.dockerignore 的作用类似 git 中的.gitignore 文件,在运行 docker build 命令时,.dockerignore 文件中列出的文件或者文件夹会从当前的构建上下文中排除,也就是说它们不会被发送到 Docker 中参与镜像的构建。

第 6 行 RUN npm run build 表示运行 npm run build 命令来编译。

这个镜像可以作为前端项目的开发环境来使用,它包含了项目编译用到的所有工具和依赖。

3. 前端多阶段构建镜像

目前构建生产环境的镜像需要分两步完成,第一步使用开发环境的镜像完成编译,第二步把编译的结果复制到最终的镜像中。如果希望把这两个步骤合并起来,可以使用多阶段构建技术实现这个目标。

多阶段构建允许在构建过程中从上一构建阶段所产生的临时镜像中复制需要的文件,这样就不会引入其他不需要的东西了,从而可以有效地降低镜像的大小。

下面是采用了多阶段构建的 Dockerfile:

```
FROM node:15 as build - stage
WORKDIR /app
COPY package * .json ./
RUN npm install
COPY ..
RUN npm run build

FROM nginx:alpine
COPY -- from = build - stage /app/dist /app
COPY docker/default.conf /etc/nginx/conf.d/default.conf
```

可以看到这个文件相当于把前面定义的两份 Dockerfile 合并起来的结果,下面讲述和之前发生变化的指令的含义。

第 1 行 FROM node as build-stage 表示把 node: 15 作为基础镜像,并且将这个构建阶段命名为 build-stage,这样可以在下一阶段的构建过程中引用这个构建阶段中的文件。

第 10 行 COPY --from=build-stage /app/dist .表示把第一阶段镜像中的/app/dist 文件夹复制到当前阶段的镜像中的/app 文件夹下。

4.5 构建镜像

到现在已经分别定义好了构建前端和后端项目的镜像所需的 Dockerfile,接下来会用到 docker build 命令来构建镜像。构建镜像类似于把程序源代码编译成可执行文件。完成构建之后，镜像就可以运行起来而成为一个容器。

4.5.1 docker build 命令

docker build 是 Docker 官方提供的构建镜像的命令，它的格式如下：

```
docker build [OPTIONS] PATH | URL | -
```

运行 docker build 需要两个东西，一个是 Dockerfile，另一个是构建上下文（Build Context）。如果使用本地的文件夹路径（也就是 PATH 参数），构建上下文就是运行 docker build 命令时 PATH 参数指定的文件夹下的所有文件。如果使用 URL 参数，一般会是一个 git 仓库的 URL 网址，这种情况下，Docker 会把这个 URL 指向的仓库下载下来，并以这个仓库中的所有文件作为构建上下文的文件。构建上文中的所有文件会被发送到 Docker 守护进程来参与镜像的构建过程。Dockerfile 中的指令使用的相对路径会以 PATH 指定的文件夹路径为基准。Dockerfile 中引用的本地文件必须是在构建上下文中存在的文件。

docker build 命令比较常用的参数有-t 参数（也可以写成--tag 形式），它的作用是给镜像命名，给镜像起一个有意义的名字之后，可以在运行容器的时候通过这个名字来方便地引用这个镜像。

Docker 规定镜像名不能包含某些特殊字符，如果镜像名由多个部分组成，各部分之间可以用斜杠分隔。-t 参数的格式是“名字:标签”，标签是可选的，标签可以包含小写字母、大写字符、数字、下画线（_）、句点（.）和连字符（-）。如果没有指定标签，Docker 则会提供默认的标签，叫作 latest，在实践中，使用默认的 latest 标签不利于对镜像版本的控制和追踪，所以最好提供一个语义化的标签，如 v1.0.1。最后一个参数句点（.）的意思是以当前文件夹作为此次构建的上下文。

docker build 会按照 Dockerfile 中的指令顺序从上到下依次执行构建指令，构建完成后，镜像会保存在本地的镜像列表中，要查看本地的镜像列表，可以用命令 docker images。

4.5.2 镜像缓存

在构建镜像过程中，Docker 会缓存产生的镜像文件系统层级，这样可以加快镜像构建的过程，对于许多具有相同基础镜像的镜像，它们的文件系统层级中有一部分是完全相同的，在这种情况下，利用 Docker 的缓存可以极大地提升构建速度。Dockerfile 中指令的顺序可能会影响构建镜像时 Docker 是否能有效利用上一次构建过程中对镜像中文件系统层级的缓存。如果某个指令会在它对应的文件系统层级中频繁地修改数据（例如把源码复制到镜像中的 COPY 指令），就应该考虑把这条指令放在整个指令序列中靠后的位置，因为这条指令会引起后续指令对应的文件系统层级的缓存失效，这样就会使下一次构建过程因无法利用缓存而相对变慢。

2min

有些情况下不希望利用缓存功能，可以在运行 docker build 命令时加上参数--no-cache=true。如果既希望利用缓存功能又希望某些指令对应的缓存失效，可以使用打破缓存的技

术(Cache Busting),举个例子,Dockerfile中有以下的指令:

```
RUN apt - get update
RUN apt - get install  - y curl
```

当执行完一次构建后,希望安装另一个软件vim,于是可把第2行指令变更为:

```
RUN apt - get install  - y curl vim
```

在下次构建镜像时可能会遇到安装失败的问题及无法升级的问题,这是因为第一行指令没有被修改,所以第一行指令的结果会被Docker缓存起来,在再次构建的过程中并不会重新运行apt-get update命令,从而导致下一步安装vim时失败。

如果希望Docker每次都重新执行apt-get update指令而不是利用之前缓存的结果,就需要把指令修改为如下形式:

```
RUN apt - get update && apt - get install  - y curl vim
```

这样还带来另一个好处,也就是减少了镜像中文件系统的层级,从而减小了镜像的体积。

Dockerfile中的每一条RUN指令都会使镜像中的文件系统产生新的层级,这样就加大了镜像的体积,所以可以利用shell的特性,把多条RUN指令中执行的命令合并到一条RUN指令中,但是这样会降低Dockerfile的可读性。

另一个可以有效降低镜像体积的办法是采用多阶段构建,利用多阶段构建的方法还可以写出可读性更高,更容易维护的Dockerfile。

4.5.3 构建用于开发环境的后端镜像

首先构建编译后端项目所需的镜像,执行的命令如下:

```
docker build  - t accounts: dev  - f Dockerfile.build .
```

命令中的-t参数表示把镜像命令定义为accounts: dev。

接下来运行accounts: dev镜像,以此来编译后端项目accounts,执行命令如下:

```
docker run  -- rm \
    - v  $PWD/dist:/go/src/github.com/bitmyth/accounts/dist \
   accounts:dev\
   ./script/make.sh
```

命令中的-v参数表示把宿主机的当前文件夹下的dist文件夹挂载到容器中的/go/src/github.com/bitmyth/accounts/dist文件夹下,这样宿主机上的dist文件夹和容器中的dist文件夹就绑定到了一起。不管是宿主机对dist文件夹中的内容做了修改还是容器对它的

dist 文件夹做了修改，对方都可以实时看到修改后的内容。Dist 文件夹用来保存项目编译后生成的可执行文件，通过这种方式就可以在宿主机上得到在容器中编译后的文件。

在上面的命令中最后的参数./script/make.sh 是会在容器中执行的命令。这个文件是一个 bash 脚本，它的内容如下：

```
#!/usr/bin/env bash
set -e

if [ -z "$VERSION" ]; then
    VERSION=$(git rev-parse HEAD)
fi

if [ -z "$CODENAME" ]; then
    CODENAME=cheddar
fi

if [ -z "$DATE" ]; then
    DATE=$(date -u '+%Y-%m-%d_%I:%M:%S%p')
fi

GOOS=Linux CGO_ENABLED=0 GOGC=off  go build -v -ldflags "\
-X github.com/bitmyth/accounts/src/app/version.Version=$VERSION \
-X github.com/bitmyth/accounts/src/app/version.Codename=$CODENAME \
-X github.com/bitmyth/accounts/src/app/version.BuildTime=$DATE " \
-a -installsuffix nocgo -o dist/accounts ./src/server
```

这个脚本的功能是使用 go build 命令来完成编译，因为编译时使用了较多的参数，所以把这个过程写在了 bash 脚本文件中。在文件中 go build 命令使用了参数-o，指定把编译结果写入 dist/accounts 文件中，所以运行完这个命令以后，会看到宿主机上的 dist 文件夹下也出现了完全相同的 accounts 文件。

使用 docker images 命令可以查看刚刚构建的 accounts：dev 镜像，输出的类似信息如下：

```
$ docker images
REPOSITORY          TAG       IMAGE ID         CREATED           SIZE
accounts            dev        52f5585c680b    18 seconds ago    1GB
```

accounts：dev 镜像可以用来提供后端项目的开发环境，在开发环境中会频繁修改代码，每次修改完后可以在 accounts:dev 容器中运行 go run src/server/main.go 命令来启动。

4.5.4 构建用于生产环境的后端镜像

下面开始构建 accounts 项目的镜像，在项目根目录下执行如下命令：

```
docker build -t accounts:v1.0.0 .
```

命令中-t 参数表示把镜像命令定义为 accounts：v1.0.0，运行命令可以看到类似如下输出：

```
[ + ] Building 0.1s (5/5) FINISHED
 => [internal] load build definition from Dockerfile                    0.0s
 => => transferring dockerfile: 36B                                     0.0s
 => [internal] load .dockerignore                                       0.0s
 => => transferring context: 2B                                         0.0s
 => [internal] load build context                                       0.0s
 => => transferring context: 61B                                        0.0s
 => CACHED [1/1] COPY dist/accounts /                                   0.0s
 => exporting to image                                                  0.0s
 => => exporting layers                                                 0.0s
 => => writing image sha256:fd53df793c7cc99ef2136317ae4e               0.0s

 => => naming to docker.io/library/accounts:v1.0.0                      0.0s
```

输出的信息表示已经完成了构建。

使用 docker images 命令可以查看刚刚构建的镜像，输出的类似信息如下：

```
$ docker images
REPOSITORY          TAG           IMAGE ID        CREATED               SIZE
accounts             v1.0.0        fb0d955758a6   About a minute ago    17.7MB
```

可以看到输出信息中出现了刚刚构建的 accounts 镜像，Docker 给 accounts 镜像加了默认的"latest"标签，IMAGE ID 这一列的值 fb0d955758a6 是 Docker 分配的唯一的标识。可以用它来引用这个镜像。CREATED 表示镜像创建的时间。SIZE 表示镜像的体积大小。

相比于用于开发环境的镜像 1GB 的体积，此次构建的生产环境镜像只有 17.7MB，这是非常显著的区别。

4.5.5 构建数据迁移镜像

现在构建用于数据迁移的镜像，执行的命令如下：

```
docker build -t goose -f Dockerfile.goose .
```

命令中的-t 参数表示把镜像命名为 goose。

可以运行 docker images 命令查看镜像，输出的类似信息如下（只截取了部分信息）：

```
REPOSITORY    TAG       IMAGE ID        CREATED          SIZE
goose         latest    e7705b1ad4f0   15 minutes ago   519MB
```

4.5.6 构建用于开发环境的前端镜像

现在使用 docker build 命令构建用于开发环境的 accounts-frontend 前端项目镜像，需要用到前端项目根目录下的 Dockerfile.build 文件。因为没有使用 Docker 默认的 Dockerfile 文件名，需要使用-f 参数指定 Dockerfile 的位置，命令如下：

```
docker build -t account-frontend-build -f Dockerfile.build .
```

构建成功后，运行下面的命令便可把容器中的 dist 文件夹复制到宿主机上：

```
docker run --rm \
    --name asset \
    -v $PWD:/tmp \
   account-frontend-build \
   cp -r /app/dist /tmp
```

这样就得到了在容器中编译生成的前端资源文件。其他人在用同样的方式完成编译后，会得到完全一样的前端资源文件。

4.5.7 构建用于生产环境的前端镜像

下面开始构建用于生产环境的 accounts-frontend 项目镜像，在项目根目录下执行的命令如下：

```
docker build -t accounts-frontend: v1.0.0 .
```

运行 docker images 命令查看镜像，输出的类似信息如下（只截取了部分信息）：

```
REPOSITORY          TAG        IMAGE ID       CREATED          SIZE
account-frontend    v1.0.0     540063784657   5 minutes ago    26.2MB
```

4.6 启动镜像

镜像是静态的，镜像启动之后会生成一个容器，容器中会运行制作镜像时打包进去的应用程序。每个容器中的进程都会对应宿主机上的一个进程。

4.6.1 启动 MySQL 镜像

考虑到后端项目 accounts 会依赖 MySQL 数据库，所以在启动后端项目的镜像之前，需要先启动一个 MySQL 数据库。同样用容器的方式运行 MySQL，这样 accounts 应用就变成了多容器应用。启动 MySQL 容器的命令如下：

```
docker run \
    --name mysql \
    -p 3306:3306 \
```

```
    -e MYSQL_ROOT_PASSWORD=123 \
    -e MYSQL_DATABASE=accounts \
   mysql:5.7 \
    --character-set-server=utf8mb4 \
    --default-time-zone=+08:00
```

下面解释命令中的参数。

命令中的参数--name mysql是把要启动的MySQL容器命令成mysql。

-e参数用来设置容器中的环境变量，这里设置的MYSQL_ROOT_PASSWORD环境变量的值将会被MySQL用来设置成root用户的访问密码。

MYSQL_DATABASE环境变量指定了MySQL启动时自动创建的数据库。这样当容器启动后，在MySQL数据库中就会看到名字叫accounts的数据库。

--character-set-server=utf8mb4参数用于设置数据库默认使用的字符集。

--default-time-zone=+08:00参数指定了容器默认的时区。

以这样的方式运行MySQL容器不方便进行数据备份。如果这个容器被删除，则MySQL数据库中的数据就丢失了。为了更好地进行数据备份，可以把数据通过挂载的方式保存在宿主机或者Docker数据卷中。默认的Docker数据卷保存在宿主机上由Docker管理的文件夹下，不需要关心它的具体路径。

现在改进上面的命令，用Docker数据卷来保存容器中的MySQL数据库所产生的数据。修改后的命令如下：

```
Docker run \
    --name mysql \
    -p 3306:3306 \
    -e MYSQL_ROOT_PASSWORD=123 \
    -e MYSQL_DATABASE=accounts \
    -v mysql:/var/lib/mysql \
   mysql:5.7 \
    --character-set-server=utf8mb4 \
    --default-time-zone=+08:00
```

命令中新加的-v mysql:/var/lib/mysql参数表示声明一个叫作mysql的数据卷并挂载到容器中的/var/lib/mysql文件夹下，这个位置是MySQL保存数据的地方。这样容器中/var/lib/mysql文件夹下的内容就会被保存到声明的mysql数据卷中。下次启动MySQL容器时只需用同样的方式挂载这个数据卷，就会拥有和之前一样的数据库环境，数据库中会包含相同的数据。

Docker中数据卷默认的驱动是本地存储，也就是保存在宿主机上，当为数据卷指定支持云端存储的驱动后，数据就可以直接备份到云端。

4.6.2 数据迁移

启动 MySQL 之后,可利用数据迁移镜像来创建数据库中的表,命令如下:

```
docker run --rm \
    -w /migrations \
    -v $PWD/src/database/migrations/:/migrations \
    -e GOOSE_DRIVER=mysql \
    -e GOOSE_DBSTRING="root:123@tcp(mysql:3306)/accounts" \
    --link mysql goose goose up
```

下面解释命令中的参数。

命令中的--rm 参数表示这个容器运行结束后会自动被删除。

-w 参数用于将容器中的工作目录设置为/migrations,这样在容器中执行的命令,如果使用的是相对路径,就会以这个目录作为基准。

-e 参数用于设置容器中的环境变量。其中 GOOSE_DRIVER 的值用来指定 goose 连接的数据库类型,GOOSE_DBSTRING 用来指定 goose 连接数据库用到的用户名、密码、数据库 IP 和监听的端口等信息。

--link 参数用来连接 mysql 容器,这样 goose 才能连接到运行在另一个容器中的数据库。采用--link 的连接方式已经被 Docker 废弃,后面会介绍更好的通过自定义 bridge 网络实现容器互连的方式。

运行命令后会看到以下输出:

```
2021/03/05 03:44:52 OK    20210304203736_create_table_user.sql
2021/03/05 03:44:52 goose: no migrations to run. current version: 20210304203736
```

表示数据迁移已经成功。可以用下面的命令在数据库容器中运行 SQL,可以通过查询来验证:

```
docker exec mysql mysql accounts -uroot -p123 -e "show tables"
```

命令输出如下:

```
mysql: [Warning] Using a password on the command line interface can be insecure.
Tables_in_accounts
goose_db_version
users
```

在 accounts 数据库中出现了 users 表和用于跟踪迁移状态的 goose_db_version 表。

现在数据库中已经有了 accounts 项目所需要的表结构,接下来就可以运行 accounts 项目了。

4.6.3 启动后端镜像

在开发环境中用 accounts:dev 镜像来运行后端项目,输入的命令如下:

```
docker run --rm \
     --name accounts \
     -v $PWD:/go/src/github.com/bitmyth/accounts \
     -p 8081:8081 \
    accounts:dev \
    go run src/server/main.go
```

命令中通过-v 参数把当前项目的根目录挂载到了容器中,这样当修改代码之后,容器中也会有最新的代码。通过 go run 命令直接就能运行项目,不需使用 go build 命令进行手动编译,这样可以加快调试和开发。

在生产环境中使用镜像 accounts:v1.0.0 来运行后端项目,输入的命令如下:

```
docker run --rm \
     --name accounts \
     -v $PWD/config:/config \
    accounts:v1.0.0
```

命令中参数--name accounts 的作用是把容器命名为 accounts,参数--rm 的意思是在容器退出后自动删除这个容器,-v 参数用于把宿主机本地的 $PWD/config 文件夹下的文件挂载到容器中的/config 文件夹下,这样容器中的程序便可以从 config 文件夹下的 plain.yaml 文件中读取配置。运行命令会看到输出的信息如下:

```
[error] failed to initialize database, got error dial tcp 127.0.0.1:3306: connect:
connection refused
```

此信息提示无法和 MySQL 建立连接。这是因为 accounts 容器和 mysql 容器具有不同的 IP 地址,accounts 容器中通过本地回环地址 127.0.0.1 访问的是 accounts 容器自身,由于 accounts 容器中并没有运行 MySQL 数据库,所以自然就无法建立连接。

要解决这个问题,最直接的办法就是把 mysql 容器的 IP 地址填到配置文件 host 配置项中。

需要修改 config/plain.yaml 文件中的 database.host 配置项,代码如下:

```
database:
  host: 127.0.0.1
```

为了找到 MySQL 容器的 IP 地址,可以使用下面的命令:

```
docker inspect mysql
```

输出信息如下(这里只截取了部分信息):

```
"NetworkSettings": {
  ...
"Gateway": "172.17.0.1",
"GlobalIPv6Address": "",
"GlobalIPv6PrefixLen": 0,
"IPAddress": "172.17.0.2",
"IPPrefixLen": 16,
"IPv6Gateway": "",
"MacAddress": "02:42:ac:11:00:02",
"Networks": {
"bridge": {
          ...
```

输出信息中可以看到"IPAddress":"172.17.0.2"这一行,表示容器的IP地址是172.17.0.2,把这个IP地址写入配置文件config/plain.yaml中,代码如下:

```
database:
  host: 172.17.0.2
```

现在重新运行accounts容器,代码如下:

```
docker run --rm \
     --name accounts \
     -v $PWD/config:/config accounts
    accounts:v1.0.0
```

可以看到输出信息是Server listen on port:8081,证明容器中应用程序启动成功了。

可以使用命令docker ps来查看正在运行的所有容器。演示代码如下:

```
$ docker ps
CONTAINER ID   IMAGE                    COMMAND
5fb66b6d7980   accounts:v1.0.0"/accounts"
fac98a271ac3   mysql:5.7                "docker-entrypoint.s…"
```

为了方便查看,只截取了部分信息,在输出的信息中可以看到刚刚启动的容器。

为什么accounts容器和mysql容器可以通过IP通信呢?这个问题需要对Docker的容器网络有基本的了解。

4.6.4 bridge网络

Docker默认会对容器的网络和宿主机的网络进行隔离,Docker中的网络模块通过可插拔的网络驱动实现。Docker预置了多种类型的网络驱动,包括bridge、host和overlay

等，其中 bridge 是容器默认选择的网络驱动。同时预置了 3 个网络，分别是 bridge、host 和 none。

可以使用 docker network 命令来查看 Docker 中与网络相关的信息，例如列出当前 Docker 中的所有网络可以用的命令：

```
$ docker network ls
NETWORK ID     NAME      DRIVER    SCOPE
14daf60e2e7a   bridge    bridge    local
fca889ae2d7c   host      host      local
27790ed24c8f   none      null      local
```

bridge 网络所使用的驱动是 bridge，它们名字是一样的，但一个是网络名，另一个是驱动名，需要注意区分。host 网络会使容器直接使用宿主机的网络栈，这样容器的网络就不会和宿主机隔离了。none 网络没有使用任何网络驱动，连接到 none 网络的容器将会失去网络通信能力。

Docker 会把没有对网络进行显式设置的容器加入 bridge 网络中。在前面启动容器的命令中并没有使用--network 参数设置容器所在的网络，所以它们都处在同一个网络 bridge 下，这样就可以通过 IP 来相互通信了。

可使用下面的命令查看 Docker 中 bridge 网络的信息：

```
docker network inspect bridge
```

可以看到输出的信息如下：

```
[
    {
"Name": "bridge",
"Id": "14daf60e2e7a7f86ae9671d054d1b593ce308ccb8a91f66da93164a5015bfad4",
"Created": "2021 - 02 - 26T01:13:56.761604873Z",
"Scope": "local",
"Driver": "bridge",
"EnableIPv6": false,
"IPAM": {
"Driver": "default",
"Options": null,
"Config": [
                {
"Subnet": "172.17.0.0/16",
"Gateway": "172.17.0.1"
                }
            ]
        },
"Internal": false,
```

```
"Attachable": false,
"Ingress": false,
"ConfigFrom": {
"Network": ""
        },
"ConfigOnly": false,
"Containers": {
"5fb66b6d79806b3dfd4c9d621820734760c5dde00f9c4af81cf51cd27fbf6845": {
"Name": "accounts",
"EndpointID": "041e274e06b2b6b6e7ac29368802163 5cc534464f0704a428de2f30317540d96",
"MacAddress": "02:42:ac:11:00:03",
"IPv4Address": "172.17.0.3/16",
"IPv6Address": ""
            },
"7d327c5478ab46c981fe4994190fd48b22394826619989dfb6c8897607c02bb7": {
"Name": "mysql",
"EndpointID": "1c643cdabea2d2a5f9771461a7b609502aa7e4c8474d322d70fbde0c4810fe3b",
"MacAddress": "02:42:ac:11:00:02",
"IPv4Address": "172.17.0.2/16",
"IPv6Address": ""
            }
        },
"Options": {
"com.docker.network.bridge.default_bridge": "true",
"com.docker.network.bridge.enable_icc": "true",
"com.docker.network.bridge.enable_ip_masquerade": "true",
"com.docker.network.bridge.host_binding_ipv4": "0.0.0.0",
"com.docker.network.bridge.name": "docker0",
"com.docker.network.driver.mtu": "1500"
        },
"Labels": {}
    }
]
```

输出信息显示 bridge 网络的范围是 172. 17. 0. 0/16，网关是 172. 17. 0. 1。在 Containers 字段下显示了当前连接到 bridge 网络中的所有容器，可以看到前面创建的 account 容器和 mysql 容器都出现在这里，证明这两个容器在同一个网络中。这就解释了为什么 accounts 容器可以通过 mysql 容器的 IP 地址访问 mysql 服务。

到这里已经把多容器应用运行起来了，注意到容器之间通信需要知道对方的准确 IP 地址，但是手动查找容器的 IP 有些不太方便，有什么办法来改进呢？DNS 可以解决这个问题。

4.6.5 自定义网络

Docker 支持用户自定义网络，在自定义的网络中，可以通过容器名来解析到容器所在的 IP 地址。这个功能是通过 Docker 中的 DNS 服务实现的。

从 Docker 1.10 版本开始，Docker 实现了内置的 DNS 服务器。不过它只对处于自定义网络中并且创建时指定了 name 或 net-alias 及 link 参数(已经废弃)的容器生效。

现在使用 docker network create 命令创建一个自定义的网络：

```
docker network create account - net
a4765cd1937678d48627b97d060f5d1a5f00986daca8e279d12b0597a2b00572
```

这样就创建了一个叫 account-net 的网络，可以使用下面的命令查看新建的网络：

```
docker network ls
NETWORK ID      NAME            DRIVER     SCOPE
a4765cd19376    account - net   bridge     local
14daf60e2e7a    bridge          bridge     local
fca889ae2d7c    host            host       local
27790ed24c8f    none            null       local
```

可以看到自定义的网络 account-net 出现在网络列表中了，它使用的是 bridge 驱动。接下来分别把 accounts 容器和 mysql 容器都连接到 account-net 网络中。

对于正在运行的 mysql 容器，可以使用 docker network connect 命令把它连接到 account-net 网络：

```
docker network connect account - net mysql
```

docker network connect 命令可以把一个运行中的容器连接到网络。相应地，使用 docker network disconnect 命令可以把容器和网络断开。

为了验证容器是否成功连接到网络，可以使用的命令如下：

```
docker inspect mysql
```

会看到类似的输出信息如下(部分截取)：

```
"Networks": {
"account - net": {
"IPAMConfig": {},
"Links": null,
"Aliases": [
"7d327c5478ab"
                ],
"NetworkID": "a4765cd1937678d48627b97d060f5d1a5f00986daca8e279d12b0597a2b00572",
"EndpointID": "0661cbb883da8aed381c810bec3276fe89a7b839456598e4f1037a8101eca3e3",
"Gateway": "172.20.0.1",
"IPAddress": "172.20.0.2",
"IPPrefixLen": 16,
```

```
"IPv6Gateway": "",
"GlobalIPv6Address": "",
"GlobalIPv6PrefixLen": 0,
"MacAddress": "02:42:ac:14:00:02",
"DriverOpts": {}
                },
"bridge": {
"IPAMConfig": null,
"Links": null,
"Aliases": null,
"NetworkID": "14daf60e2e7a7f86ae9671d054d1b593ce308ccb8a91f66da93164a5015bfad4",
"EndpointID": "1c643cdabea2d2a5f9771461a7b609502aa7e4c8474d322d70fbde0c4810fe3b",
"Gateway": "172.17.0.1",
"IPAddress": "172.17.0.2",
"IPPrefixLen": 16,
"IPv6Gateway": "",
"GlobalIPv6Address": "",
"GlobalIPv6PrefixLen": 0,
"MacAddress": "02:42:ac:11:00:02",
"DriverOpts": null
```

输出信息中的 Networks 字段显示 mysql 容器连接到了 bridge 和 account-net 两个网络，证明操作成功了。

之前创建 account 容器和 mysql 容器时都指定了 name 参数，所以只要把它们都连接到自定义网络 account-net，就可以通过容器名访问对方了，这样就能解决手工查找 IP 的问题。

accounts 容器在启动时会尝试连接 MySQL，如果在启动的时候没有连接到 accounts-net 网络，accounts 就无法连接 MySQL。下面修改启动它的命令，实现 accounts 容器启动时就连接到 account-net 网络。

首先停止运行中的 accounts 容器，命令如下：

```
$ docker stop accounts
```

把配置文件中的 host 参数修改成 mysql 容器的名字 mysql：

```
database:
  host: mysql
```

这样就关闭了 accounts 容器，然后修改 docker run 命令，加上参数--network：

```
$ docker run --rm \
    --name accounts \
    -v $PWD/config:/config \
```

```
    --network account-net
  accounts:v1.0.0
```

这样后端容器就可以通过 mysql 访问数据库容器了。

4.6.6 启动前端镜像

接下来运行前端的项目 account-frontend：

```
$ docker run --rm \
    --name account-frontend \
    -p8088:80 \
  account-frontend:v1.0.0
```

再次使用 docker ps 命令查看刚刚启动的容器信息：

```
CONTAINER ID  IMAGE    PORTS              NAMES
b4414089e21baccount-frontend0.0.0.0:8088->80/tcpaccount-frontend
```

第 1 章介绍的身份认证应用需要用到 3 个容器，具有依赖关系的容器需要按依赖顺序有序地启动才可以正常启动并工作。在实践中多个容器相互协作的应用场景十分常见，手动启动多个需要相互协作的容器不仅麻烦而且容易出错，在后续章节会使用更好的方式来处理这种多个容器相互协作的问题。

现在应用就可以通过浏览器访问了，打开网址 http://localhost:8088 会看到首页。

4.7 发布镜像

镜像制作完成后，可以把镜像发布到镜像仓库，镜像仓库可以是公开的，也可以是私有的。这样别人就可以通过下载开发者发布的镜像运行开发者的程序。通过发布镜像实现了软件交付，这样的交付方式让用户不再为环境问题所困扰，避免了许多因为环境不兼容导致的程序无法正常运行的问题。用户将从这种交付方式中获得前所未有的便捷和愉悦。

4.7.1 Docker Hub

使用 docker push 命令上传镜像时，默认会上传到 Docker 官方的公开镜像仓库 Docker Hub 中。它同时也是使用 docker pull 命令下载镜像时的默认的镜像仓库。

在尝试将镜像上传到 Docker Hub 之前，需要注册一个账号。在网页 https://hub.docker.com/signup 中注册账号后，在控制台输入命令 docker login 来登录自己的账号，命令及登录信息如下：

```
$ docker login

Login with your Docker ID to push and pull images from Docker Hub. If you don't have a Docker ID,
head over to https://hub.docker.com to create one.
Username: bitmyth
Password:
Login Succeeded
```

提示 Login Succeeded 表示登录成功，如果输入了错误的账号或密码会提示 Error response from daemon: Get https://registry-1.docker.io/v2/: unauthorized: incorrect username or password。

在已经登录的情况下，重新运行命令 docker login 不会重新要求输入账号和密码，而是直接使用已有的登录凭证去验证，然后提示登录成功。

```
$ docker login
Authenticating with existing credentials...
Login Succeeded
```

完成登录后，尝试把镜像 accounts 发布到自己的 Docker 镜像仓库，在此之前需要对镜像名称进行修改，变成"账号/镜像名:标签"这样的格式，通过如下命令来完成修改，这里笔者用账号 bitmyth 来演示。

```
$ docker tag accounts:v1.0.0 bitmyth/accounts:v1.0.0
```

现在镜像有了新名字 bitmyth/accounts:v1.0.0。新生成的镜像并不会占用额外的空间，因为它和原来的镜像具有相同的文件系统层级，并且两个镜像具有相同的镜像 ID。

接着运行 docker push bitmyth/accounts 命令来发布镜像，命令如下：

```
$ docker push bitmyth/accounts:v1.0.0

The push refers to repository [docker.io/bitmyth/accounts]
c344074aa03d: Pushed
latest: digest: sha256: cff3230826f7d6838d416a1fbce2c631e19a3e38016fa77223a7b2bfab231b5c
size: 527
```

镜像发布后，在世界任何地方就都可以使用 docker pull 命令下载这个镜像并通过 docker run 命令运行它。

除了公有的 Docker Hub 镜像仓库，还可以使用各大云厂商提供的镜像仓库，也可以利用 Docker 提供的 registry 镜像来自建镜像仓库。

4.7.2 私有镜像仓库

首先利用 registry 镜像在本地搭建一个镜像仓库来熟悉它的工作流程。

Docker Registry 需要 Docker Engine 的版本不低于 1.6.0。运行下面的命令来启动 registry 镜像：

```
docker run -d -p 5000:5000 --name registry registry:2
```

这样就拥有了一个本地的镜像仓库。下面尝试把镜像上传到这个本地的仓库。

首先下载 alpine 镜像，命令如下：

```
docker pull alpine
```

接着把 alpine 镜像按照“仓库地址/镜像名”的格式重命名：

```
docker tag alpine localhost: 5000/alpine
```

命令中的 localhost:5000 就是本地仓库的访问地址。默认情况下 Docker 访问的镜像仓库是 Docker Hub，它的地址是 docker.io。例如运行的 docker pull alpine 命令本质上是 docker pull docker.io/library/alpine:latest 的简写，所以为了区别默认的仓库，需要把镜像命名为 localhost:5000/alpine。这样在操作这个镜像时，Docker 就会使用 localhost:5000 这个位置的仓库。

现在把 localhost:5000/alpine 镜像上传到本地仓库，命令如下：

```
docker push localhost: 5000/alpine
```

为了测试下载镜像的功能，先把 localhost:5000/alpine 镜像从 Docker 删除，命令如下：

```
docker rmi localhost: 5000/alpine
```

现在从本地仓库下载它，命令如下：

```
docker pull localhost: 5000/alpine
```

到这里已经演示完本地镜像仓库的上传和下载镜像的功能。下面搭建可用于生产环境的远程镜像仓库。生产环境需要解决访问控制和安全问题。

4.8 CI/CD

CI/CD，即持续集成/持续部署，是一种加速软件安全交付的方法。通过使用各种工具将软件生命周期中的集成和部署环节自动化，集成过程中产生的错误得以及时反馈，开发者可以第一时间对错误进行修改。CI/CD 有效地提高了软件的生产效率，成为现代 DevOps 领域的重要组成部分。常见的 CI/CD 工具有 GitHub Actions、GitLab Runner、Jenkins、

Circle CI、Zuul 等。

采用 CI/CD 以后，代码发生变更后会自动触发 CI/CD 流程。实现对软件频繁地进行集成和部署，因为每次只发布小的改动，从而可以有效降低软件发布的风险。代码发布以后如果出现故障，也有利于及时定位代码的缺陷位置。

CI 阶段除了会执行编译通常还会执行软件测试，实现了测试的强制化执行。这对于提前发现软件缺陷有重大意义。

容器化应用的生产流程一般包括开发、测试、构建镜像、发布镜像等环节。除了开发以外，其他大部分环节是可以自动化的，如果每次提交代码后都手工进行单元测试、构建镜像、发布镜像等环节将需要大量的重复劳动。这一节讲述如何在第 1 章介绍的用户认证项目中使用 GitHub Actions 这个 CI/CD 工具来提升生产流程自动化水平，从而提高生产效率，解放人力。

2min

4.8.1 后端引入 CI/CD

后端项目 accounts 是 Go 项目，Go 项目一般会使用 go test 进行测试，以及使用 go build 来编译。后端项目使用了 GitHub Actions 作为 CI/CD 工具。

项目中的.github/workflows/go.yaml 文件是一个 GitHub Actions 工作流定义文件，内容如下：

```
name: Go

on:
  push:
    branches: [ main ]
    tags:
      - "v*.*.*"
  pull_request:
    branches: [ main ]

jobs:
  build:
    runs-on: Ubuntu-latest
    steps:
    - uses: actions/checkout@v2

    - name: Set up Go
      uses: actions/setup-go@v2
      with:
        go-version: 1.15

    - name: Build
      run: go build -v ./...
```

```
    - name: Test
      run: go test -v ./...
```

下面解释文件中指令的含义。

定义工作流的名字是 Go,代码如下:

```
name: Go
```

定义触发执行任务的事件,这里定义了 2 个事件,一个是 push 到 main 分支或者提交了新的标签并且标签的格式满足 v*.*.*,例如 v1.0.0 就是一个满足条件的标签。另一个事件是提交了 pull request,也就是请求和主分支 main 合并代码,代码如下:

```
on:
  push:
    branches: [ main ]
    tags:
      - "v*.*.*"
  pull_request:
    branches: [ main ]
```

定义执行的任务。这里只定义了一个任务,名字是 build。build 任务的运行环境是 ubuntu-latest。

```
jobs:
  build:
    runs-on: ubuntu-latest
```

定义执行 build 任务的具体步骤,代码如下:

```
steps:
```

定义 build 任务的第 1 步,用来把项目源代码检出(checkout)到工作目录,代码如下:

```
- uses: actions/checkout@v2
```

定义 build 任务的第 2 步,用来准备 Go 环境,使用 Go 1.15 版本,代码如下:

```
- name: Set up Go
  uses: actions/setup-go@v2
  with:
    go-version: 1.15
```

定义 build 任务的第 3 步,在项目根目录执行 go build -v ./...命令来编译项目,代码如下:

```
- name: Build
  run: go build -v ./...
```

定义 build 任务的第 4 步，在项目根目录执行 go test -v ./...命令来运行项目中的 Go 测试用例。

```
- name: Test
  run: go test -v ./...
```

当任务中有任意一个步骤失败时就会中断任务，不会继续执行后续的步骤。

引入这个工作流之后，在满足触发条件的事件发生后，GitHub Action 会自动执行以上任务来及时给予开发者反馈。当然推荐开发者在本地开发环境进行测试和构建，但是这里定义的这个工作流依然有必要。开发者有时会遗忘测试环节，或者由于某种原因没有严格执行测试。这种情况下，CI/CD 中定义的任务会形成一种对开发流程的强制约束。以最大可能提前发现质量缺陷，从而提高软件质量。

除了进行基本的测试和构建任务，还需要把构建 Docker 镜像，以及发布镜像也实现自动化。在项目中还有一个文件.github/workflows/docker-image.yaml，这个文件定义了自动完成构建和发布镜像的工作流程，内容如下：

```
name: CI to Docker Hub

on:
branch
  push:
    tags:
      - "v*.*.*"
  pull_request:
    branches: [ main ]

  workflow_dispatch:

jobs:

  build:

    runs-on: Ubuntu-latest

    steps:
      - name: Cache Docker layers
        uses: actions/cache@v2
        with:
```

```
          path: /tmp/.buildx-cache
          key: ${{ runner.os }}-buildx-${{ GitHub.sha }}
          restore-keys: |
            ${{ runner.os }}-buildx-

      - name: Check-out repo
        uses: actions/checkout@v2

      - name: Login to Docker Hub
        uses: docker/login-action@v1
        with:
          username: ${{ secrets.DOCKER_HUB_USERNAME }}
          password: ${{ secrets.DOCKER_HUB_ACCESS_TOKEN }}

      - name: Set up Docker Buildx
        id: buildx
        uses: docker/setup-buildx-action@v1

      - name: Get latest tag
        uses: olegtarasov/get-tag@v2.1
        id: tagName

      - name: Show tag
        run: echo $GIT_TAG_NAME
      - name: Build and push
        id: docker_build
        uses: docker/build-push-action@v2
        with:
          context: ./
          file: ./Dockerfile.multi
          builder: ${{ steps.buildx.outputs.name }}
          push: true
          tags: bitmyth/accounts:${{ steps.tagName.outputs.tag }}
          cache-from: type=local,src=/tmp/.buildx-cache
          cache-to: type=local,dest=/tmp/.buildx-cache

      - name: Show Image digest
        run: echo ${{ steps.docker_build.outputs.digest }}
```

下面介绍文件中指令的作用。

定义工作流的名字是 CI to Docker Hub,代码如下：

```
name: CI to Docker Hub
```

定义触发执行任务的事件,这里定义了 3 个事件,一个是 push 到 main 分支或者提交了新的标签并且标签的格式满足 v＊.＊.＊，例如 v1.0.0 就是一个满足条件的标签。另一个事件是提交了 pull request，也就是请求和主分支 main 合并代码。还有一个 workflow_dispatc 是直接在 GitHub Actions 页面单击任务执行按钮后触发，代码如下：

```
on:
branch
  push:
    tags:
      - "v*.*.*"
  pull_request:
    branches: [ main ]

  workflow_dispatch:
```

定义使用本地文件来缓存 Docker 的文件系统层级。GitHub 托管的运行器每次都会在全新的虚拟环境中启动，这样每次构建 Docker 镜像都无法利用上次构建产生的缓存。这会使任务运行时间变长。为了更快地构建，可使用 GitHub Actions 的 cache 操作，代码如下：

```
- name: Cache Docker layers
  uses: actions/cache@v2
  with:
    path: /tmp/.buildx-cache
    key: ${{ runner.os }}-buildx-${{ GitHub.sha }}
    restore-keys: |
      ${{ runner.os }}-buildx-
```

定义了把项目源代码检出(checkout)到工作目录的操作，代码如下：

```
- name: Check-out repo
  uses: actions/checkout@v2
```

下面的代码定义的操作是登录到 Docker Hub。用到的认证凭据是从 GitHub secret 中取出的，代码如下：

```
- name: Login to Docker Hub
  uses: docker/login-action@v1
  with:
    username: ${{ secrets.DOCKER_HUB_USERNAME }}
    password: ${{ secrets.DOCKER_HUB_ACCESS_TOKEN }}
```

下面的代码定义的操作是设置 Docker Buildx。它用来构建 Docker 镜像的工具，代码如下：

```
- name: Set up Docker Buildx
  id: buildx
  uses: docker/setup-buildx-action@v1
```

下面的代码定义的操作是获取最新的代码标签。后续会用这个标签来命名 Docker 镜像,代码如下:

```
- name: Get latest tag
  uses: olegtarasov/get-tag@v2.1
  id: tagName
```

下面的代码定义的操作是输出 $GIT_TAG_NAME 环境变量,它是由上一步操作设置的,代码如下:

```
- name: Show tag
  run: echo $GIT_TAG_NAME
```

下面的代码定义的操作是构建 Docker 镜像并上传。file 属性指定的./Dockerfile.multi 会用来作为镜像定义文件。tags 属性用于指定镜像名字。cache-from 和 cache-to 属性用来指定缓存文件存取的位置,代码如下:

```
- name: Build and push
  id: docker_build
  uses: docker/build-push-action@v2
  with:
    context: ./
    file: ./Dockerfile.multi
    builder: ${{ steps.buildx.outputs.name }}
    push: true
    tags: bitmyth/accounts:${{ steps.tagName.outputs.tag }}
    cache-from: type=local,src=/tmp/.buildx-cache
    cache-to: type=local,dest=/tmp/.buildx-cache
```

下面的代码定义的操作是显示构建产生镜像的摘要。

```
- name: Show Image digest
  run: echo ${{ steps.docker_build.outputs.digest }}
```

4.8.2 前端引入 CI/CD

后端项目 accounts-frontend 是 Vue 项目,Vue 项目可以使用 npm 进行测试,使用 npm run build 来编译。前端项目同样使用了 GitHub Actions 作为 CI/CD 工具,代码如下:

```
name: CI to Docker Hub

on:
  push:
    tags:
      - "v*.*.*"
  workflow_dispatch:

jobs:
  build:
    # The type of runner that the job will run on
    runs-on: Ubuntu-latest

    steps:
      - name: Check-out repo
        uses: actions/checkout@v2

      - name: Login to Docker Hub
        uses: docker/login-action@v1
        with:
          username: ${{ secrets.DOCKER_HUB_USERNAME }}
          password: ${{ secrets.DOCKER_HUB_ACCESS_TOKEN }}

      - name: Set up Docker Buildx
        id: buildx
        uses: docker/setup-buildx-action@v1

      - name: Set env
        id: tagName
        run: |
          echo "GIT_TAG_NAME=${GITHUB_REF#refs/*/}">> $GITHUB_ENV
          echo ::set-output name=tag::${GITHUB_REF#refs/*/}
          echo $GIT_TAG_NAME

      - name: Build and push
        id: docker_build
        uses: docker/build-push-action@v2
        with:
          context: ./
          file: ./Dockerfile
          push: true
          tags: bitmyth/accounts-frontend:${{ steps.tagName.outputs.tag }}

      - name: Show Image digest
        run: echo ${{ steps.docker_build.outputs.digest }}
```

第5章 容器编排

5.1 容器编排简介

首先理解编排两个字的含义，编排在计算机领域是指对系统的自动配置和管理。像Ansible这样的工具就是用来对服务器进行配置和管理的编排工具，而容器编排是指容器的配置、部署和管理等任务的自动化。容器编排可以把开发者从复杂且重复的容器管理工作中拯救出来。

如果只需管理一个容器，则大可不必使用容器编排，但现实是一个容器往往无法承载日益复杂的应用，如果把应用所有的组件都放到一个容器中运行，就丧失了使用容器的诸多优势，无法完成快速扩容缩容、无法对某个组件进行单独部署和升级及无法灵活地调度到其他服务器。现在微服务的概念已经深入人心，越来越多的应用以松耦合的形式来设计架构。对这样的应用容器化后，一定是以多容器的形式来运行的。

从4.6节手动运行和管理容器的过程中可以体会到多容器的管理比较复杂。手工管理不仅非常低效，而且极易出错，在容器数量增长到一定量级的情况下，使用容器编排的必要性就更加显著。尤其是对于稳定性有极高要求的生产环境，最大可能减少手工操作就能降低出错的风险。

中大型的应用一般会由多个组件构成，例如一个常见Web应用程序可能需要同时用到Web服务器容器、数据库容器及用作缓存的Redis容器，后续可能会引入消息队列容器。把应用容器化之后，每个组件都会运行在独立的容器中。这样应用就需要多个容器紧密配合才能运行。为了让多个容器正确协作，需要保证容器按一定的顺序启动，需要维持运行中的容器数量，需要控制各个容器的状态，通过容器编排技术，可以自动化完成这些任务，使多个容器组成的整体达到期望的状态。

容器编排是使用容器的实践中比较复杂的一个课题。Docker提供了一定的容器编排的能力，而作为容器编排领域事实标准的Kubernetes提供了更全面、更强大的编排能力。

5.2 Docker Compose

Docker提供了Compose组件，用于实现简单的容器编排。Docker Compose会从文件中读取应用所需的全部容器的定义，这个文件默认的文件名是docker-compose.yaml，可以把它称为 Compose 文件。有了这个文件以后，只需运行 docker-compose up 命令就可以把应用所需的所有容器启动，把所需的网络和存储准备就绪。Compose 文件相当于把项目的运行环境文档化。实现了用一个命令就完成了烦琐启动多个容器的任务。提供给项目一个完整且隔离的运行环境。

Docker Compose 只能管理单节点上的容器，所以更适合在开发和测试环境下使用。

下面使用 Docker Compose 对第 1 章中介绍的应用作容器编排。

5.2.1 Compose 文件

在后端项目的根目录下定义一个 docker-compose.yaml 文件，内容如下：

```
version: "3.9"

services:
  web:
    image: accounts-frontend:v1.0.0
    depends_on:
      - api
    ports:
      - "8088:80"

api:
    depends_on:
      - mysql
    image: accounts:v1.0.0
    ports:
      - "8081:80"
    volumes:
      - ./config:/config
    restart: always

  mysql:
    image: mysql:5.7
    volumes:
      - mysql:/var/lib/mysql
    restart: always
    environment:
      MYSQL_ROOT_PASSWORD: 123
MYSQL_DATABASE: accounts
    command: --character-set-server=utf8mb4 --default-time-zone=+08:00

db-migration:
    image: goose
    depends_on:
```

```
        - mysql
      working_dir: /migrations
      volumes:
        - ./src/database/migrations:/migrations
      command: 'goose up'
      environment:
        GOOSE_DRIVER: mysql
        GOOSE_DBSTRING: root:123@tcp(mysql)/accounts

volumes:
  mysql:
    external: true
```

下面分析一下这个文件的内容。

下面一行定义了 Compose 文件格式的版本，推荐使用最新的版本，但是也要注意和 Docker 的版本兼容，代码如下：

```
version: "3.9"
```

从这一行开始定义应用需要的所有服务。服务是对多个运行相同任务的容器的抽象。因为 Compose 支持扩展运行某个服务的实例数，所以服务会运行在一个或多个容器实例中，代码如下：

```
services:
```

定义前端项目 accounts-frontend 提供的服务，并把它命名为 web，web 这个名字将成为这个服务中容器的网络别名，也就是说，同一网络中的其他容器可以使用 web 作为主机名解析到这个容器。image 字段用于指定 web 容器的镜像。depends_on 字段用于指定 web 容器依赖的其他所有容器。depends_on 字段决定了容器启动的顺序，它会使依赖的容器先行启动，然后启动 web 容器。ports 字段定义了容器端口的映射，会把宿主机的 TCP 8080 端口映射到容器中的 TCP 80 端口，代码如下：

```
web:
  image: accounts - frontend:v1.0.0
  depends_on:
    - api
  ports:
    - "8088:80"
```

定义后端项目 accounts 提供的服务，并把它命名为 api。depends_on 字段用于指定这个服务依赖于 mysql 服务。ports 字段把宿主机的 TCP 8081 端口映射到容器中的 TCP 80 端口。volumes 字段用于将宿主机上与这个 docker-compose.yaml 文件同级的 config 文件挂载到容器中的/config 文件夹下。restart 字段用于指定容器在中止运行后重启的策略，always 表示在任何情况下中止运行后都会重启，代码如下：

```
api:
    depends_on:
      - mysql
    image: accounts:v1.0.0
    ports:
      - "8081:80"
    volumes:
      - ./config:/config
    restart: always
```

定义 MySQL 服务并把它命名为 mysql。image 字段指定了容器的镜像是 mysql:5.7。volumes 字段用于把名为 mysql 的数据卷挂载到容器中的/var/lib/mysql 文件夹下。

environment 字段用于定义容器中的环境变量。这里定义了 MYSQL_ROOT_PASSWORD，用来设置 root 用户的密码。MYSQL_DATABASE 环境变量用来设置 MySQL 启动时自动创建的数据库名字，当指定的数据库已经存在时，不会影响现有的数据。

command 中的--character-set-server=utf8mb4 指定了 MySQL 启动时的默认字符集为 utf8mb4，不设置默认字符集会默认为 latin1，--default-time-zone=+08:00 将默认时区设置为东八时区。由于 mysql 容器的默认执行的命令是 mysqld，代码如下：

```
mysql:
    image: mysql:5.7
    volumes:
      - mysql:/var/lib/mysql
    restart: always
    environment:
      MYSQL_ROOT_PASSWORD: 123
MYSQL_DATABASE: accounts
    command: --character-set-server=utf8mb4 --default-time-zone=+08:00
```

定义运行数据迁移的服务并把它命名为 db-migration。image 字段指定了容器的镜像是 goose，代码如下：

```
db-migration:
    image: goose
    depends_on:
      - mysql
    working_dir: /migrations
    volumes:
      - ./src/database/migrations:/migrations
```

```
    command: 'goose up'
    environment:
      GOOSE_DRIVER: mysql
      GOOSE_DBSTRING: root:123@tcp(mysql)/accounts
```

定义应用中需要用到的数据卷。第 1 个数据卷 mysql 正是上面定义的 mysql 容器，此容器用来挂载数据卷，规定了它的属性 external 是 true，表示这个数据卷是在外部的 Docker 中已有的，无须由 Compose 创建。

```
volumes:
  mysql:
    external: true
```

5.2.2　Compose 环境变量

目前的 Compose 文件中没有使用环境变量，当需要改变文件中的某些值的时候需要修改 Compose 文件，这样不太方便。如果镜像版本发生了变化，就需要在文件中改动。

Compose 支持使用环境变量替代 Compose 文件中的值。例如可以把文件中的这部分代码进行替换：

```
image: accounts-frontend: v1.0.0
```

替换成使用环境变量的格式：

```
image: ${ACCOUNTS_FRONTEND_IMAGE}
```

这样当 accounts-frontend 镜像发生变化的时候，通过改动环境变量 ACCOUNTS_FRONTEND_IMAGE 就可以避免修改 Compose 文件。例如当镜像版本升级到 v1.1.0 时，把 ACCOUNTS_FRONTEND_IMAGE 环境变量设置为 accounts-frontend: v1.1.0，这样 Compose 就会在启动后自动把 Compose 文件中相应的变量替换成 accounts-frontend: v1.1.0。如果 Compose 文件中用到的环境变量没有定义，则最终对应的位置会变成一个空字符串。

也可以把环境变量写到一个叫作.env 的文件中，这会使环境变量的管理变得方便。Compose 在启动时会自动读取 Compose 文件所在文件夹下的.env 文件。在.env 文件中写入以下内容：

```
ACCOUNTS_FRONTEND_IMAGE=accounts-frontend: v1.0.0
```

为了验证环境变量配置是否正确，预览一下经过替换变量处理后的 Compose 文件，命令如下：

```
docker-compose config
```

在输出的信息中会看到变量 ACCOUNTS_FRONTEND_IMAGE 替换后的结果如下：

```
image: accounts-frontend: v1.0.0
```

如果环境变量文件的位置和 Compose 文件不在同一个文件夹下或者文件名并不是默认的.env，就可以使用--env-file 参数来告诉 Compose 它的具体位置。利用这个功能，可以把不同环境下用到的环境变量写到不同的文件中，例如把集成环境用到的环境变量写到.env.ci 文件中，把开发环境用到的环境变量写到.env.dev 中。之后在运行 Compose 的时候把对应的文件通过--env-file 参数传递到 Compose。

如果 Compose 文件中的某些值重复出现在多个位置，则使用环境变量还可以使修改值的过程变得更简单。不再需要在文件中多处修改同一个值，只需修改一次环境变量。

下面是 Compose 文件引入更多环境变量后最终的版本：

```
version: "3.9"

services:
  web:
    image: ${ACCOUNT_FRONTEND_IMAGE}
    depends_on:
      - api
    ports:
      - "8088:80"

api:
    depends_on:
      - mysql
    image: ${ACCOUNT_IMAGE}
    ports:
      - "8081:80"
    volumes:
      - ./config:/config
    restart: always
    command: go run src/server/main.go

  mysql:
    image: mysql:5.7
    volumes:
      - mysql:/var/lib/mysql
    restart: always
    environment:
      MYSQL_ROOT_PASSWORD: ${MYSQL_PASSWORD}
```

```
      MYSQL_DATABASE: ${MYSQL_DB}
    command: --character-set-server=utf8mb4 --default-time-zone=+08:00

  db-migration:
    image: goose
    depends_on:
      - mysql
    working_dir: /migrations
    volumes:
      - ./src/database/migrations:/migrations
    command: goose up
    environment:
      GOOSE_DRIVER: mysql
      GOOSE_DBSTRING: root:${MYSQL_PASSWORD}@tcp(mysql:3306)/${MYSQL_DB}
volumes:
  mysql:
    external: true
```

可以看到引入环境变量之后，降低了 Compose 文件中值的耦合。相应地，需要在.env 文件中增加新引入的环境变量，修改后的代码如下：

```
ACCOUNTS_FRONTEND_IMAGE=accounts-frontend:v1.0.0
ACCOUNT_IMAGE=accounts:v1.0.0
MYSQL_PASSWORD=123
MYSQL_DB=accounts
```

5.2.3 Compose 运行应用

现在 docker-compose.yaml 文件已经准备好了，下面用 Docker Compose 来运行应用。命令如下：

```
docker-compose up
```

运行命令后会看到输出信息中包含以下部分：

```
Creating network "accounts_default" with the default driver
Creating accounts_mysql_1 ... done
Creating accounts_api_1 ... done
Creating accounts_db-migration_1 ... done
Creating accounts_web_1          ... done
```

输出日志比较多，并且都显示在终端，如果希望在后台运行，则可以在命令中加上-d 参数。如果以-d 参数运行，则查看日志需要使用 docker-compose logs 命令。

输出信息第一行表示 Docker Compose 自动创建了一个基于默认驱动器（也就是

bridge)的网络,网络名为 accounts_default。这样就为应用提供了网络隔离。

用 docker network 命令进行验证,命令如下:

```
docker network ls
NETWORK ID      NAME               DRIVER    SCOPE
f64a9cafa09e    accounts_default   bridge    local
```

发现多了一个 accounts_default 网络。

通过 Docker Compose 运行起来的容器同样也可以用 docker ps 命令看到。为了方便观察,这里对 docker ps 命令输出的信息做了适当截取,命令如下:

```
docker ps
CONTAINER ID   IMAGE                       NAMES
77a16cd7ac42   accounts - frontend:v1.0.0   accounts_web_1
7a5e6fc79984   accounts:v1.0.0 accounts_api_1
608fbe95fc17   mysql:5.7                   accounts_mysql_1
```

下面介绍一下 Docker Compose 对容器命名的方式。以 accounts_web_1 容器为例,accounts_web_1 中的 accounts 部分是项目名,默认为用项目根目录的文件夹名字命名,如果希望自己指定项目名,则可以在启动时加上参数-p,web 是在 Compose 文件中定义的容器名或者叫服务名,最后的数字 1 表示 web 容器的个数,也就是它一共运行了多少个实例。

也可以用 Compose 提供的命令来查看容器的运行情况,命令如下:

```
docker - compose ps
```

输出如下:

```
Name                        Command                        State          Ports
---------------------------------------------------------------------------------------
accounts_api_1   /accounts                              Up     0.0.0.0:8081 -> 80/tcp
accounts_db - migration_1   goose up                          Exit 1
accounts_mysql_1   docker - entrypoint.sh mysqld          Up        3306/tcp, 33060/tcp
accounts_web_1    /docker - entrypoint.sh ngin ...        Up     0.0.0.0:8088 -> 80/tcp
```

和 docker ps 命令不同的是:这条命令会只显示 docker-compose.yaml 文件中有关的容器。这样就可以方便观察。

在输出的信息中可以发现 accounts_db-migration_1 的状态是 Exit 1,证明容器中产生了某种错误而退出了。

查看 Compose 输出的日志,会发现下面的信息:

```
db - migration_1    | 2021/03/05 12:47:25 goose run: dial tcp 172.18.0.2:3306: connect:
connection refused
```

表明 db-migration_1 连接 MySQL 数据库失败。这个容器用来运行 goose 进行数据迁移，如果无法与数据库时连接，就无法正常运行。

尝试单独运行 db-migration，使用的命令如下：

```
docker - compose run db - migration
```

运行命令会看到以下输出：

```
Creating accounts_db - migration_run ... done
2021/03/05 12:56:50 OK     20210304203736_create_table_user.sql
2021/03/05 12:56:50 goose: no migrations to run. current version: 20210304203736
```

这次 db-migration 运行成功了，说明之前出现连接失败是因为 MySQL 没有准备就绪。

虽然这个 Compose 文件中定义了 db-migration 与 mysql 之间的依赖关系，但是这只能使 Compose 按照依赖关系的顺序运行容器，并不保证容器就绪的顺序也严格遵从这个依赖关系。只要被依赖的容器开始运行，Compose 就会开始创建依赖它的容器。容器进入运行状态不一定表示它已经准备就绪，有的容器如 mysql 从开始运行到就绪需要一段时间来初始化，mysql 就绪的标志是在日志中输出[Note] mysqld：ready for connections。从日志中可以看出 mysql 容器从启动到就绪所用的时间远大于 db-migration，当 db-migration 尝试连接 mysql 的时候，mysql 并没有准备就绪。这就是为什么 db-migration 在启动后连接 mysql 会失败的原因。

同样地，api 服务的容器也会由于这个原因在启动过程中连接数据库失败，但是 api 会在连接数据库失败后不断重试，在 mysql 就绪以后，重试连接就成功了，所以不需要人工介入 accounts 也能在最后进入正常状态。

服务就绪顺序与依赖顺序不一致问题的一个很好的解决方案就是提高对依赖的服务处于未就绪状态的容忍度，也就是增加服务的弹性。

到这里应用就已经通过 Compose 这种方式正常运行起来了。打开网址 http://localhost：8088 可以看到前端的页面可正常显示。

当前版本的 docker-compose.yaml 文件引用了一个外部的 mysql 数据卷，增加了在其他机器上复现环境的复杂性，在其他机器上要用这个文件运行 docker-compose up，也要先手动创建数据卷。在不需要引用外部数据卷的场景下，可以把文件中的 volumes 部分做如下修改：

```
volumes:
    mysql:
```

去掉了属性 external:true。这样，mysql 数据卷就会由 Compose 来创建和管理了。

为了观察修改后的文件所带来的变化，用 Compose 重新运行应用。重启之前需要关闭整个应用。命令如下：

```
docker - compose down
```

这个命令会把与应用相关的所有容器都中止，和手动管理每个容器相比，Docker Compose 把多个容器作为整体进行管理，带来了很大的便捷。

接着重新运行 docker-compose up，会看到输出信息中相较之前多了一行这样的文本：

```
Creating volume "accounts_mysql" with default driver
```

表明 Compose 创建了一个名为 accounts_mysql 的数据库，可以用 docker volume 命令验证，命令如下：

```
$ docker volume ls
DRIVER      VOLUME NAME
local       accounts_mysql
```

本地的数据卷中出现了新创建的 accounts_mysql 数据卷。证明 Compose 会根据文件中的定义自动创建所需要的数据卷。

Compose 会保留应用使用的数据卷，下次启动应用时会把上次运行产生的数据卷复制到新的容器中，这样就不会使数据丢失了。

要关闭应用，需要用到下面的命令：

```
docker - compose stop
```

运行命令会看到以下输出信息：

```
Stopping accounts_web_1      ... done
Stopping accounts_api_1 ... done
Stopping accounts_mysql_1    ... done
```

可以看到 docker-compose stop 命令会把与应用相关的所有运行中的容器都中止。

再次启动容器时可以用的命令如下：

```
docker - compose start
```

如果需要中止应用并且把应用产生的所有容器和网络都清除，则可以运行的命令如下：

```
docker - compose down
```

运行命令后输出的信息如下：

```
Stopping accounts_web_1      ... done
Stopping accounts_api_1 ... done
```

```
Stopping accounts_mysql_1      ... done
Removing accounts_web_1                ... done
Removing accounts_api_1       ... done
Removing accounts_db-migration_1 ... done
Removing accounts_mysql_1              ... done
Removing network accounts_default
```

在 Compose 文件中声明了命名数据卷的情况下，为数据安全起见，使用 docker-compose down 命令关闭应用并不会清除它创建的数据卷。需要加上--volume 参数才会把数据卷也清除。重新执行 docker-compose up 启动应用，然后输入的命令如下：

```
docker-compose down --volume
```

运行这条命令会看到下面的输出信息：

```
Stopping accounts_web_1         ... done
Stopping accounts_api_1 ... done
Stopping accounts_mysql_1      ... done
Removing accounts_web_1                ... done
Removing accounts_db-migration_1 ... done
Removing accounts_api_1       ... done
Removing accounts_mysql_1              ... done
Removing network accounts_default
Removing volume accounts_mysql
```

可以看到加了--volume 参数后，输出信息中出现了下面这一行：

```
Removing volume accounts_mysql
```

说明把数据卷 accounts_mysql 删除了。

5.2.4 Compose 更新应用

当更新应用代码以后，会构建出新版本的容器镜像，接下来应如何对运行中的应用进行升级呢？不需要手动中止旧容器，然后运行新容器来完成升级，只需修改 Compose 文件中相应的镜像版本，修改完 Compose 文件后，重新运行 docker-compose up，Compose 就会自动中止旧容器，创建新容器。这样就完成了对应用的更新。

5.3 Docker Swarm

与 Docker Compose 只能管理本地单个节点上的容器不同的是 Docker Swarm 可以管理多个节点上的容器，也就是具有了管理 Docker 集群的能力。学习 Swarm 可以熟悉运维和管理容器集群过程中会遇到的一些常见问题。

使用 Swarm 要求 Docker Engine 的版本不低于 1.12。

Swarm 集群中的节点分为 manager 节点和 worker 节点。节点即 Docker 主机，也就是运行 Docker Engine 的机器，Swarm 支持在多个节点上运行，也支持在单节点上运行。manager 节点也可以执行 worker 节点的任务，当然也可以设置成 manager 节点只执行 manager 的任务。Swarm 中的节点必须以 Swarm 模式运行。

在分布式环境中，节点之间需要确保安全的通信，不仅需要对通信进行加密，还需对身份进行验证。Swarm 集群节点之间通过双向 TLS 实现安全通信，当运行 docker swarm init 命令时，这个 manager 节点上会生成一个新的 root CA 和一对公私钥。同时 manager 节点还会生成 2 个 token，一个是 manager token 用于其他 manager 节点加入集群，另一个是 worker token，用于其他 worker 节点加入集群。每个 token 都包含 root CA 证书的摘要信息还有一个随机生成的密钥。当节点加入集群时，节点就可以根据 token 中的 root CA 摘要来验证 manager 的真实性，而 manager 节点用 token 中的密钥来验证待加入节点的有效性。

当节点加入集群时，还会在节点生成一对公私钥及 root CA 证书。文件保存在/var/lib/docker/swarm/certificates 下。其中公钥证书的文件名是 swarm-node. crt，私钥的文件名为 swarm-node. key。root CA 的证书名为 swarm-root-ca. crt。swarm-root-ca. crt 文件内容和集群中第 1 个 manager 节点上的 root CA 的文件内容完全一致。

manager 节点负责管理 worker 节点，具有 leader 身份的 manager 节点负责将任务分配到 worker 节点。manager 节点之间使用 Raft 算法来达成共识。所有的 manager 节点都参与维护 Swarm 集群内部状态，使其在不同的 manager 节点保持一致。

Swarm 集群中的 manager 节点数量决定了集群的容错能力，如果集群节点数是 n，则集群最多可以容忍($n-1/2$)个节点失败。例如 3 节点的集群，最多可以容忍 1 个节点失败，4 节点的集群，同样也只能容忍 1 个节点失败，所以最好使用奇数个 manager 节点。偶数个节点和比它小 1 的奇数相比并不能提高集群的容错能力。为了更佳的可用性，可以把 manager 节点分布到不同的 IDC 区域。

虽然增加 manager 数量可以提高集群容错能力，但是也会影响 Swarm 的性能。Docker 官方建议使用的 manager 节点数不超过 7 个。

worker 节点只负责运行服务而不负责管理集群。所有的 worker 节点上都运行一个代理，用来通知 master 节点自己的状态。这样 master 节点就可以根据通知来对 worker 节点进行管理。在 worker 上的任务失败后，manager 可以把这个任务调度到其他节点执行。

5.3.1 创建 Swarm 集群

下面演示 Swarm 集群功能，准备 3 台 Linux 服务器，1 台作为 manager 节点，其余 2 台作为 worker 节点。所有服务器都安装了 Docker。3 台服务器的操作系统都是 Ubuntu 20.04。

在第 1 台服务器上创建 Swarm 集群的 manager 节点,命令如下:

```
docker swarm init --advertise-addr 54.255.243.130
```

命令中的--advertise 参数用来指定 manager 节点的 IP 地址。运行命令输出的信息如下:

```
Swarm initialized: current node (f2hkxhks4g4msvann1ehargyy) is now a manager.

To add a worker to this swarm, run the following command:

    docker swarm join --token \
    SWMTKN-1-4loyfwaqwwoqdoutbscaw0t877h4o3pnoa0yytf6t5mstd5uwj-e97svc3o150sm7xp8qx18ad5p
54.255.243.130:2377

To add a manager to this swarm, run 'docker swarm join-token manager' and follow the instructions.
```

表示集群中的 manager 节点启动了。输出信息中提示 worker 节点加入这个集群的命令如下:

```
docker swarm join --token \
    SWMTKN-1-4loyfwaqwwoqdoutbscaw0t877h4o3pnoa0yytf6t5mstd5uwj-e97svc3o150sm7xp8qx18ad5p
54.255.243.130:2377
```

之后会在 worker 节点上运行此命令来加入集群。

worker 加入集群用到的 worker token 后续还可以单独查询,命令如下:

```
docker swarm join-token worker
```

运行命令后输出的信息如下:

```
To add a worker to this swarm, run the following command:

    docker swarm join --token \
    SWMTKN-1-4loyfwaqwwoqdoutbscaw0t877h4o3pnoa0yytf6t5mstd5uwj-e97svc3o150sm7xp8qx18ad5p
54.255.243.130:2377
```

如果希望其他节点以 manager 角色加入集群,则需要获取 manager token,命令如下:

```
docker swarm join-token manager
```

运行命令后输出的信息如下:

```
To add a manager to this swarm, run the following command:

    docker swarm join -- token \
    SWMTKN - 1 - 55zvv6p3lqogypvzwmbo76sd9c90wx214a74o4myqgs3cq4ht8 - btap3xq3znef742sageey94yt
172.31.30.250:2377
```

创建集群后,查看一下目前集群中的所有节点,命令如下:

```
docker node ls
```

运行命令后输出类似如下信息:

```
ID          HOSTNAME STATUS AVAILABILITY MANAGER STATUS ENGINE VERSION
r29f.. *  ip - 172 - 31 - 30 - 250 ReadyActive Leader   20.10.5
```

在输出信息中对 ID 列做了截取,STATUS 列是 Active,表示 manager 节点运行正常。

下面在第 2 台服务器(作为 worker 节点)运行命令如下:

```
docker swarm join -- token \
SWMTKN - 1 - 4loyfwaqwwoqdoutbscaw0t877h4o3pnoa0yytf6t5mstd5uwj - e97svc3o150sm7xp8qx18ad5p
54.255.243.130:2377
```

运行命令后输出的信息如下:

```
This node joined a swarm as a worker.
```

表示这台服务器作为 worker 节点成功加入了 Swarm 集群中。

同样地,在第 3 台服务器也执行同样的操作,命令如下:

```
docker swarm join -- token \
SWMTKN - 1 - 4loyfwaqwwoqdoutbscaw0t877h4o3pnoa0yytf6t5mstd5uwj - e97svc3o150sm7xp8qx18ad5p
54.255.243.130:2377
```

同样地,会输出如下信息:

```
This node joined a swarm as a worker.
```

现在重新查看集群中的节点,命令如下:

```
docker node ls
```

运行命令后输出类似如下信息(前 4 列):

```
ID                              HOSTNAME              STATUS    AVAILABILITY
f2hkxhks4g4msvann1ehargyy *    ip-172-31-30-250     Ready     Active
iobh29mwpmnq0fu3w1q85qt73      ip-172-31-32-49      Ready     Active
t1e9y94rrw6rrrxz16og1djt5      ip-172-31-35-212     Ready     Active
```

输出信息中后2列的内容如下：

```
MANAGER STATUS    ENGINE VERSION
Leader              20.10.5
                    20.10.5
                    20.10.5
```

表示3台服务器节点都成功加入了一个Swarm集群。输出信息中的STATUS列全部是Ready，表示集群中的节点都处于正常状态。AVAILABILITY列全部是Active，表示节点都可以用于运行服务所包含的任务，也就是说可以把运行任务的容器调度到这些节点上。

管理Swarm集群时需要在manager节点运行命令，如果在worker节点运行管理Swarm集群相关的命令，则提示如下：

```
Error response from daemon: This node is not a swarm manager. Worker nodes can't be used to view
or modify cluster state. Please run this command on a manager node or promote the current node to
a manager.
```

目前集群中有了3个节点，1个manage，2个worker，为了提升集群的容错能力，需要至少3个manager节点，可以在节点加入集群时使用manager token来成为manager节点，也可以把现在集群中的2个worker节点升级为manager节点，命令如下：

```
docker node promote ip-172-31-32-49
```

运行命令后输出的信息如下：

```
Node ip-172-31-32-49 promoted to a manager in the swarm.
```

同样地，将另外一个worker节点也升级为manager节点，命令如下：

```
docker node promote ip-172-31-35-212
```

重新查看集群节点信息，命令如下：

```
docker node ls
```

输出信息如下：

```
ID                          HOSTNAME           STATUS   AVAILABILITY
r29f4jyze4zlaydw5gqvljb2w *  ip-172-31-30-250   Ready    Active
uks3x7m5xth4efe6ywn5q4ost   ip-172-31-32-49    Ready    Active
a4odmh76r7983qskgix939os9   ip-172-31-35-212   Ready    Active
```

后两列信息如下：

```
MANAGER STATUS   ENGINE VERSION
Leader            20.10.5
Reachable         20.10.5
Reachable         20.10.5
```

同样支持把 manager 节点降级为 worker 节点，命令如下：

```
docker node demote ip-172-31-32-49
```

运行命令后输出的信息如下：

```
Manager ip-172-31-32-49 demoted in the swarm.
```

如果需要把节点从集群中移除，可以在节点上运行命令 docker swarm leave。

5.3.2 将样例服务部署到 Swarm 集群

Swarm 中运行的任务称为服务。1 个服务可以有多个容器实例，可部署到 1 个或多个节点上。Swarm 中服务的部署模式分为两种，一种是 replicated，另一种是 global。replicated 是默认的部署模式，在这种模式下，服务可以运行多个副本，但是不能保证服务在每个节点都有容器实例在运行，而在 global 模式下，集群中的所有节点都会部署 1 个服务实例。新加入集群的节点也会由 Swarm 自动启动 1 个服务实例。通常需要以 global 模式部署的服务，包括日志收集、监控系统及其他需要在每个节点都运行的服务。

下面演示如何在 Swarm 集群部署服务。登录到已经搭建好的集群中的第 1 台服务器，也就是 manager 节点，然后创建一个简单的服务，命令如下：

```
docker service create --replicas 1 --name helloworld alpine ping docker.com
```

命令会生成一个 alpine 镜像的容器，在容器中运行 ping 命令。--replicas 参数用来设置同时运行多少个容器实例，--replicas 1 表示只运行一个容器实例。运行命令后输出的信息如下：

```
ew7ri8clfc34fmtrhq00t2o9h
overall progress: 1 out of 1 tasks
1/1: running   [==================================================>]
verify: Service converged
```

如果输出信息,则表示服务创建成功了。

下面查看 Docker 中的容器信息,命令如下:

```
docker ps
```

运行命令后输出的信息如下:

```
CONTAINER ID   IMAGE           COMMAND              CREATED          STATUS
fe9661fe7f93   alpine:latest   "ping docker.com"    8 seconds ago    Up 7 seconds
```

还可以用 docker service 命令查看 Swarm 集群中的服务列表,命令如下:

```
docker service ls
```

运行命令后输出的信息如下:

```
ID             NAME         MODE         REPLICAS    IMAGE           PORTS
ew7ri8clfc34   helloworld   replicated   1/1         alpine:latest
```

输出信息中的 REPLICAS 列显示的 1/1 表示服务一共有 1 个容器在运行,期望运行的实例数是 1。

查看服务的详细信息,命令如下:

```
docker service inspect --pretty helloworld
```

运行命令后输出的信息如下:

```
ID:             ew7ri8clfc34fmtrhq00t2o9h
Name:           helloworld
Service Mode:Replicated
 Replicas:   1
Placement:
UpdateConfig:
 Parallelism:   1
 On failure:   pause
 Monitoring Period: 5s
 Max failure ratio: 0
 Update order:        stop-first
RollbackConfig:
 Parallelism:   1
 On failure:   pause
 Monitoring Period: 5s
 Max failure ratio: 0
```

```
    Rollback order:     stop - first
    ContainerSpec:
     Image:
        alpine:latest@sha256:a75afd8b57e7f34e4dad8d65e2c7ba2e1975c795ce1ee22fa34f8cf46f96a3be
     Args:              ping docker.com
     Init:              false
    Resources:
    Endpoint Mode:      vip
```

5.3.3 伸缩样例服务

Swarm 中的服务可以很方便地进行伸缩，首先登录到 manager 节点，把 helloworld 服务扩展到 2 个容器实例，命令如下：

```
docker service scale helloworld = 2
```

运行命令后输出的信息如下：

```
helloworld scaled to 2
overall progress: 2 out of 2 tasks
1/2: running   [ ==================================================>]
2/2: running   [ ==================================================>]
verify: Service converged
```

表示 helloworld 服务扩展成功，接着查看 helloworld 服务的容器实例的信息，命令如下：

```
docker service ps helloworld
```

运行命令后输出的信息如下(前 5 列)：

```
ID              NAME            IMAGE            NODE                  DESIRED STATE
84361iq8n17e    helloworld.1    alpine:latest    ip - 172 - 31 - 30 - 250    Running
7bjpfhxppb1q    helloworld.2    alpine:latest    ip - 172 - 31 - 32 - 49     Running
```

后 3 列信息如下：

```
CURRENT STATE              ERROR     PORTS
Running 2 minutes ago
Running 30 seconds ago
```

从输出信息中的 NODE 列可以看出 Swarm 把 helloworld 服务的一个容器实例调度到了一个 worker 节点(ip-172-31-32-49)。把另一个实例调度到了 manager 节点(ip-172-31-

30-250)。

实验结束,删除 helloworld 服务,命令如下:

```
docker service rm helloworld
```

5.3.4　更新样例服务

这一节演示如何对 Swarm 中的服务升级,首先部署 3 个使用 redis:3.0.6 镜像的容器实例,然后升级到 redis:3.0.7 镜像。Swarm 具有支持滚动更新的特性,也就是按照指定的并发度,依次对节点中服务的容器实例进行更新。

下面创建服务 redis,命令如下:

```
docker service create \
    --replicas 3 \
    --name redis \
    --update-delay 10s \
   redis:3.0.6
```

命令中的--replicas 3 表示运行 3 个容器实例,--update-delay 10s 表示依次更新每个容器的时间间隔为 10s。时间单位可以是小时、分钟和秒,分别用 h、m、s 表示。可以搭配起来使用,例如 1h5m30s 表示 1 小时 5 分钟 30 秒。Swarm 调度器默认在同一时刻只会更新 1 个容器实例,如果想改变同一时刻更新容器的数量,则需要使用参数--update-parallelism。例如指定--update-parallelism 3 就会同时对 3 个容器实例进行更新。

运行命令后输出的信息如下:

```
5rf7xzeu2rmw2e8j65mk4spm7
overall progress: 3 out of 3 tasks
1/3: running   [==================================================>]
2/3: running   [==================================================>]
3/3: running   [==================================================>]
verify: Service converged
```

输出信息表示已经成功运行了 3 个容器,查看 Swarm 中的服务列表,命令如下:

```
docker service ls
```

运行命令后输出的信息如下:

```
ID              NAME     MODE         REPLICAS   IMAGE          PORTS
5rf7xzeu2rmw    redis    replicated   3/3        redis: 3.0.6
```

进一步查看运行 redis 服务的容器信息和所在节点,命令如下:

```
docker service ps redis
```

运行命令后输出的信息如下(前 4 列)：

```
ID              NAME      IMAGE         NODE                    DESIRED STATE
z78tjmschmpa    redis.1   redis:3.0.6   ip-172-31-35-212        Running
0ftx5ta6mq7t    redis.2   redis:3.0.6   ip-172-31-32-49         Running
etui5zcr9xt9    redis.3   redis:3.0.6   ip-172-31-30-250        Running
```

后 3 列信息如下：

```
CURRENT   STATE   ERROR     PORTS
Running   49      seconds   ago
Running   48      seconds   ago
Running   48      seconds   ago
```

从 NODE 列可以看出 3 个容器实例运行在 3 个不同的节点上。

查看 redis 服务的详细信息，命令如下：

```
docker service inspect --pretty redis
```

--pretty 参数表示不使用 JSON 格式输出。运行命令后输出的信息如下：

```
ID:     5rf7xzeu2rmw2e8j65mk4spm7
Name:     redis
Service Mode:Replicated
 Replicas:   3
Placement:
UpdateConfig:
 Parallelism:   1
 Delay:     10s
 On failure:   pause
 Monitoring Period: 5s
 Max failure ratio: 0
 Update order:       stop-first
RollbackConfig:
 Parallelism:   1
 On failure:   pause
 Monitoring Period: 5s
 Max failure ratio: 0
 Rollback order:     stop-first
ContainerSpec:
 Image:
    redis:3.0.6@sha256:6a692a76c2081888b589e26e6ec835743119fe453d67ecf03df7de5b73d69842
```

```
Init:              false
Resources:
Endpoint Mode:      vip
```

在 UpdateConfig 条目下有一项 Parallelism:1,表示更新服务的时候最多只能同时更新一个容器实例。也就是更新操作的并行度,可以想象,这个值越大,更新全部服务所用的时间就会越短,但是为了系统的稳定性,通常会选择增量更新。

下面演示把 redis 服务的镜像升级到 3.0.7,同样需要在 manager 节点上操作。命令如下:

```
docker service update --image redis:3.0.7 redis
```

运行命令后输出的信息如下:

```
redis
overall progress: 3 out of 3 tasks
1/3: running   [==================================================>]
2/3: running   [==================================================>]
3/3: running   [==================================================>]
verify: Service converged
```

重新查看 redis 服务的详细信息,命令如下:

```
docker service inspect --pretty redis
```

运行命令后输出的信息如下:

```
ID:     5rf7xzeu2rmw2e8j65mk4spm7
Name:     redis
Service Mode:   Replicated
 Replicas:   3
UpdateStatus:
 State:      updating
 Started:   55 seconds ago
 Message:   update in progress
Placement:
UpdateConfig:
 Parallelism:   1
 Delay:      10s
 On failure:   pause
 Monitoring Period: 5s
 Max failure ratio: 0
 Update order:        stop-first
```

```
RollbackConfig:
 Parallelism:   1
 On failure:   pause
 Monitoring Period: 5s
 Max failure ratio: 0
 Rollback order:     stop-first
ContainerSpec:
 Image:
    redis:3.0.7@sha256:730b765df9fe96af414da64a2b67f3a5f70b8fd13a31e5096fee4807ed802e20
 Init:     false
Resources:
Endpoint Mode:   vip
```

5.3.5 维护 Swarm 节点

当 Swarm 集群中的节点需要维护时，需要把节点上运行的任务调度到正常工作的节点，在把节点设置为维护状态时，Swarm 会自动完成任务调度。

下面演示维护节点时 Swarm 自动调度任务的过程。

首先登录到 manager 节点，创建一个服务，命令如下：

```
docker service create --replicas 3 --name redis --update-delay 10s redis:3.0.6
```

运行命令，以便创建 3 个运行 redis:3.0.6 镜像的容器作为 redis 服务。

查看运行 redis 服务的节点信息，命令如下：

```
docker service ps redis
```

运行命令后输出的信息如下(前 5 列)：

```
ID             NAME      IMAGE         NODE                  DESIRED STATE
hjrzup5y7gxj   redis.1   redis:3.0.6   ip-172-31-32-49        Running
xegd5ktg3yo5   redis.2   redis:3.0.6   ip-172-31-30-250       Running
se1tz60os1u1   redis.3   redis:3.0.6   ip-172-31-35-212       Running
```

后 3 列信息如下：

```
CURRENT STATE                ERROR      PORTS
Running 39 seconds ago
Running 39 seconds ago
Running 39 seconds ago
```

从输出信息中的 NODE 列可以看出 3 个容器实例运行在 3 个不同的节点上。

下面把节点 ip-172-31-32-49 设置成维护状态，命令如下：

```
docker node update --availability drain ip-172-31-32-49
```

命令的--availability drain 表示把节点 ip-172-31-32-49 设置为 drain，也就是维护状态。drain 在英文中是抽干、排干的意思，设置成 drain 状态的节点上的容器实例会被 Swarm 结束运行，然后调度到其他正常节点上继续运行。相当于把节点上的容器抽出去，就像把泳池中的水排走一样。

现在再次查看运行 redis 服务的节点信息，命令如下：

```
docker service ps redis
```

运行命令后输出的信息如下（前 5 列）：

```
ID           NAME   IMAGE    NODE     DESIRED STATE
eo1jkk73u2q9   redis.1       redis:3.0.6   ip-172-31-30-250   Running
hjrzup5y7gxj    \_ redis.1   redis:3.0.6   ip-172-31-32-49    Shutdown
xegd5ktg3yo5   redis.2       redis:3.0.6   ip-172-31-30-250   Running
se1tz60os1u1   redis.3       redis:3.0.6   ip-172-31-35-212   Running
```

后 3 列信息如下：

```
CURRENT STATE                ERROR     PORTS
Running 1 second ago
Shutdown 2 seconds ago
Running about a minute ago
Running about a minute ago
```

输出信息中显示节点 ip-172-31-32-49 运行 redis 服务的容器实例被关闭了，Swarm 把这个关闭的任务调度到了节点 ip-172-31-30-250 上继续运行，并且保持了原来的实例名字 redis.1。这样在节点 ip-172-31-30-250 上就有了 2 个运行 redis 服务的容器实例。

可以查看一个节点 ip-172-31-30-250 上的 Docker 容器的进程，命令如下：

```
docker ps
```

运行命令后输出的信息如下（部分截取）：

```
CONTAINER ID    IMAGE          COMMAND                    CREATED
7e3db2f9b28e    redis:3.0.6   "/entrypoint.sh redi..."About a minute
agod3e4344a4aaf    redis:3.0.6    "/entrypoint.sh redi..."2 minutes ago
```

输出信息显示有 2 个 redis:3.0.6 镜像的容器。

现在节点 ip-172-31-32-49 处于不可用于调度的状态，再次查看节点信息，命令如下：

```
docker node ls
```

运行命令后输出的信息如下(前 4 列)：

```
ID                             HOSTNAME            STATUS    AVAILABILITY
r29f4jyze4zlaydw5gqvljb2w *    ip-172-31-30-250    Ready     Active
6zxgzwnpaurtlcd874qozn181      ip-172-31-32-49     Ready     Drain
a4odmh76r7983qskgix939os9      ip-172-31-35-212    Ready     Active
```

后 2 列信息如下：

```
MANAGER STATUS   ENGINE VERSION
Leader            20.10.5
                  20.10.5
                  20.10.5
```

可以看到节点 ip-172-31-32-49 对应的 AVAILABILITY 列显示的状态为 Drain。

单独查看节点 ip-172-31-32-49 的具体信息，命令如下：

```
docker node inspect --pretty ip-172-31-32-49
```

运行命令后输出的信息如下(部分截取)：

```
ID:                      6zxgzwnpaurtlcd874qozn181
Hostname:                ip-172-31-32-49
Joined at:               2021-03-10 02:27:10.221061622 +0000 utc
Status:
 State:                  Ready
 Availability:           Drain
 Address:                172.31.32.49
```

同样可以看到节点 ip-172-31-32-49 的 Availability 属性是 Drain。

节点维护结束后，再把它恢复为正常状态。命令如下：

```
docker node update --availability active ip-172-31-32-49
```

恢复为 active，也就是活跃状态后，重新查看节点 ip-172-31-32-49 的信息，命令如下：

```
docker node inspect --pretty ip-172-31-32-49
```

运行命令后输出的信息如下(部分截取)：

```
ID:                      6zxgzwnpaurtlcd874qozn181
Hostname:                ip-172-31-32-49
Joined at:               2021-03-10 02:27:10.221061622 +0000 utc
```

```
Status:
 State:              Ready
 Availability:             Active
 Address:     172.31.32.49
```

可以看到节点 ip-172-31-32-49 的 Availability 属性恢复成了 Active。这时 Swarm 又可以调度任务到它上面运行了。

5.3.6 Swarm 路由网格

Swarm 集群中的服务一般会部署到多个节点上运行，当 Swarm 中的服务通过某个网络端口对外提供服务时，Swarm 的 Routing Mesh(路由网格)功能使集群中任何一个节点都能在服务暴露的端口上接受请求，包括没有运行服务实例的节点，也可以在服务暴露的端口接受请求。Swarm 路由网格会把节点收到的请求自动转发到运行服务的容器实例中。这将有利于提高服务的可用性，例如当某个节点上运行服务的容器实例意外中止后，这个节点上收到的请求会被转发到正常的节点上进行处理。

启用 Swarm 路由网格功能需要使 Swarm 集群中的所有节点都开放以下端口：

TCP 或 UDP 协议的 7946 端口用于查找网络中的容器，称为容器网络发现，UDP 协议的 4789 端口用于容器入口网络。

在保证节点间可以互相访问上述端口的情况下，创建 Swarm 集群，这样才能开启 Swarm 的路由网格功能。这需要设置节点的防火墙或节点使用的网络安全组中的入站策略。

下面演示 Swarm 路由网格的功能，首先登录到 Swarm 集群中的 manager 节点。在 manager 节点创建一个名为 nginx 的服务，命令如下：

```
docker service create \
   --name nginx \
   --publish published=8080,target=80 \
   --replicas 2 \
  nginx
```

运行命令后输出的信息如下：

```
zljsb4htvze5vm7mgj5nb2zfc
overall progress: 2 out of 2 tasks
1/2: running   [==================================================>]
2/2: running   [==================================================>]
verify: Service converged
```

现在查看 nginx 服务的容器实例的节点分布情况，命令如下：

```
docker service ps nginx
```

运行命令后输出的信息如下(前 5 列):

```
ID            NAME     IMAGE          NODE                DESIRED STATE
jhp6u085n5rh  nginx.1  nginx:latest   ip-172-31-30-250    Running
9hdrifxm9xf3  nginx.2  nginx:latest   ip-172-31-35-212    Running
```

后 3 列信息如下:

```
CURRENT STATE              ERROR     PORTS
Running 4 minutes ago
Running 4 minutes ago
```

输出信息表示 nginx 服务现在有两个容器实例,分别运行在节点 ip-172-31-30-250 和节点 ip-172-31-35-212 上。

在节点 ip-172-31-30-250 上验证服务是否正常,命令如下:

```
curl localhost: 8080
```

运行命令后输出的信息如下:

```
<!DOCTYPE html>
<html>
<head>
<title>Welcome to Nginx!</title>
...
```

查看目前集群中的所有节点,命令如下:

```
docker node ls
```

运行命令后输出的信息如下(前 4 列):

```
ID                             HOSTNAME          STATUS    AVAILABILITY
r29f4jyze4zlaydw5gqvljb2w *    ip-172-31-30-250  Ready     Active
6zxgzwnpaurtlcd874qozn181      ip-172-31-32-49   Ready     Active
a4odmh76r7983qskgix939os9      ip-172-31-35-212  Ready     Active
```

表明 ip-172-31-32-49 节点上没有 nginx 服务的容器实例,登录到 ip-172-31-32-49 节点上查看 Docker 进程信息来验证一下,命令如下:

```
Ubuntu@ip-172-31-32-49:~$ docker ps
```

运行命令后输出的信息如下:

```
CONTAINER ID   IMAGE     COMMAND   CREATED   STATUS     PORTS     NAMES
```

输出信息中没有出现任何运行中的容器，而Swarm路由网格功能可以使集群中没有运行服务实例的节点也可以在服务暴露的端口接受请求。下面验证ip-172-31-32-49节点是否监听了nginx服务暴露的8080端口，命令如下：

```
sudo lsof -i:8080
```

这个命令会打印出所有监听8080端口的进程列表，输出的信息如下：

```
COMMAND PID USER   FD   TYPE DEVICE SIZE/OFF NODE NAME
dockerd 577 root   20u  IPv6  69103       0t0  TCP *:http-alt (LISTEN)
```

可以看到ip-172-31-32-49节点虽然没有运行nginx的服务容器实例，但是依然监听了nginx服务暴露的8080端口。

尝试访问ip-172-31-32-49节点上的8080端口，命令如下：

```
curl localhost:8080
```

运行命令后输出的信息如下：

```
<!DOCTYPE html>
<html>
<head>
<title>Welcome to Nginx!</title>
...
```

证明Swarm的路由网络功能工作正常。

还会对请求做负载均衡，为了验证这一点，在ip-172-31-32-49节点上多次访问nginx服务，构造一个特殊的地址访问，命令如下：

```
curl localhost:8080?node=ip-172-31-32-49
```

接着登录到ip-172-31-30-250，查看nginx服务对应的容器的日志，查看日志命令如下：

```
docker logs -f 1f00fef00d2e
```

其中1f00fef00d2e是运行nginx服务的容器ID。容器ID可通过docker ps命令查询。在输出的日志底部会看到类似的请求日志：

```
10.0.0.12 - - [10/Mar/2021:06:37:41 +0000] "GET /?node=ip-172-31-32-49 HTTP/1.1"
200 612 "-" "curl/7.68.0" "-"
```

在 ip-172-31-35-212 节点上进行同样的操作，同样会看到类似的请求日志。说明 Swarm 路由网络具有负载均衡的功能。

在节点 ip-172-31-30-250 上查看 nginx 服务日志还可以使用命令 docker service logs nginx。这是因为该节点是 manager 节点。

5.3.7 开发环境 Swarm 部署

1. 准备 Swarm 环境

与 Docker Compose 类似，Swarm 也支持在描述应用的清单文件中读取容器的所有配置，包括镜像、端口、存储和网络等，并且文件格式和 Compose 是兼容的。在 Swarm 中称这个文件为 stack 文件。

这一节演示把用户认证项目部署到本地开发环境中的 Swarm。本地运行单节点的 Swarm 集群。Docker Desktop 默认只支持单节点的 Swarm 集群。如果需要在本地运行多节点的 Swarm，则需要搭配使用 Docker Machine。

在使用 Swarm 之前，需要让 Docker Engine 以 Swarm 模式运行。需运行的命令如下：

```
docker swarm init
```

2. 定义 stack 文件

下面是 accounts 项目根目录下 docker-compose.yaml 文件中的内容：

```
version: "3.9"

services:
  web:
    image: ${ACCOUNT_FRONTEND_IMAGE}
    depends_on:
      - api
    ports:
      - "8088:80"

api:
    depends_on:
      - mysql
    image: ${ACCOUNT_IMAGE}
    ports:
      - "8081:80"
    volumes:
      - ./config:/config
command: go run src/server/main.go
    restart: always

  MySQL:
```

```
    image: mysql:5.7
    volumes:
      - mysql:/var/lib/mysql
    restart: always
    environment:
      MYSQL_ROOT_PASSWORD: ${MYSQL_PASSWORD}
      MYSQL_DATABASE: ${MYSQL_DB}
    command: --character-set-server=utf8mb4 --default-time-zone=+08:00

  db-migration:
    image: goose
    depends_on:
      - mysql
    working_dir: /migrations
    volumes:
      - ./src/database/migrations:/migrations
    command: goose up
    environment:
      GOOSE_DRIVER: mysql
      GOOSE_DBSTRING: root:${MYSQL_PASSWORD}@tcp(mysql:3306)/${MYSQL_DB}
volumes:
  mysql:
```

把 docker-compose.yaml 复制为 stack.yaml 文件，下面会用 stack.yaml 文件作为 Swarm stack 文件。

Swarm 不支持 restart，所以需要修改 stack.yaml 文件，把 restart:always 删除掉。

Swarm 中的服务在退出后会被自动重启，而 db-migration 服务在完成数据库迁移后会自动退出，不需要重启。修改 db-migration 服务的重启策略，代码如下：

```
db-migration:
    deploy:
      restart_policy:
        condition: none
```

其他保持不变，添加了 deploy 字段。用于指定与服务部署相关的配置。deploy 字段下的 restart_policy 用于指定服务的重启策略。restart_policy 下的 condition 用于指定触发重启的条件。默认值是 any，也就是任何情况下，只要应用中止运行都会被重启。这里将 condition 指定为 none，意思是任何情况下都不会自动重启。

3. 部署应用

现在就可以用 Swarm 来部署应用 accounts 了，在后端项目的根目录下运行的命令如下：

```
docker stack deploy --compose-file stack.yaml accounts
```

可以看到以下输出信息：

```
Ignoring unsupported options: restart

Creating network accounts_default
Creating service accounts_mysql
Creating service accounts_web
Creating service accounts_api
```

这个 deploy 子命令拥有别名 up，也就是说可以把上面命令中的 docker stack deploy 换成 docker stack up。命令中--compose-file 用于指定 Compose 文件的位置。最后一个参数 accounts 是为应用所包括的所有容器的集合起的名字。

如何理解 docker stack 命令名字的含义呢？英文 stack 的意思是一堆，引申含义是大量、许多。应用包含了多个服务，所以把多服务组成的整体称为 stack。

4. 查看服务状态

确认应用包含的容器已经在正常运行，命令如下：

```
docker stack services accounts
```

运行命令后输出的信息如下(前 4 列)：

```
ID              NAME                       MODE            REPLICAS
2kxwbk7xo6lk    accounts_api               replicated      1/1
fdyzh2ojm1h6    accounts_db-migration      replicated      0/1
s9lg0nf74b9r    accounts_mysql             replicated      1/1
237i5hbprlz7    accounts_web               replicated      1/1
```

后 2 列信息如下：

```
IMAGE                        PORTS
accounts:dev                 *:8081->80/tcp
goose:latest
mysql:5.7
accounts-frontend:latest     *:8088->80/tcp
```

输出信息的 REPLICAS 字段显示 api、mysql 和 web 服务都已经正常运行。

5. 关闭应用

关闭整个应用可以使用的命令如下：

```
docker stack rm accounts-stack
```

运行命令后输出的信息如下：

```
Removing service accounts - stack_accounts
Removing service accounts - stack_mysql
Removing service accounts - stack_web
Removing network accounts - stack_default
```

5.3.8 生产环境 Swarm 部署

集群环境下，如果服务可以在任意节点运行将有利于提高服务的可用性。当运行服务的某个节点失败后，可以快速地在其他节点重新启动服务。另外因为这样的服务对节点不挑剔，所以调度成功的概率会更高。

5.3.7 节中的 docker-compose.yaml 文件中的 api 服务声明需要挂载本地的配置文件./config，代码如下：

```
api:
    volumes:
      - ./config:/config
```

依赖本地文件就需要在服务启动前在所有节点准备好依赖的文件，这样显然比较烦琐。Docker Swarm 支持把配置信息存储在 Docker configs 中，这样就不需要把配置文件挂载到容器中了。

Docker configs 一般用来存储非机密的配置信息，如果需要存储密码、API 密钥及证书等机密信息，最好使用 Docker secrets。

需要注意的是 Docker configs 只对 Swarm 中的服务生效，对于使用 docker run 命令创建的独立容器和 Docker Compose 生成的容器则无法使用 configs。

当添加新的 configs 时，configs 中的信息会被安全地发送到 Swarm 中的 manager 节点，然后存储在 Raft 日志中。Raft 日志会被复制到集群中的所有 manager 节点中。这样保证了 configs 数据的高可用性。

服务使用 configs 后，相应的 configs 数据会在运行服务的容器中以文件的形式出现。对于 Linux 环境下的 Docker 容器，这个文件的默认位置是根目录/<配置名字>下。

1. 定义 stack 文件

保留用于开发环境的 docker-compose.yaml 文件，新建 stack.yaml 文件。把 docker-compose.yaml 文件的内容复制到 stack.yaml 文件中，下面开始修改新创建的文件。

在新创建的文件 stack.yaml 中加入 configs 定义，代码如下：

```
configs:
  plain_config:
    file: ./config/plain.yaml
```

接着对 api 服务相应的部分进行修改，代码如下：

```
api:
    configs:
      - source: plain_config
        target: /config/plain.yaml
        mode: 0440
```

configs的定义有两种语法，一种是长语法，另一种是短语法。这里使用的是长语法，长语法提供了更细粒度的参数。source:plain_config表示配置来源是plain_config，也就是引用全局定义的plain_config。target:/config/plain.yaml表示将配置文件挂载到容器中的/config/plain.yaml文件。mode:0440表示将容器中配置文件的读写权限设置为0440。

这样配置文件./config/plain.yaml中的内容就会通过configs的方式出现在api服务的容器中的/config/plain.yaml文件。

配置文件config/plain.yaml中包含了数据库密码，可以把密码等机密信息放到单独的文件，以便分开管理。这样当将应用部署到Swarm中时，可以利用Docker secret来存取数据库密码。

下面把配置文件config/plain.yaml中的数据库密码放到单独的文件config/secret.yaml中，新建文件secret.yaml，在文件中写入的内容如下：

```
database:
  password: 123
```

这时plain.yaml文件的内容如下：

```
server:
  port: 80
database:
  host: mysql
  port: 3306
  username: root
  schema: accounts
```

在stack.yaml中加入configs定义，代码如下：

```
secrets:
  secret_config:
    file: ./config/secret.yaml
```

接着在api服务中添加secrets定义，修改后代码如下：

```
api:
    secrets:
```

```
    - source: secret_config
      target: /config/secret.yaml
      mode: 0440
```

修改完成后的 stack.yaml 内容如下：

```
version: "3.9"

services:
  web:
    image: ${ACCOUNT_FRONTEND_IMAGE}
    depends_on:
      - api
    ports:
      - "8088:80"

  api:
    depends_on:
      - mysql
    image: ${ACCOUNT_IMAGE}
    ports:
      - "8081:80"
    configs:
      - source: plain_config
        target: /config/plain.yaml
        mode: 0440
    secrets:
      - source: secret_config
        target: /config/secret.yaml
        mode: 0440

  mysql:
    image: mysql:5.7
    volumes:
      - mysql:/var/lib/mysql
    environment:
      MYSQL_ROOT_PASSWORD: ${MYSQL_PASSWORD}
      MYSQL_DATABASE: ${MYSQL_DB}
    command: --character-set-server=utf8mb4 --default-time-zone=+08:00

  db-migration:
    image: goose
    depends_on:
      - mysql
    working_dir: /migrations
```

```
    volumes:
      - ./src/database/migrations:/migrations
    command: goose up
    environment:
      GOOSE_DRIVER: mysql
      GOOSE_DBSTRING: root:${MYSQL_PASSWORD}@tcp(mysql)/${MYSQL_DB}
    deploy:
      restart_policy:
        condition: none

volumes:
  mysql:

configs:
  plain_config:
    file: ./config/plain.yaml

secrets:
  secret_config:
    file: ./config/secret.yaml
```

接下来登录到 manager 节点进行操作。

2. 设置环境变量

登录到 manager 节点以后，把 accounts 项目复制到当前工作目录下，因为 stack.yaml 中使用了环境变量，所以需要把.env 文件复制到 accounts 项目根目录下。

.env 文件内容如下：

```
ACCOUNT_FRONTEND_IMAGE=bitmyth/accounts-frontend:v1.0.0
ACCOUNT_IMAGE=bitmyth/accounts:v1.0.0
MYSQL_PASSWORD=123
MYSQL_DB=accounts
```

.env 文件中出现的镜像都必须从镜像仓库下载。

当使用 stack.yaml 文件把应用部署到 Swarm 时，Swarm 并不会像 Compose 一样自动读取.env 文件中配置的环境变量，需要手动设置用到的环境变量。命令如下：

```
export $(cat .env)
```

这样就可以使.env 文件中定义的全部环境变量在当前的 shell 进程中生效。

3. 部署应用

完成环境变量设置后，将应用部署到生产环境的 Swarm 集群中，命令如下：

```
docker stack deploy --compose-file stack.yaml accounts
```

运行命令后输出的内容如下：

```
Ignoring unsupported options: restart

Creating network accounts_default
Creating secret accounts_secret_config
Creating config accounts_plain_config
Creating service accounts_api
Creating service accounts_mysql
Creating service accounts_db-migration
Creating service accounts_web
```

输出信息表示所有的服务都已经开始创建了，并且创建了一个 config，名字叫 accounts_plain_config。另外创建了一个 secret，名字叫 accounts_secret_config。

用 docker config 命令可查看 config 列表，命令如下：

```
docker config ls
```

运行命令后输出的信息如下：

```
ID                          NAME                    CREATED          UPDATED
pte1p0uyrb7021wioxhrqnt64   accounts_plain_config   21 seconds ago   21 seconds ago
```

继续查看 accounts_plain_config 的详细内容，命令如下：

```
docker config inspect accounts_plain_config
```

运行命令后输出的信息如下：

```
[
    {
"ID": "pte1p0uyrb7021wioxhrqnt64",
"Version": {
"Index": 39458
        },
"CreatedAt": "2021-03-11T01:31:52.595487186Z",
"UpdatedAt": "2021-03-11T01:31:52.595487186Z",
"Spec": {
"Name": "accounts_plain_config",
"Labels": {
"com.docker.stack.namespace": "accounts"
            },
"Data": "c2VydmVyOgogIHBvcnQ6IDgwCmRhdGFiYXNlOgogIGhvc3Q6IG15c3FsCiAgcG9ydDogMzMwNgogIHVz
ZXJuYW1lOiByb290CiAgcGFzc3dvcmQ6IDEyMwogIHNjaGVtYTogYWNjb3VudHMK"
        }
```

```
    }
]
```

输出信息中显示的Data字段就是配置的具体内容，这是base64编码后的结果。可以用命令base64来解码查看，命令如下：

```
base64 -d <(echo c2VydmVyOgogIHBvcnQ6IDgwCmRhdGFiYXNlOgogIGhvc3Q6IG15c3FsCiAgcG9ydDogMzMw
NgogIHVzZXJuYW1lOiByb290CiAgcGFzc3dvcmQ6IDEyMwogIHNjaGVtYTogYWNjb3VudHMK)
```

运行命令后输出的解码的内容如下：

```
server:
  port: 80
database:
  host: mysql
  port: 3306
  username: root
  password: 123
  schema: accounts
```

对比这个输出内容和./config/plain.yaml文件，发现它们是一致的。

下面查看secret列表以确认secret是否创建成功，命令如下：

```
docker secret ls
```

运行命令后输出的信息如下：

```
ID                          NAME                     DRIVER    CREATED         UPDATED
sicei6vjehoh8hj4b64ia2esi   accounts_secret_config             2 minutes ago   2 minutes ago
```

继续查看accounts_secret_config的详细内容，命令如下：

```
docker inspect accounts_secret_config
```

运行命令后输出的信息如下：

```
[
    {
"ID": "sicei6vjehoh8hj4b64ia2esi",
"Version": {
"Index": 74589
        },
"CreatedAt": "2021-03-11T07:02:22.774283061Z",
"UpdatedAt": "2021-03-11T07:02:22.774283061Z",
```

```
"Spec": {
"Name": "accounts_secret_config",
"Labels": {
"com.docker.stack.namespace": "accounts"
            }
        }
    }
]
```

查看 Swarm 中的服务列表，命令如下：

```
docker service ls
```

运行命令后输出的内容如下（前 4 列）：

```
ID              NAME                    MODE          REPLICAS
9iqolp4cb3jj    accounts_api            replicated    1/1
rwptwtkorrhl    accounts_db-migration   replicated    0/1
0v6re1vdchf4    accounts_mysql          replicated    1/1
sx7b95d9au5q    accounts_web            replicated    1/1
```

后两列内容如下：

```
IMAGE                               PORTS
bitmyth/accounts:v1.0.0             *:8081->80/tcp
goose:latest
mysql:5.7
bitmyth/accounts-frontend:v1.0.0    *:8088->80/tcp
```

4. 查看服务状态

accounts_api 服务启动后，可以请求它所提供的注册接口，以此来验证是否工作正常，命令如下：

```
curl -v  localhost:8081/v1/register -d'{"name":"test"}'
```

如果服务正常运行，则会收到类似响应，响应内容如下：

```
{"token":"eyJhbGciOiJIUzI1NiIsInR5cCI6IkpXVCJ9.eyJ1aWQiOjQsInVzZXIiOnRydWUsImV4cCI6MTYxNTU2MjA5NywiaWF0IjoxNjE1Mzg5Mjk3LCJpc3MiOiJCaWthc2gifQ.USttmL2zRAzlLzNb30dBu7txcWn5NLHzMh88d4-ybBc","user":{"ID":4,"Name":"test","Password":"","Email":"","Phone":"","Avatar":"","CreatedAt":"2021-03-10T15:14:57.870701764Z","UpdatedAt":"2021-03-10T15:14:57.870701764Z","DeletedAt":null}}
```

accounts_api 服务是无状态的，所以可以很方便地把 accounts_api 服务实例扩展到

3个，命令如下：

```
docker service scale accounts_api = 3
```

运行命令后输出的内容如下：

```
accounts_api scaled to 3
overall progress: 3 out of 3 tasks
1/3: running   [==================================================>]
2/3: running   [==================================================>]
3/3: running   [==================================================>]
verify: Service converged
```

以上内容表明全部启动成功，说明 Docker configs 可以让不同的节点共享。

查看 accounts_api 服务的容器实例分布在哪些节点，命令如下：

```
docker service ps accounts_api
```

运行命令后输出的内容如下(前4列)：

```
ID              NAME              IMAGE                       NODE
jd0p13e9s39r    accounts_api.1    bitmyth/accounts:v1.0.0    ip-172-31-30-250
7zy6mq1vfkrz    accounts_api.2    bitmyth/accounts:v1.0.0    ip-172-31-35-212
tdf2qiithyz4    accounts_api.3    bitmyth/accounts:v1.0.0    ip-172-31-32-49
```

后4列内容如下：

```
DESIRED STATE  CURRENT STATE             ERROR      PORTS
Running         Running 9 minutes ago
Running         Running 4 minutes ago
Running         Running 4 minutes ago
```

5. 关闭应用

关闭整个应用，命令如下：

```
docker stack rm accounts
```

运行命令后输出的信息如下：

```
Removing service accounts_api
Removing service accounts_db-migration
Removing service accounts_mysql
Removing service accounts_web
Removing secret accounts_secret_config
Removing config accounts_plain_config
Removing network accounts_default
```

5.3.9 约束服务调度

在生产环境中 Swarm 集群通常有多个节点。在调度下，服务会在多节点间转移。这样依赖本地数据卷的服务在调度到其他节点继续运行时，就会面临数据丢失的情况。例如 MySQL 服务，在前面定义的 stack.yaml 文件中，MySQL 使用了一个本地数据卷。MySQL 的数据会保存到这个数据卷中。如果 MySQL 被调度到了其他节点，则在其他节点上就无法找到之前保存的数据。

这个问题可以通过指定 MySQL 服务运行在固定的节点来解决。也就是在调度过程中限制 MySQL 服务，使它只能调度到满足要求的节点。这个功能在 Swarm 中称为放置约束(Placement Constraints)。顾名思义就是对把服务放到哪个节点运行进行约束。

放置约束条件通过对节点属性进行筛选实现。支持通过匹配节点的标签(node.labels)实现，也可以通过匹配节点的角色(node.role)、节点 ID(node.id)及节点的主机名(node.hostname)等实现。

如果使用标签实现放置约束，则需要定义标签并为节点设置标签，然后在服务定义中加上需要匹配的节点标签作为限制条件，这样服务就会调度到和标签匹配的节点上。

只有节点拥有的标签和服务放置约束中的指定的标签匹配。服务才会被调度成功。指定了放置限制的服务，如果在调度时没有节点拥有匹配的标签，则服务将无法调度，会一直处于待处理状态。如果节点无法满足服务的其他放置约束条件，则服务同样无法调度。

Swarm 集群中节点的标签格式为 key=value。可以为节点指定一个或多个标签。

下面使用标签作为 MySQL 服务的放置约束。假设 MySQL 服务可以运行的节点需要有标签 role=db，修改 stack.yaml 文件中 MySQL 服务的定义，代码如下：

```
mysql:
    image: mysql:5.7
    deploy:
      placement:
        constraints:
          - "node.labels.role == db"
```

image 字段下添加的 deploy 字段用于指定服务部署相关的配置。deploy 字段下的 placement 用于指定服务放置配置。placement 下的 constraints 用于指定放置的限制条件，这里指定了标签匹配条件。约束条件为 node.labels.role==db，表示部署到拥有 role=db 标签的节点。注意条件中的操作符是 2 个等号“==”。约束条件中用“==”表示匹配，用“!=”表示排除。

同样地，db-migration 服务因为挂载了本地的数据迁移文件，也需要在固定的节点执行，所以要为它指定放置约束。修改后的 db-migration 服务的定义如下：

```
  db-migration:
    image: goose
    depends_on:
      - mysql
    working_dir: /migrations
    volumes:
      - ./src/database/migrations:/migrations
    command: goose up
    environment:
      GOOSE_DRIVER: mysql
      GOOSE_DBSTRING: root:${MYSQL_PASSWORD}@tcp(mysql)/${MYSQL_DB}
    deploy:
      restart_policy:
        condition: none
      placement:
        constraints:
          - "node.labels.role == db"
```

假设希望 MySQL 运行的节点是 ip-172-31-30-250，首先为节点 ip-172-31-30-250 设置标签 role=db，命令如下：

```
docker node update --label-add role=db ip-172-31-30-250
```

验证标签设置是否生效，命令如下：

```
docker node inspect ip-172-31-30-250
```

在运行命令后输出的信息中会看到 Spec 字段，字段如下：

```
"Spec": {
"Labels": {
"role": "db"
            },
```

表示节点 ip-172-31-30-250 已经拥有了标签 role=db。

完成修改后，重新部署应用，命令如下：

```
docker stack deploy --compose-file stack.yaml accounts
```

运行命令后等待部署结束，然后查看 MySQL 服务是否被调度到节点 ip-172-31-30-250 运行，命令如下：

```
docker service ps accounts_mysql
```

运行命令后输出信息中的 NODE 列的内容如下：

```
NODE
ip-172-31-30-250
```

说明放置约束生效了。MySQL 服务运行在包含 role＝db 标签的节点 ip-172-31-30-250 上。

5.3.10　日志收集

1. 简介

日志对于观测应用运行情况，定位和修复故障及系统运行状态监控等起着重要作用。Swarm 中运行的服务如果有多个容器实例，则查看服务所产生的日志可以使用 docker service logs 命令。它会把服务中所有容器的日志集中起来显示，但是使用 docker service logs 命令查看日志的方式在搜索时不方便，并且需要登录到服务器上操作。

常用的解决方案是用专门的收集工具把容器日志转发到搜索引擎。下面介绍如何使用 EFK(Elasticsearch＋Fluentd＋Kibana)工具栈实现日志收集和处理。

2. Fluentd

Fluentd 是开源的日志数据收集工具，使用简单并且性能优异。支持对日志进行自定义格式处理和转发到自定义的日志处理服务。Fluentd 对 Docker 的支持很好，是 Docker 内置的日志驱动之一。Fluentd 是 CNCF(云原生计算基金会)已完成项目之一。

Fluentd 同时具有良好的扩展性和灵活性，支持使用自定义插件来满足个性化需求。

EFK 技术栈中 Fluentd 需要把收集的日志转发到 Elasticsearch，所以要为 Fluentd 安装 Elasticsearch 插件，而 Fluentd 官方的镜像没有安装 Elasticsearch 插件，需要基于官方镜像构建自定义的镜像。下面开始构建 Fluentd 镜像，所需的 Dockerfile 如下：

```
FROM fluent/fluentd:v1.12.0-debian-1.0

USER root

RUN ["gem", "install", "fluent-plugin-elasticsearch", "--no-document", "--version", "4.3.3"]

USER fluent
```

在这个 Dockerfile 同级目录下运行命令，以便构建镜像，命令如下：

```
docker build -t bitmyth/fluentd-es:v1.0.0 .
```

这样就拥有了安装了 Elasticsearch 插件的 Fluentd 镜像。接着把这个镜像上传到 Docker Hub。这样集群中的节点就可以下载这个镜像了，命令如下：

```
docker push bitmyth/fluentd-es:v1.0.0 .
```

使用 Fluentd 收集 Swarm 集群中的所有容器的日志，需要将 Fluentd 设置为 Swarm 全局服务。修改 stack.yaml 文件，增加 Fluentd 服务的定义，代码如下：

```
fluentd:
    image: bitmyth/fluentd-es:v1.0.0
    environment:
      FLUENTD_CONF: 'fluent.conf'
    configs:
      - source: fluentd_config
        target: /fluentd/etc/fluent.conf
    ports:
      - "24224:24224"
      - "24224:24224/udp"
    deploy:
      mode: global
```

下面介绍指令的含义。

environment 字段中定义了 FLUENTD_CONF 环境变量，把它的值设置为 fluent.conf。这个环境变量用于指定 Fluentd 读取配置的文件名，代码如下：

```
environment:
  FLUENTD_CONF: 'fluent.conf'
```

configs 字段中定义了 fluentd_conf，target：/fluentd/etc/fluent.conf 表示把配置 fluentd_config 挂载到容器中的/fluentd/etc/fluent.conf 文件。Fluentd 默认的配置文件夹是/fluentd/etc/，这样 Fluentd 就可以读取到自定义的 fluent.conf 文件中，代码如下：

```
configs:
  - source: fluentd_config
    target: /fluentd/etc/fluent.conf
```

公开 Fluentd 监听的 TCP 端口 24224 和 UDP 端口 24224，代码如下：

```
ports:
  - "24224:24224"
  - "24224:24224/udp"
```

服务将以 global 模式部署。也就是在所有节点都会启动一个 Fluentd 容器实例。这样 Fluentd 就可以收集到全部节点上运行的容器所产生的日志。

```
deploy:
  mode: global
```

上述 Fluentd 定义中用到的 fluentd_config 定义如下：

```
configs:
  fluentd_config:
    file: ./config/fluentd/fluent.conf
```

表明 fluentd_config 配置项的内容将从./config/fluentd/fluent.conf 文件中读取。Fluentd 默认将文件的编码配置为 UTF-8 或 ASCII。

./config/fluentd/fluent.conf 文件内容如下：

```
<source>
  @type forward
  port 24224
  bind 0.0.0.0
</source>

<match *.**>
  @type copy

<store>
    @type elasticsearch
    host elasticsearch
    port 9200
    logstash_format true
    logstash_prefix fluentd
    logstash_dateformat %Y%m%d
    include_tag_key true
    type_name access_log
    tag_key @log_name
    flush_interval 1s
</store>

<store>
    @type stdout
</store>
</match>
```

下面介绍配置文件中的指令含义。

定义日志的输入源，也就是数据的来源。Fluentd 配置文件中支持定义多个 source。每个 source 都必须指定@type 参数来表明使用什么输入插件。@type forward 表示处理输入的插件类型是转发。除了 forward，Fluentd 还支持 http 插件。http 插件会提供 HTTP 服务，接收以 HTTP 发送的数据，而 forward 提供 TCP 服务，接收以 TCP 数据包的形式发送的数据。当然这两个输入插件可以同时启用。如果内置的输入插件无法满足需求，Fluentd 支持开发自定义输入插件。

指令 port 24224 表示 Fluentd 通过 TCP 端口 24224 接收日志数据，bind 0.0.0.0 表示监听任意的网卡设备。

Fluentd 中接收的数据会被包装成一个事件。事件包含了 tag 属性、time 属性和 record 属性。tag 属性用于识别日志数据的来源。它是用英文句点“.”分隔的字符串。Fluentd 推荐在 tag 中使用的字符是小写的字母、数字及下画线，用正则表达式表示为^[a-z0-9_]+$。事件的 time 属性是 UNIX 时间格式，由输入插件指定它的值。事件的 record 属性是一个 JSON 对象，包含了实际输入的数据，代码如下：

```
<source>
  @type forward
  port 24224
  bind 0.0.0.0
</source>
```

指令 match 用于控制 Fluentd 如何匹配和输出数据。match *.** 表示匹配拥有 *.** 格式的 tag(标签)的事件。每个 match 指令都必须指定一个匹配模式和一个@type 参数。匹配模式用于匹配事件的标签。match 中的@type 用于指定使用什么输出插件。

这里的@type copy 指定了 copy 插件。它是 Fluentd 内置的一个标准输出插件。copy 会把事件复制到多个输出。

指令 match 下包含的 store 指定了存储输出的目的地。match 指令中至少需要指定一个 store，代码如下：

```
<match *.**>
  @type copy
```

store 指令下的@type elasticsearch 用于指定使用 elasticsearch 插件。host 用于指定 elasticsearch 的 IP 地址，因为 Swarm 中的服务可以通过 DNS 来解析它的 IP 地址，所以这里使用 elasticsearch 服务名作为 host 字段的值。port 9200 指定 elasticsearch 服务的 TCP 端口号为 9200。如果需要发送到多个 elasticsearch 服务，则可以通过 hosts 指定，例如：hosts:host1:port1,host2:port2。通过示例中这种格式指定了多个 elasticsearch 服务的情况下，指定单个 elasticsearch 服务的 host 和 port 字段会被忽略，代码如下：

```
<store>
    @type elasticsearch
    host elasticsearch
    port 9200
    logstash_format true
    logstash_prefix fluentd
    logstash_dateformat %Y%m%d
    include_tag_key true
    type_name access_log
```

```
    tag_key @log_name
    flush_interval 1s
</store>
```

采用与 Logstash 兼容的格式把数据写入索引，代码如下：

```
logstash_format true
```

将索引的前缀设置为 fluentd，代码如下：

```
logstash_prefix fluentd
```

将索引名字中的日期格式设置为%Y%m%d，例如 20210325，代码如下：

```
logstash_dateformat %Y%m%d
```

将 Fluentd 的 tag 加入 Elasticsearch 数据中，代码如下：

```
include_tag_key true
```

指定写入 Elasticsearch 中的文档类型的名字。type_name 参数在 Elasticsearch 7 中会设置为固定值_doc，而 Elasticsearch 8 会忽略这个参数，代码如下：

```
type_name access_log
```

指定写入 Elasticsearch 中的 Fluentd tag 的键名为@log_name，代码如下：

```
tag_key @log_name
```

指定将缓存数据写入目的地的时间间隔为 1s，代码如下：

```
flush_interval 1s
```

3. Elasticsearch

Elasticsearch 是非常流行的开源搜索引擎，在 EFK 技术栈中的角色是保存和索引 Fluentd 收集的日志数据。下面定义 Elasticsearch 服务，修改 stack.yaml 文件，代码如下：

```
elasticsearch:
    image: docker.elastic.co/elasticsearch/elasticsearch:7.10.2
    environment:
      - "discovery.type=single-node"
    ports:
      - "9200:9200"
    deploy:
```

```
      resources:
        limits:
          cpus: '2'
          memory: 2G
        reservations:
          cpus: '1'
          memory: 1.5G
```

由于 Elasticsearch 服务需要使用比较多的内存，所以在上述定义中增加了对 CPU 和内存资源使用限制的声明。

限制 Elasticsearch 最多使用 2 个 CPU 和 2GB 的内存。在不限制容器使用内存的情况下，如果容器使用了过多的系统内存，则会被内核强制中止运行，代码如下：

```
limits:
  cpus: '2'
  memory: 2G
```

为 Elasticsearch 服务预留 1 个 CPU 和 1.5GB 内存。如果没有节点可以满足指定的计算资源需求，则服务将无法调度成功，代码如下：

```
reservations:
  cpus: '1'
  memory: 1.5G
```

4. Kibana

Kibana 为使用 Elasticsearch 提供了友好的 UI 界面，降低了使用 Elasticsearch 的难度。方便用户查询 Elasticsearch 中的数据。

下面定义 Kibana 服务，修改 stack.yaml 文件，代码如下：

```
kibana:
  image: kibana:7.10.1
  depends_on:
    - "elasticsearch"
  ports:
    - "5601:5601"
```

depends_on 声明的 Kibana 服务依赖于 Elasticsearch，当 Elasticsearch 启动后才会启动 Kibana。

5. 使用 Fluentd 日志驱动

现在修改第 1 章介绍的用户认证应用中包含的服务所用的日志驱动。首先修改后端项目，也就是 api 服务，设置它的日志驱动，代码如下：

```
api:
    logging:
driver: "fluentd"
      options:
        fluentd-address: localhost:24224
        tag: api.{{.Name}}
```

其他部分保持不变，增加了 logging 设置。将日志驱动设置为 fluentd，日志驱动选项通过 options 定义。options 下的 fluentd-address 定义了 fluentd 监听的网络地址。由于 Fluentd 是以 global 模式部署的，所以每个节点都可以通过 localhost 这个地址访问它。options 下的 tag 定义了日志的名字。{{.Name}}会被替换成容器的名字。最后 api.{{.Name}}会变成类似 api.accounts_api.1.gvq4cxpd99fdt7grxkr38zle4 这样的字符串。

同样地，可为前端项目（Web 服务）设置日志驱动，修改后的代码如下：

```
web:
    logging:
      driver: "fluentd"
      options:
        fluentd-address: localhost:24224
        tag: web.{{.Name}}
```

Web 服务定义的其他部分保持不变，增加了 logging 设置。

6. 部署

增加与日志相关的设置后，重新将应用部署到 Swarm 集群。首先设置环境变量，命令如下：

```
export $(cat .env.prod)
```

这样就可以使在.env 文件中定义的全部环境变量在当前的 shell 进程中生效。完成环境变量设置后，再将应用部署到生产环境的 Swarm 集群中，命令如下：

```
docker stack deploy --compose-file stack.yaml accounts
```

运行命令后输出的内容如下：

```
Creating network accounts_default
Creating secret accounts_secret_config
Creating config accounts_fluentd_config
Creating config accounts_plain_config
Creating service accounts_fluentd
Creating service accounts_elasticsearch
Creating service accounts_kibana
Creating service accounts_web
```

```
Creating service accounts_api
Creating service accounts_mysql
Creating service accounts_db-migration
```

查看 Fluentd 服务的状态，命令如下：

```
docker service ps accounts_fluentd
```

运行命令后输出信息的后 5 列如下：

```
NODE                DESIRED STATE   CURRENT STATE          ERROR     PORTS
ip-172-31-35-212    Running         Running 3 minutes ago
ip-172-31-32-49     Running         Running 3 minutes ago
ip-172-31-30-250    Running         Running 3 minutes ago
```

可以看到 Swarm 集群中所有的节点运行着 Fluentd 服务。

接下来打开浏览器访问 Kibana，设置要查看的索引的模式。默认首页如图 5-1 所示。

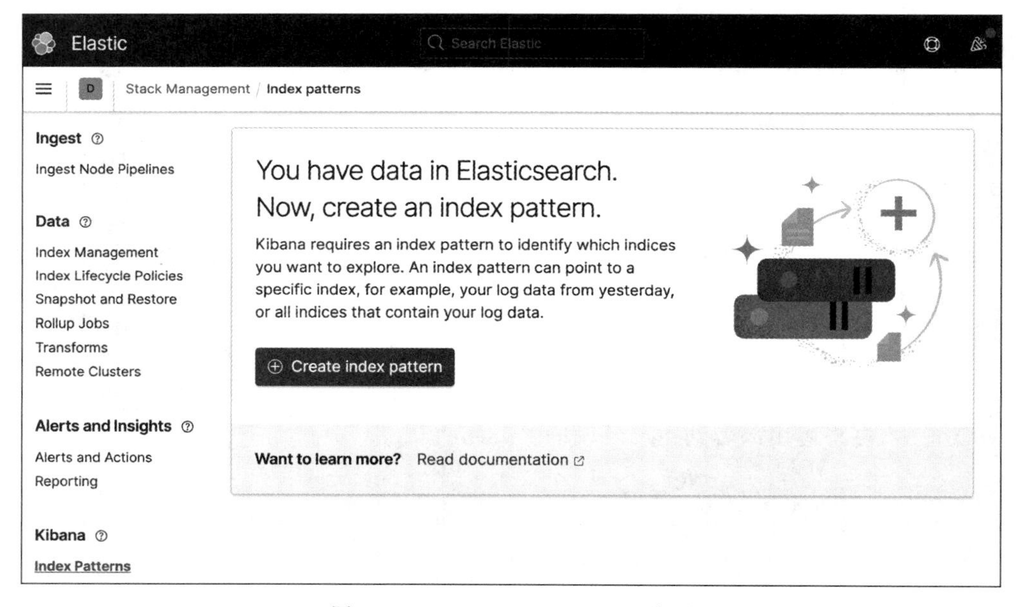

图 5-1　Kibana Index patterns 首页

单击页面中的 Create index pattern 按钮，会跳转到如图 5-2 所示页面。

在页面中 Index pattern name 输入框中输入 fluentd-*，如图 5-3 所示。

然后单击 Next step 按钮。在新页面中选择 Time field 下拉列表框中的@timestamp，之后单击 Create index pattern 按钮，如图 5-4 所示。

Create index pattern

An index pattern can match a single source, for example, `filebeat-4-3-22` , or **multiple** data sources, `filebeat-*` .

Read documentation

Step 1 of 2: Define an index pattern

Index pattern name

index-name-*

Next step

Use an asterisk (*) to match multiple indices. Spaces and the characters \, /, ?, ", <, >, | are not allowed.

Include system and hidden indices

Your index pattern can match your 1 source.

fluentd-20210325 Index

Rows per page: 10

图 5-2 创建 Kibana index patterns 页

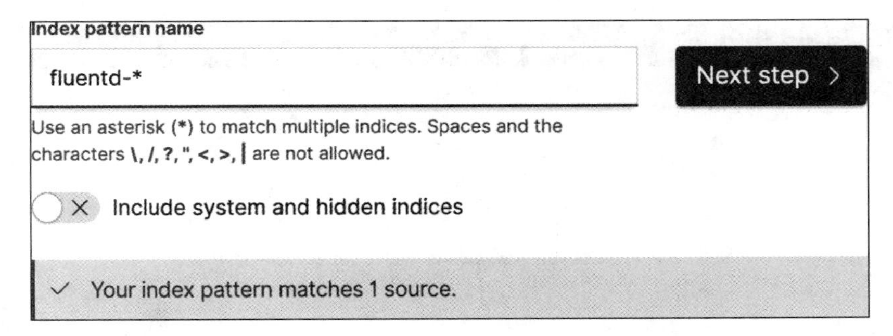

图 5-3 输入 Index pattern name

Step 2 of 2: Configure settings

Specify settings for your **fluentd-*** index pattern.

Select a primary time field for use with the global time filter.

Time field Refresh

@timestamp

Show advanced settings

Back Create index pattern

图 5-4 选择 Time field

接下来单击页面左上角菜单，在展开的菜单中选中 Discover，如图 5-5 所示。

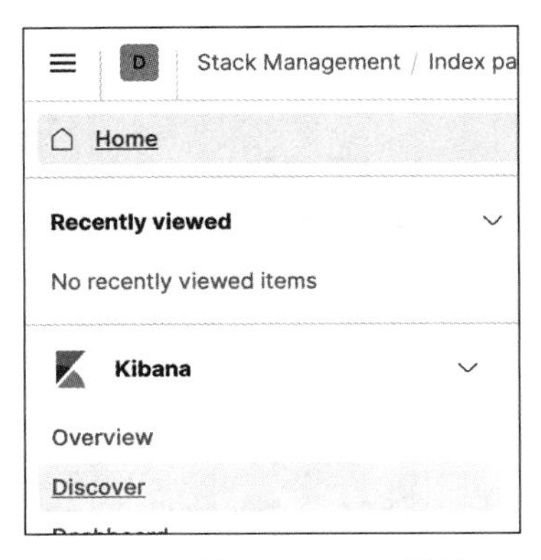

图 5-5　单击 Discover 链接

现在就可以在页面中观察日志了。访问前端项目，完成注册和登录，这样可以产生一些日志数据，然后回到 Kibana 中查看这些日志数据。Kibana 会展示最新的日志，如图 5-6 所示。

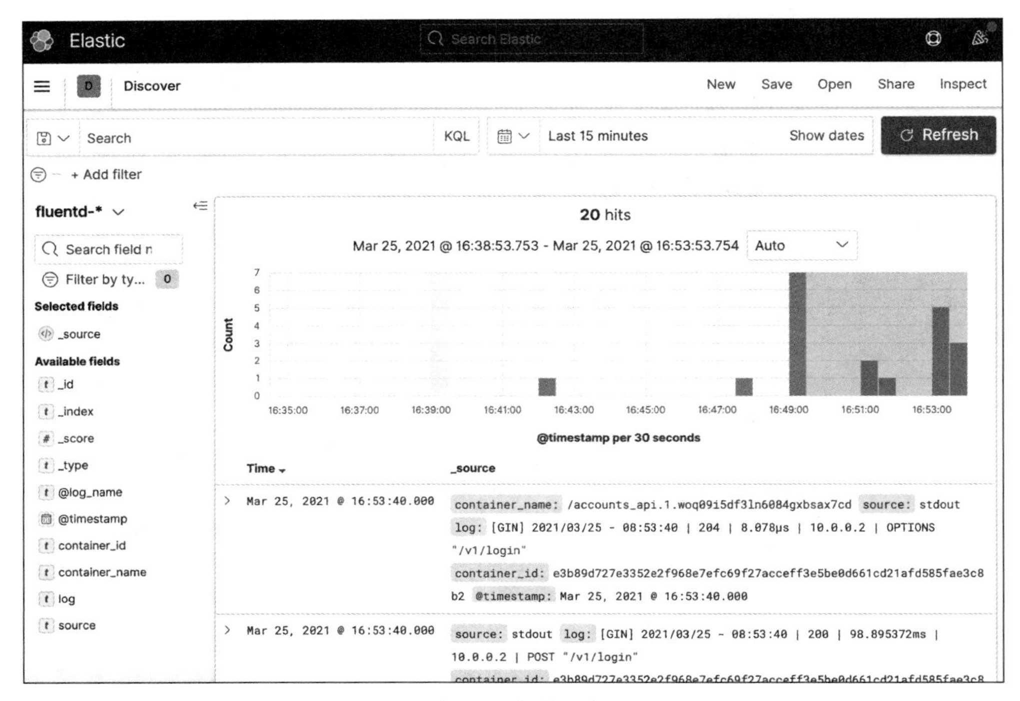

图 5-6　查看日志

在日志中可以看到前端项目和后端项目的日志。以前端项目产生的日志为例，如图 5-7 所示。

图 5-7　前端日志示例

日志记录的字段如表 5-1 所示。

表 5-1　前端日志字段表

字　段　名	字　段　值
@log_name	web. accounts_web. 1. gvq4cxpd99fdt7grxkr38zle4
@timestamp	Mar 25，2021 @ 16：51：04. 000
_id	RaCUaHgBvNUrEt1ffJ79
_index	fluentd-20210325
_score	-
_type	_doc
container_id	747e096cabfd7f94e05c734776d3eac03396e865c0cc9d07c5d1df24729f8bd7
container_name	/accounts_web. 1. gvq4cxpd99fdt7grxkr38zle4
log	10. 0. 0. 2 - - [25/Mar/2021:08:51:04 +0000] "GET / HTTP/1. 1" 200 904 "-""-""-"
source	stdout

可以从 container_name 字段判断出日志产生自哪个服务，通过 container_id 可以准确定位产生日志的容器。

7. 总结

基于 EFK 技术栈实现的日志收集系统可以实时地收集日志，并且可以方便地检索日志，以及定位故障，从而实现集群分布式环境下容器日志的统一处理。

第6章 云原生软件生产流程

软件生产由传统方式到云原生方式的转变体现在软件生产的整个流程，包括开发、测试、交付、升级和维护等。云原生的生产方式可以实现软件的快速开发、稳定交付。在大规模部署和升级的场景下，云原生可以节省许多时间和人力成本。

在设计和开发阶段就为软件在云上运行而做考虑，从而生产出可以高效利用云计算资源的软件称为云原生的软件。

6.1 云原生简介

云原生理念的诞生和云计算的发展密切相关。云计算让IT基础实施实现了随用随取。云计算的成熟是云原生发展的基石，使云原生的许多概念得以落地。例如快速部署、快速扩容缩容、快速故障排除等。

云原生应用一般有以下特点：

在应用的设计和开发阶段就为部署到云上做适配。例如尽量把应用中的服务设计成无状态的形式，无状态的特性使编排工具可以很方便地把服务从一个失败的节点调度到其他正常的节点执行，同时也方便扩容和缩容。

应用由多个松耦合的小模块构成而不是一个庞大的单体项目。这种模式称为微服务。小的模块更容易进行容器化，因为小的模块依赖较少，复杂度低。这样构建的容器镜像体积也会较小。微服务的拆分粒度需要视实际情况而定，拆分过细会导致微服务数量过多，从而带来更多的管理难度。微服务容器可以进行独立开发、部署和伸缩，具有极高的灵活性。不同的开发团队可以负责不同的微服务模块，这样可以更专注地进行开发。微服务可以独立地部署，这提高了部署的频率，缩短了每个功能点的发布周期，从而提高了软件生产的敏捷性，同时由于每个微服务只专注于应用的某个模块，即使出现故障，影响范围也是可控的。这样就降低了发布的风险。

通过容器来交付和发布应用，应用代码中会加入容器化需要的文件，例如镜像定义文件，容器编排工具需要的文件等。代码的变更和环境的变更都会及时反映到构建产生的容器镜像中，使运行环境和代码实时同步到镜像中。容器化应用可以通过容器编排工具实现自动部署、自动伸缩、自动故障修复等，这将极大减少运维的工作。让团队有更多精力投入核心业务上而不是运维上。

容器化使应用可以不再依赖于具体的底层平台，在需要迁移的时候可以很方便地整体迁移到公有云、私有云或者混合云。

云原生应用的基础设施是由云计算提供的，提供云计算服务的各大云厂商一般有不同的服务规格和接入方式，所以云原生应用在具体落地过程中需要根据云厂商的产品说明文档进行适配和调整。虽然接入方式有差异，但是不同云厂商提供的功能大体一致。目前头部的云厂商如 AWS(Amazon Web Service)、Microsoft Azure 和阿里云都提供了容器服务和 Kubernetes 集群等产品。

云原生既是一个技术理念，又是一个技术生态，目前已经产生了许多与云原生相关的开源项目。这离不开云原生领域的一个著名的基金会 Cloud Native Computing Foundation (CNCF)，中文名字为云原生计算基金会。CNCF 致力于培育和维护一个厂商中立的开源生态系统。它是非营利性质的 Linux 基金会的一部分。CNCF 托管和孵化了许多重要的与云原生技术相关的开源项目，包括 Kubernetes、Prometheus、Envoy 等重要的项目。目前已经有超过 10 万开发者在为 CNCF 项目贡献代码。CNCF 还运营着世界上最大的开源软件开发者会议。

6.2　云计算的能力

云计算弱化了传统 IT 硬件概念，革命性地降低了企业在 IT 基础设施上的建设成本。在 21 世纪初期，自建机房还是企业常规的解决方案，但是现在越来越多的企业不再选择自建机房，而是直接使用更便捷更稳定的云上资源。包括对数据安全有极高要求的政企项目和银行项目也开始在云上部署。这就像 19 世纪诞生了稳定的电力系统之后，工厂不再需要自己购置发电机来发电一样。

机房的建设不仅耗时耗力而且价格不菲，合格的机房需要满足各种严苛条件下的高可用性要求。机房内需要常年保持稳定的温度和湿度，具备良好的通风、防尘条件及稳定的供电系统。机房的运维需要专业的团队和成熟的经验，这些都是巨大的成本。

云厂商拥有更成熟的机房建设和运维经验，而且随着用户规模的增长，规模效益愈加明显。这就形成了良性循环，用户规模越大，云计算的单位成本就越低，形成了共赢的局面。得益于云厂商的极高规格的数据中心和强大的异地多活容灾策略，云厂商可向用户提供极高的产品可用性保证。

自建机房无法满足对计算资源、网络和存储等的弹性需求。当急需增加计算资源时，从购置新服务器到安装环境再到交付使用往往有很长流程，无法及时响应需求。当对计算资源的需求减小时，部分计算资源又会进入闲置状态，产生了大量浪费。

云计算实现了弹性获取计算资源，用户可以按需使用，按时计费，减少了大量资源浪费。

用户只需要在网页按所需的计算资源进行配置，单击几次鼠标就能轻松获取所需的计算资源。在使用完毕后,可以随时释放不再需要的计算资源,避免资源闲置和浪费。这一方面节省了时间成本和IT成本,另一方面云计算对IT基础设施的快速扩展能力提高了用户系统响应的及时性。

除了提供计算资源、网络、存储等基础设施,云厂商还提供了许多托管的软件服务,例如常用的数据库像MySQL、Redis,还有常用的负载均衡、日志收集等。

6.3 云原生的优势

和传统的软件生产方式相比,云原生的优势主要在于它提高了应用发布和运维的效率,显著降低了运维的复杂性。提升了应用的稳定性和用户体验。有利于各组织在公有云、私有云和混合云等新型动态环境中,构建和运行可弹性扩展的应用。

使用传统方式发布应用前需要手动配置服务器环境,然后把代码上传到服务器,之后通过将服务器加入负载均衡器的出口列表等方式把流量接入服务器,而当使用云原生的方式时,应用运行环境和待发布的代码集成在容器中,不需要在服务器上手动配置运行环境。这一方面提高了部署效率,另一方面因为容器可以提供和开发环境完全一致的环境,避免了手动配置环境可能产生的失误所带来的安全风险,也提高了部署的安全性。

云原生的容器部署方式还带来了显著的效率提升。容器从下载到运行起来通常只需数分钟时间,使软件发布不再是一件耗时又费力的事情,而使用传统方式从手动配置环境到手动运行应用通常需要数倍的时间。随着软件公司逐渐提高应用的发布频率,在部署环节使用的时间会占据越来越大的比重。这样使用云原生方式就会节省大量的时间和成本,从而带来更大的竞争优势。每个创新都能更快地落地,更快地触达用户。

对于发布后需要紧急回滚的情况,容器部署的方式可以在数秒到数分钟内完成。只需停止当前运行中的容器,重新启动上个版本的镜像,而传统方式在这种情况下很可能因为时间紧急而无法精心重建上一版本软件的环境,只能优先把代码回滚到上一版本,当上一版本和当前版本的软件所需的运行环境有差异时,例如新版本所需的环境中对某个系统库进行了升级,而回滚后的代码并不支持升级后的某个系统库,这将导致回滚后的代码无法正常工作。虽然大多情况下环境的变更不会像软件变更一样频繁,但是在这种需要回滚的情况下,无法快速重建环境毫无疑问是个潜在的巨大风险,而且回滚时间越长,造成的损失一般也会更大,影响的用户范围也会越广。容器方式可以将回滚时间降到最短以降低损失。

在运维方面,传统方式高度依赖于人工来解决故障。当硬盘写满后,需要人工配置新的硬盘并挂载到服务器。或者将服务器上的应用整体迁移到更高配置的新服务器上。如果使用了负载均衡技术,则还需要在恢复之前把产生故障的服务器节点从负载均衡中剔除,这都需要大量的时间。使用云原生的方式可以借助编排工具自动完成失败任务的调度,并且把进入失败节点的流量自动定向到其他正常节点。

云原生应用可以高效利用云计算资源。相比传统方式开发的应用具备更好的弹性。在用户请求流量剧烈波动的情况下,传统方式开发的应用无法敏捷地实施扩容和缩容来应对系统压力的波动。在流量波峰到来时，极易造成系统因耗尽计算资源而崩溃，而如果提前冗余过多的计算资源，又会造成资源闲置和浪费。云原生应用在面对同样的情况时，可以通过编排工具快速完成扩容和缩容，既能满足对流量波动的及时响应，又避免了资源的闲置。

云原生显著改变了软件的开发方式，增加了开发的复杂性，但是这一切都是值得的。

6.4　云原生的劣势

云原生方式采用的容器化部署相比传统方式的裸机部署会有性能损耗，这是使用Docker引入的资源虚拟化成本。

云原生应用采用的微服务的架构会增加服务间通信的成本。传统方式下单体应用的服务之间通过函数调用通信，几乎没有延迟，而分布式的微服务架构下的服务间通过网络通信势必带来较高的延迟，并且由于网络的波动还增加了通信的失败概率。

在传统方式下开发应用的流程中没有使用容器，也就省去了使用容器带来的开发和管理成本，而云原生应用的开发需要精心设计运行应用的容器。还需要对镜像进行管理和存储，这一定程度上增加了开发的复杂性。此外，对开发者也提出了更高的要求。

第7章 云原生基础设施

7.1 Kubernetes 是什么

Kubernetes 是开源的容器编排平台,支持集群环境下部署和管理容器化应用。目前已经是容器编排领域的事实标准,成为云原生时代的操作系统。Kubernetes 一词来自希腊语,是舵手的意思,Kubernetes 就像是舵手驾驶着一艘载满容器的船。Kubernetes 的图标是船舵的样子,也呼应了这一点。

Kubernetes 也叫 K8s,Kubernetes 的首字母是 K,尾字母是 s 中间有 8 个字母,所以缩写为 K8s。

Kubernetes 是 CNCF(云原生计算基金会)的第 1 个完成的项目,目前已拥有庞大的生态和蓬勃发展的社区。在软件领域的企业中普及率快速提高。越来越多的公司开始在生产环境中使用 Kubernetes,运行成千上万的容器。诸多的云厂商也提供了云上的 Kubernetes 环境,支持一键创建 Kubernetes 集群,方便用户快速使用 Kubernetes 来部署和管理他们的应用。

Kubernetes 由谷歌公司于 2013 年开源。它的设计思想来源于谷歌内部运行多年的 Borg 系统,吸收了谷歌多年管理大规模生产集群的经验,也包含了开源社区不断贡献的智慧。

Kubernetes 可以使复杂系统变得可控和稳定。可以显著提升系统的灵活性和可用性,从而提高运维管理的效率。利用 Kubernetes 高效而强大的编排能力,可以实现使用一个简单的命令就可以把应用部署到数量庞大的容器中,当计算资源不足的时候,可以快速地扩容,在资源空闲的时候,又可以快速地进行缩容,节省了使用计算资源的成本。

和 Docker Swarm 相比,Kubernetes 提供了更强大的功能。

Kubernetes 是一个相对复杂的系统,由许多组件构成。对它们的整体了解有助于理解 Kubernetes 的设计理念,为后续深入学习打下基础。

Kubernetes 集群中包含两种节点,一种是 Control Plane 节点,也叫 master 节点,运行 Control Plane(控制面),另一种是 worker 节点。

Kubernetes 中的容器运行在 Pod 中,Pod 运行在 Node 中,默认情况下 Control Plane

不会运行用户提交的 Pod，所以每个集群至少有 1 个 worker 节点，否则 Pod 将无法调度和运行。

通常情况下集群中的 Control Plane 会在多台服务器上运行，集群中也会有多个 Node 来运行 Pod。这样会提高集群的容错性和保障高可用性。

Control Plane 是 Kubernetes 中的容器编排层，通过它提供的 API，可以定义和部署容器及管理容器的整个生命周期。Control Plane 控制着整个集群。

Control Plane 中的组件包括 API server、etcd、Scheduler、Controller manager、Cloud controller manager。这些组件可以运行在集群中的任意一台服务器上，但是为简单起见，初始脚本会在一台服务器上启动它们，并且不会把用户提交的容器调度到这台服务器上运行。下面分别进行介绍。

API server 提供 Kubernetes API 访问服务。当使用客户端工具 kubectl 和集群交互时，API server 作为服务器端负责响应。它的具体实现是 kube-apiserver。

etcd 是一致、高可用的键值存储，用来保存 Kubernetes 集群的数据。

Scheduler 负责监控新创建的 Pod，如果 Pod 没有指定 Node，Scheduler 则会把它调度到合适的 Node 上运行。Scheduler 的具体实现通过 kube-scheduler 实现。

Controller manager 负责管理一系列控制器，包括 Node controller、Job controller、Endpoints controller、Service Account & Token controllers。这些控制器会一直监视集群状态并根据情况做出对应操作，以维持集群处于期望状态。理论上，每个控制器都在单独的进程中运行，但是为了降低复杂性，它们被编译到一个二进制文件中，并且运行在同一个进程中。Controller manager 的具体实现通过 kube-controller-manager 实现。

Cloud controller manager 用于和云厂商提供的 API 交互。Cloud controller manager 会针对具体的云厂商运行特定的控制器。如果没有用到云厂商，例如在本地搭建的 Kubernetes 集群，则集群中不会存在 Cloud controller manager。

7.2　客户端工具 kubectl

7.2.1　kubectl 简介

kubectl 是 Kubernetes 的命令行工具。用户可以通过 kubectl 和 Kubernetes 集群交互、完成部署和管理应用等任务。kubectl 和 Kubernetes 之间的关系类似于 Docker CLI 和 Docker Engine 的关系。

为了使 kubectl 正常工作，kubectl 的版本和 Kubernetes 的版本只允许相差一个 minor version（小版本号）。例如 v1.2 版本的 kubectl 和 v1.1、v1.2、v1.3 版本的 Kubernetes 搭配会正常工作。

7.2.2　使用 HomeBrew 安装 kubectl

如何使用 macOS 操作系统，并且安装了 HomeBrew 包管理器，则可以使用 brew 命令

安装 kubectl,命令如下:

```
brew install kubectl
```

安装完成后验证是否安装成功,命令如下:

```
kubectl version --client
```

如果 kubectl 安装成功,则运行命令后输出的类似信息如下:

```
Client Version: version.Info{Major:"1", Minor:"19", GitVersion:"v1.19.7", GitCommit:
"1dd5338295409edcfff11505e7bb246f0d325d15", GitTreeState:"clean", BuildDate:"2021-01-
13T13:23:52Z", GoVersion:"go1.15.5", Compiler:"gc", Platform:"darwin/amd64"}
```

7.2.3 使用 apt 安装 kubectl

在 Linux 环境下,可以使用包管理器安装 kubectl。在基于 Debian 的操作系统环境下可以使用 apt 安装 kubectl。下面介绍安装步骤。

首先更新 apt,并且安装 Kubernetes apt 库会用到的包。命令如下:

```
sudo apt-get update
sudo apt-get install -y apt-transport-https ca-certificates curl
```

接着添加 Kubernetes 官方的 GPG key,命令如下:

```
sudo curl -fsSLo /usr/share/keyrings/kubernetes-archive-keyring.gpg \
https://packages.cloud.google.com/apt/doc/apt-key.gpg
```

然后添加 Kubernetes 的 apt 库。

```
echo "deb [signed-by=/usr/share/keyrings/kubernetes-archive-keyring.gpg] \
https://apt.kubernetes.io/ kubernetes-xenial main" \
|sudo tee /etc/apt/sources.list.d/kubernetes.list
```

现在再次更新 apt,并且安装 kubectl,命令如下:

```
sudo apt-get update
sudo apt-get install -y kubectl
```

安装完成后验证是否安装成功,命令如下:

```
kubectl version --client
```

7.2.4 使用 curl 安装 kubectl

使用 curl 命令下载 kubectl 二进制文件，需要根据具体平台选择下载链接，下面下载的是适用于 Linux 平台的最新版本的 kubectl，命令如下：

```
curl -LO "https://dl.k8s.io/release/$(curl -L -s \
https://dl.k8s.io/release/stable.txt)/bin/linux/amd64/kubectl"
```

如果是 Mac 平台，则运行命令如下：

```
curl -LO "https://dl.k8s.io/release/$(curl -L -s \
https://dl.k8s.io/release/stable.txt)/bin/darwin/amd64/kubectl"
```

命令中访问的 https://dl.k8s.io/release/stable.txt 文件的内容是最新的 kubectl 稳定版本号，例如 v1.20.5。如果需要安装指定版本的 kubectl，则需要把命令中的 $(curl -L -s https://dl.k8s.io/release/stable.txt)替换成指定的版本，如 v1.20.0。

下载结束后，下载 kubectl 校验文件用于验证，kubectl 校验文件中的版本号需要和上一步下载的 kubectl 二进制文件的版本号匹配，命令如下：

```
curl -LO "https://dl.k8s.io/$(curl -L -s \
https://dl.k8s.io/release/stable.txt)/bin/linux/amd64/kubectl.sha256"
```

使用 sha256sum 命令校验，命令如下：

```
echo "$(<kubectl.sha256) kubectl" | sha256sum --check
```

若校验通过，则输出的信息如下：

```
kubectl: OK
```

现在开始安装 kubectl，通过 install 命令将 kubectl 复制到/usr/local/bin/kubectl，并且将 kubectl 文件的权限设置为 0755，所属用户为 root，用户组为 root。用户和用户组参数需要根据实际情况设置，命令如下：

```
sudo install -o root -g root -m 0755 kubectl /usr/local/bin/kubectl
```

查看一下/usr/local/bin/kubectl 文件，命令如下：

```
Ls -alF /usr/local/bin/kubectl
```

运行命令后看到的输出如下：

```
-rwxr-xr-x 1 root root 40263680 Mar 26 09:28 /usr/local/bin/kubectl*
```

安装完成后验证是否安装成功，命令如下：

```
kubectl version --client
```

7.2.5 设置 kubectl 命令自动补全

为了方便使用 kubectl 命令，可以设置 kubectl 命令的自动补全功能。在设置前，需要确保已经安装了 bash-completion。如果没有安装 bash-completion，则可以使用命令 apt-get install bash-completion 或者 yum install bash-completion 安装。

在 bash 脚本环境下，使用命令 kubectl completion bash 可以获取命令补全脚本。在 zsh 环境下，可以使用命令 kubectl completion zsh。把补全脚本写入文件，使用 source 命令使其生效，下面以 bash 为例演示如何设置自动补全功能，命令如下：

```
echo 'source <(kubectl completion bash)' >>~/.bashrc
```

另一种方式是把 kubectl 命令的补全脚本写入/etc/bash_completion.d/中，命令如下：

```
kubectl completion bash >/etc/bash_completion.d/kubectl
```

如果对 kubectl 命令设置了别名，例如 k，则这种情况需要额外设置。命令如下：

```
echo 'alias k=kubectl' >>~/.bashrc
echo 'complete -F __start_kubectl k' >>~/.bashrc
```

zsh 环境下设置 kubectl 命令补全功能与上述步骤类似。

7.3 本地启动 Kubernetes

Docker Desktop 内置了 Kubernetes。利用它可以在本地启动单节点的 Kubernetes 集群。启动 Docker Desktop 之后，单击它的图标，在弹出的菜单中单击 Dashboard，如图 7-1 所示。

然后在 Dashboard 界面左侧单击菜单 Kubernetes。在右侧选中 Enable Kubernetes 选项，如图 7-2 所示。

选中 Enable Kubernetes 之后，单击右下角的 Apply&Restart 按钮使配置生效。上述截图中使用 macOS 系统作演示。Windows 系统下的操作与此类似。

这样就开启了本地的 Kubernetes。等待数分钟后 Kubernetes 便会准备就绪，可以用 kubectl 命令查看 Kubernetes 信息。命令如下：

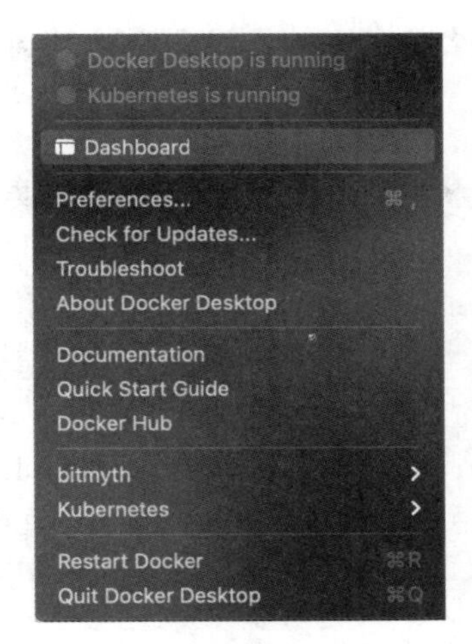

图 7-1　打开 Docker Desktop Dashboard

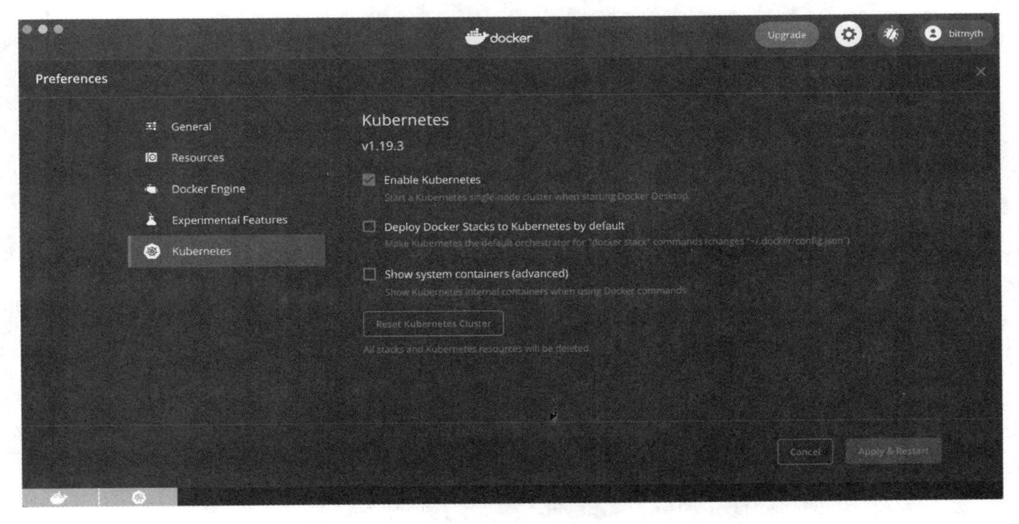

图 7-2　Docker Desktop 开启 Kubernetes

```
kubectl cluster - info
```

如果 Kubernetes 已准备就绪，运行命令则输出的信息如下：

```
Kubernetes master is running at https://kubernetes.docker.internal:6443
KubeDNS is running at \
https://kubernetes. docker. internal: 6443/api/v1/namespaces/kube - system/services/kube -
dns:dns/proxy

To further debug and diagnose cluster problems, use 'kubectl cluster - info dump'.
```

除了可以使用 Docker Desktop 内置的 Kubernetes,还可以通过 minikube 或者 kind 启动本地的 Kubernetes。

7.4 使用 kubeadm 创建 Kubernetes 集群

生产环境可以使用 kubeadm 工具部署 Kubernetes 集群。

7.4.1 环境要求

使用 Linux 服务器。Kubernetes 最佳的运行平台是 Linux 发行版,例如 Ubuntu 和 CentOS。

集群中的每台服务器都至少有 2 GB 的内存空间和 2 个 CPU。如果计算资源没有满足最低要求,在运行 kubeadm init 命令时会报错,将会看到类似信息如下:

```
[init] Using Kubernetes version: v1.21.0
[preflight] Running pre - flight checks
error execution phase preflight: [preflight] Some fatal errors occurred:
 [ERROR NumCPU]: the number of available CPUs 1 is less than the required 2
 [ERROR Mem]: the system RAM (978 MB) is less than the minimum 1700 MB
[preflight] If you know what you are doing, you can make a check non - fatal with ' -- ignore -
preflight - errors = ...'
To see the stack trace of this error execute with -- v = 5 or higher
```

服务器之间的网络需要互通。每台服务器都必须有唯一的主机名、MAC 地址及 product_uuid。

需要禁用 Swap,这样 kubelet 才可以正常工作。

需要保证 Kubernetes 用到的端口可以访问。这可能需要对防火墙或者服务器的安全组策略进行配置。

Kubernetes 中的 Control plane(控制面)需要用到的端口如表 7-1 所示。

表 7-1 Kubernetes 中的 Control plane 需要用到的端口

协议	方向	端口范围	用途	使用方
TCP	入	4443 *	Kubernetes API Server	全部
TCP	入	2379—2380	etcd 客户端 API	kube-apiserver、etcd
TCP	入	10250	kubelet API	自身、Control plane
TCP	入	10251	kube-scheduler	自身
TCP	入	10252	kube-controller-manager	自身
TCP	入			

端口后的 * 表示该端口号配置可以被覆盖。

Kubernetes 中的 Worker nodes(工作节点)需要用到的端口如表 7-2 所示。

表 7-2　Kubernetes 中的 Worker nodes 需要用到的端口

协议	方向	端口范围	用　途	使用方
TCP	入	10250	kubelet API	自身、Control plane
TCP	入	30000—32767	NodePort Services	全部

7.4.2　安装容器运行时

Kubernetes 运行容器需要容器运行时，常用的容器运行时包括 Docker、containerd 和 CRI-O。如果没有指定容器运行时，Kubernetes 则会在启动时自动扫描常见的运行时所用到的 UNIX domain sockets 文件。如果检测到了两个及以上的运行时，Kubernetes 则会产生错误而退出。

文中使用 Docker 作为 Kubernetes 容器运行时。

默认情况下 Docker 使用 cgroupfs 作为 cgroup driver，由于后续 Kubernetes 集群会运行在 Ubuntu 系统中，主流的 Linux 发行版大部分使用 Systemd 作为 init 进程。Systemd 高度集成了 cgroup，同时使用 systemd 和 cgroupfs 不利于资源视图的一致性。需将 Docker 的 cgroup driver 设置为 systemd，在 Docker 配置文件/etc/docker/daemon.json 中写入的内容如下：

```
{
"exec-opts": ["native.cgroupdriver=systemd"]
}
```

之后重启 Docker 使配置生效，命令如下：

```
systemctl restart docker
```

查看 Docker cgroup driver 信息验证是否生效，命令如下：

```
docker info --format "{{.CgroupDriver}}"
```

如果成功将 cgroup driver 配置为 systemd，则运行命令后输出的信息如下：

```
systemd
```

7.4.3　安装 kubeadm、kubelet、kubectl

首先更新 apt，并且安装 Kubernetes apt 库会用到的包，命令如下：

```
sudo apt - get update
sudo apt - get install  - y apt - transport - https ca - certificates curl
```

接着添加 Kubernetes 官方的 GPG key,命令如下:

```
sudo curl  - fsSLo /usr/share/keyrings/kubernetes - archive - keyring.gpg \
https://packages.cloud.google.com/apt/doc/apt - key.gpg
```

然后添加 Kubernetes 的 apt 库,命令如下:

```
echo "deb [signed - by = /usr/share/keyrings/kubernetes - archive - keyring.gpg] \
https://apt.kubernetes.io/ kubernetes - xenial main" \
|sudo tee /etc/apt/sources.list.d/kubernetes.list
```

现在再次更新 apt,然后安装 kubelet、kubeadm 和 kubectl,命令如下:

```
sudo apt - get update
sudo apt - get install  - y kubelet kubeadm kubectl
```

设置 kubelet、kubeadm 和 kubectl 不被系统自动更新升级及卸载,命令如下:

```
sudo apt - mark hold kubelet kubeadm kubectl
```

可以参考网址 http://manpages.ubuntu.com/manpages/bionic/man8/apt-mark.8.html 查看关于 apt-mark 命令的更多信息。

安装完成后,可以使用 kubeadm init 命令来初始化集群。在此之前,建议检测是否可以访问 Kubernetes 初始化过程中用到的 gcr.io 等镜像仓库。检测命令如下:

```
kubeadm config images pull
```

如果访问镜像仓库没有问题,则输出的信息如下:

```
[config/images] Pulled k8s.gcr.io/kube - apiserver:v1.20.5
[config/images] Pulled k8s.gcr.io/kube - controller - manager:v1.20.5
[config/images] Pulled k8s.gcr.io/kube - scheduler:v1.20.5
[config/images] Pulled k8s.gcr.io/kube - proxy:v1.20.5
[config/images] Pulled k8s.gcr.io/pause:3.2
[config/images] Pulled k8s.gcr.io/etcd:3.4.13 - 0
[config/images] Pulled k8s.gcr.io/coredns:1.7.0
```

7.4.4 初始化

现在初始化 Kubernetes 集群,命令如下:

```
sudo kubeadm init
```

运行命令后输出的类似信息如下：

```
[init] Using Kubernetes version: v1.20.5
[preflight] Running pre-flight checks
 [WARNING IsDockerSystemdCheck]: detected "cgroupfs" as the Docker cgroup driver. The
recommended driver is "systemd". Please follow the guide at https://kubernetes.io/docs/setup/
cri/
 [WARNING SystemVerification]: this Docker version is not on the list of validated versions:
20.10.5. Latest validated version: 19.03
[preflight] Pulling images required for setting up a Kubernetes cluster
[preflight] This might take a minute or two, depending on the speed of your internet connection
[preflight] You can also perform this action in beforehand using 'kubeadm config images pull'
[certs] Using certificateDir folder "/etc/kubernetes/pki"
[certs] Generating "ca" certificate and key
[certs] Generating "apiserver" certificate and key
[certs] apiserver serving cert is signed for DNS names [ip-172-31-27-163 kubernetes
kubernetes.default kubernetes.default.svc \
kubernetes.default.svc.cluster.local] and IPs [10.96.0.1 172.31.27.163]
[certs] Generating "apiserver-kubelet-client" certificate and key
[certs] Generating "front-proxy-ca" certificate and key
[certs] Generating "front-proxy-client" certificate and key
[certs] Generating "etcd/ca" certificate and key
[certs] Generating "etcd/server" certificate and key
[certs] etcd/server serving cert is signed for DNS names [ip-172-31-27-163 localhost] and
IPs [172.31.27.163 127.0.0.1 ::1]
[certs] Generating "etcd/peer" certificate and key
[certs] etcd/peer serving cert is signed for DNS names [ip-172-31-27-163 localhost] and
IPs [172.31.27.163 127.0.0.1 ::1]
[certs] Generating "etcd/healthcheck-client" certificate and key
[certs] Generating "apiserver-etcd-client" certificate and key
[certs] Generating "sa" key and public key
[kubeconfig] Using kubeconfig folder "/etc/kubernetes"
[kubeconfig] Writing "admin.conf" kubeconfig file
[kubeconfig] Writing "kubelet.conf" kubeconfig file
[kubeconfig] Writing "controller-manager.conf" kubeconfig file
[kubeconfig] Writing "scheduler.conf" kubeconfig file
[kubelet-start] Writing kubelet environment file with flags to file "/var/lib/kubelet/
kubeadm-flags.env"
[kubelet-start] Writing kubelet configuration to file "/var/lib/kubelet/config.yaml"
[kubelet-start] Starting the kubelet
[control-plane] Using manifest folder "/etc/kubernetes/manifests"
[control-plane] Creating static Pod manifest for "kube-apiserver"
[control-plane] Creating static Pod manifest for "kube-controller-manager"
```

```
[control-plane] Creating static Pod manifest for "kube-scheduler"
[etcd] Creating static Pod manifest for local etcd in "/etc/kubernetes/manifests"
[wait-control-plane] Waiting for the kubelet to boot up the control plane as static Pods from
directory "/etc/kubernetes/manifests". This can take up to 4m0s
[kubelet-check] Initial timeout of 40s passed.
[apiclient] All control plane components are healthy after 59.502631 seconds
[upload-config] Storing the configuration used in ConfigMap "kubeadm-config" in the "kube-
system" Namespace
[kubelet] Creating a ConfigMap "kubelet-config-1.20" in namespace kube-system with the
configuration for the kubelets in the cluster
[upload-certs] Skipping phase. Please see --upload-certs
[mark-control-plane] Marking the node ip-172-31-27-163 as control-plane by adding the
labels "node-role.kubernetes.io/master=''" and \
"node-role.kubernetes.io/control-plane='' (deprecated)"
[mark-control-plane] Marking the node ip-172-31-27-163 as control-plane by adding the
taints [node-role.kubernetes.io/master:NoSchedule]
[bootstrap-token] Using token: q5aa3f.t4ps1zoixl8nidf9
[bootstrap-token] Configuring bootstrap tokens, cluster-info ConfigMap, RBAC Roles
[bootstrap-token] configured RBAC rules to allow Node Bootstrap tokens to get nodes
[bootstrap-token] configured RBAC rules to allow Node Bootstrap tokens to post CSRs in order
for nodes to get long term certificate credentials
[bootstrap-token] configured RBAC rules to allow the csrapprover controller automatically
approve CSRs from a Node Bootstrap Token
[bootstrap-token] configured RBAC rules to allow certificate rotation for all node client
certificates in the cluster
[bootstrap-token] Creating the "cluster-info" ConfigMap in the "kube-public" namespace
[kubelet-finalize] Updating "/etc/kubernetes/kubelet.conf" to point to a rotatable kubelet
client certificate and key
[addons] Applied essential addon: CoreDNS
[addons] Applied essential addon: kube-proxy

Your Kubernetes control-plane has initialized successfully!

To start using your cluster, you need to run the following as a regular user:

  mkdir -p $HOME/.kube
  sudo cp -i /etc/kubernetes/admin.conf $HOME/.kube/config
  sudo chown $(id -u):$(id -g) $HOME/.kube/config

Alternatively, if you are the root user, you can run:

  export KUBECONFIG=/etc/kubernetes/admin.conf

You should now deploy a pod network to the cluster.
Run "kubectl apply -f [podnetwork].yaml" with one of the options listed at:
  https://kubernetes.io/docs/concepts/cluster-administration/addons/
```

```
Then you can join any number of worker nodes by running the following on each as root:

kubeadm join 172.31.27.163:6443 --token q5aa3f.t4ps1zoixl8nidf9 \
    --discovery-token-ca-cert-hash sha256:8154c4ce72f3d12897d34e98d2ae64218f1baf5a2
ea270b1b6366218eaffe8e8
```

运行命令 kubeadm init 后从输出的信息中会看到以下提示：

```
Your Kubernetes control-plane has initialized successfully!
```

表示 Kubernetes 中的 control plane 成功完成初始化。

运行命令 kubeadm init 后在输出的信息的末尾包含了加入其他节点的命令，命令如下：

```
kubeadm join 172.31.27.163:6443 --token q5aa3f.t4ps1zoixl8nidf9 \
    --discovery-token-ca-cert-hash sha256:8154c4ce72f3d12897d34e98d2ae64218f1baf5a2
ea270b1b6366218eaffe8e8
```

需要把这个命令保存下来，在加入其他节点时使用。

默认情况下，kubeadm 会使用默认的网关设备 IP 设置 control plane 中的 API server 的 IP 地址，需要设置其他网卡设备的 IP，需要在 kubeadm init 命令中指定参数--apiserver-advertise-address=< ip-address >。

7.4.5 设置 kubeconfig

1. 访问本地 Kubernetes 集群

使用 Kubernetes 集群前，需要设置 kubectl 配置文件，称为 kubeconfig 文件，配置文件需提供 Kubernetes 集群地址、集群的 CA 证书、客户端证书和密钥等信息。这样 kubectl 才能正常访问 Kubernetes 集群。

默认情况下，kubectl 会在 $HOME/.kube/文件夹下查找名为 config 的文件并作为 kubeconfig 文件。也可以通过环境变量 KUBECONFIG 或者--kubeconfig 参数指定其他文件路径。

运行命令 kubeadm init 后在输出的信息中会提示以非 root 用户使用 Kubernetes 的设置方法，命令如下：

```
mkdir -p $HOME/.kube
sudo cp -i /etc/kubernetes/admin.conf $HOME/.kube/config
sudo chown $(id -u):$(id -g) $HOME/.kube/config
```

上述命令的含义分别是新建 $HOME/.kube 文件夹，然后把/etc/kubernetes/admin.conf 复制到 $HOME/.kube/config 文件，最后把 $HOME/.kube/config 文件所属用户和

用户组设置为当前用户和当前用户所属的用户组。

如果以 root 用户身份使用 Kubernetes 集群，则可以将环境变量设置为 KUBECONFIG，命令如下：

```
export KUBECONFIG=/etc/kubernetes/admin.conf
```

查看 Kubernetes 集群中的节点，命令如下：

```
kubectl get node
```

运行命令后输出的类似信息如下：

```
NAME              STATUS     ROLES                   AGE   VERSION
ip-172-31-27-142  NotReady   control-plane,master    16h   v1.20.5
```

表示集群中的 control plane 节点的状态是 NotReady，即未就绪。

查看节点详细信息，命令如下：

```
kubectl describe node ip-172-31-27-142
```

输出信息中的 Conditions 部分的提示如下：

```
RunTime network not ready: NetworkReady=false reason: NetworkPluginNotReady message: docker:
network plugin is not ready: cni config uninitialized
```

说明未就绪的原因是因为没有安装网络插件。

查看所有命名空间下的 Pod，命令如下：

```
kubectl get pod --all-namespaces
```

运行命令后输出的类似信息如下：

```
NAMESPACE     NAME                                       READY   STATUS    RESTARTS   AGE
kube-system   coredns-74ff55c5b-2ssnn                    0/1     Pending   0          16h
kube-system   coredns-74ff55c5b-glbtp                    0/1     Pending   0          16h
kube-system   etcd-ip-172-31-27-142                      1/1     Running   0          16h
kube-system   kube-apiserver-ip-172-31-27-142            1/1     Running   0          16h
kube-system   kube-controller-manager-ip-172-31-27-142   1/1     Running 0 16h
kube-system   kube-proxy-d622k   1/1   Running   0   16h
kube-system   kube-scheduler-ip-172-31-27-142            1/1     Running   0          16h
```

2. 访问远程 Kubernetes 集群

如果希望远程访问 Kubernetes 集群，则需要在运行 kubeadm init 命令创建集群时使用参数--control-plane-endpoint=< ip-address >指定集群 control plane 所在的服务器公网 IP。之后把远程 Kubernetes 集群的 control plane 所在服务器上的文件/etc/kubernetes/admin.conf 复制到本地，运行 kubectl 时需加上--kubeconfig 参数，命令如下：

```
kubectl --kubeconfig ./admin.conf get nodes
```

也可以把 admin.conf 文件移动到本地的~/.kube 文件夹下。这样在运行 kubectl 命令时就不需要指定--kubeconfig 参数了，命令如下：

```
ssh <host> sudo cat /etc/kubernetes/admin.conf >~/.kube/config
```

7.4.6 安装网络插件

Kubernetes 中的网络插件由第三方提供，网络插件都可实现 Kubernetes 的网络模型 CNI(Container Network Interface)。

这里使用 Flannel 网络插件，需要设置 podCIDR，在运行 kubeadm init 命令时，需加上参数--pod-network-cidr=10.244.0.0/16，命令如下：

```
sudo kubeadm init --pod-network-cidr=10.244.0.0/16
```

安装 Flannel 网络插件，命令如下：

```
kubectl apply -f \
https://raw.GitHubusercontent.com/coreos/flannel/master/Documentation/kube-flannel.yaml
```

运行命令后输出的信息如下：

```
podsecuritypolicy.policy/psp.flannel.unprivileged created
clusterrole.rbac.authorization.k8s.io/flannel created
clusterrolebinding.rbac.authorization.k8s.io/flannel created
serviceaccount/flannel created
configmap/kube-flannel-cfg created
daemonset.apps/kube-flannel-ds created
```

等待几分钟，便可查看节点状态，命令如下：

```
kubectl get node
```

运行命令后输出的信息如下：

```
NAME              STATUS   ROLES                  AGE     VERSION
ip-172-31-27-142  Ready    control-plane,master   7m44s   v1.20.5
```

显示节点 ip-172-31-27-142 的状态变成了 Ready。

查看 Pod 信息，命令如下：

```
kubectl get pod --all-namespace
```

运行命令后输出的信息如下：

```
NAMESPACE     NAME                                             READY   STATUS    RESTARTS AGE
kube-system coredns-74ff55c5b-7zv9c                            1/1     Running   0           7m8s
kube-system coredns-74ff55c5b-vw77b                            1/1     Running   0   7m8s
kube-system etcd-ip-172-31-27-142                              1/1     Running   0    7m25s
kube-system kube-apiserver-ip-172-31-27-142   1/1   Running   0   7m25s
kube-system kube-controller-manager-ip-172-31-27-142 1/1   Running   0 7m25s
kube-system kube-flannel-ds-ldvrb                    1/1     Running   0 70s
kube-system kube-proxy-x9vvb   1/1     Running   0           7m9s
kube-system kube-scheduler-ip-172-31-27-142          1/1     Running   0   7m25s
```

7.4.7 部署样例程序

准备好 Kubernetes 集群环境后，可以创建一个示例 Deployment 来测试功能是否正常，命令如下：

```
kubectl create deployment kubernetes-bootcamp \
--image=gcr.io/google-samples/kubernetes-bootcamp:v1
```

如果集群可以正常工作，则运行命令后输出的信息如下：

```
deployment.apps/kubernetes-bootcamp created
```

查看 Deployment，命令如下：

```
kubectl get deployment
```

运行命令后输出的信息如下：

```
NAME                   READY   UP-TO-DATE   AVAILABLE   AGE
kubernetes-bootcamp    0/1     1            0           23s
```

查看创建的 Pod 信息，命令如下：

```
kubectl describe pods
```

在输出信息中包含的 Events 字段如下：

```
Events:
  Type     Reason            Age   From                 Message
  ----     -------           ----  ----                 --------
  Warning  FailedScheduling  14m   default-scheduler  0/1 nodes are available: 1 node(s)
had taint {node-role.kubernetes.io/master: }, that the pod didn't tolerate.


  Normal   Scheduled         11m   default-scheduler   Successfully assigned default/
kubernetes-bootcamp-57978f5f5d-v5czq to ip-172-31-28-71
  Normal   Pulling           11m   kubelet             Pulling image "gcr.io/google-samples/
kubernetes-bootcamp:v1"
  Normal   Pulled            11m   kubelet             Successfully pulled image "gcr.io/google-
samples/kubernetes-bootcamp:v1" in 8.758335361s
  Normal   Created           11m   kubelet             Created container kubernetes-bootcamp
  Normal   Started           11m   kubelet             Started container kubernetes-bootcamp
```

Events 中按时间顺序展示了与 Pod 相关的事件，最新的事件在最上方，输出信息显示 Pod 调度失败了，失败原因为 0/1 nodes are available，即没有合适的 Node 来放置 Pod。

默认情况下，Pod 不会被调度到 control plane 节点运行，如果希望 control plane 节点也可以运行用户提交的 Pod，例如在单节点的 Kubernetes 集群中开发，则可以将 control plane 节点去掉标记 node-role.kubernetes.io/master，命令如下：

```
kubectl taint nodes --all node-role.kubernetes.io/master-
```

7.4.8 将 Node 添加到集群

在同一可用区创建另一台服务器，并使用同样的步骤安装 Docker、kubeadm、kubectl、kubelet。之后加入已经创建好的集群中，命令如下：

```
sudo kubeadm join 172.31.27.142:6443 --tokendv0xuq.x3atqs88re7rxyqo \
--discovery-token-ca-cert-hash \
sha256:45e89f053255ca2ad62a594e4b187a198d8289c970b9728e02ab80e3565ce886
```

运行命令后输出的信息如下：

```
[preflight] Running pre-flight checks
[WARNING SystemVerification]: this Docker version is not on the list of validated versions:
20.10.5. Latest validated version: 19.03
[preflight] Reading configuration from the cluster...
```

```
[preflight] FYI: You can look at this config file with 'kubectl -n kube-system get cm kubeadm-config -o yaml'
[kubelet-start] Writing kubelet configuration to file \
"/var/lib/kubelet/config.yaml"
[kubelet-start] Writing kubelet environment file with flags to file \
"/var/lib/kubelet/kubeadm-flags.env"
[kubelet-start] Starting the kubelet
[kubelet-start] Waiting for the kubelet to perform the TLS Bootstrap...

This node has joined the cluster:
* Certificate signing request was sent to apiserver and a response was received.
* The Kubelet was informed of the new secure connection details.

Run 'kubectl get nodes' on the control-plane to see this node join the cluster.
```

回到 control plane 节点上，查看集群中的节点，命令如下：

```
kubectl get node -o wide
```

运行命令后输出的信息如下(前 6 列)：

```
NAME                STATUS   ROLES                    AGE    VERSION   INTERNAL-IP
ip-172-31-24-124    Ready    <none>                   77s    v1.20.5   172.31.24.124
ip-172-31-27-142    Ready    control-plane,master     123m   v1.20.5   172.31.27.142
```

后 4 列信息如下：

```
EXTERNAL-IP   OS-IMAGE             KERNEL-VERSION    CONTAINER-RUNTIME
<none>        Ubuntu 20.04.2 LTS   5.4.0-1038-aws    docker://20.10.5
<none>        Ubuntu 20.04.2 LTS   5.4.0-1038-aws    docker://20.10.5
```

这样集群中就有了 worker 节点，此时查看集群中的 Pod，验证是否调度成功，命令如下：

```
kubectl get pod -o wide
```

运行命令后输出的信息如下(前 5 列)：

```
NAME                                   READY   STATUS    RESTARTS   AGE
kubernetes-bootcamp-57978f5f5d-v5czq   1/1     Running   0          3m13s
```

后 4 列信息如下：

```
IP           NODE             NOMINATED NODE   READINESS GATES
10.244.1.2   ip-172-31-28-71  <none>           <none>
```

输出信息显示 Pod 调度成功,已经在运行。

运行 proxy,命令如下:

```
kubectl proxy&
```

调用接口,命令如下:

```
curl http://localhost: 8001/version
```

输出的信息如下:

```
{
"major": "1",
"minor": "20",
"gitVersion": "v1.20.5",
"gitCommit": "6b1d87acf3c8253c123756b9e61dac642678305f",
"gitTreeState": "clean",
"buildDate": "2021 - 03 - 18T01:02:01Z",
"goVersion": "go1.15.8",
"compiler": "gc",
"platform": "Linux/amd64"
}
```

输出信息说明 Pod 中的应用运行成功。

7.5　创建托管的 Kubernetes 集群

许多云服务商提供了托管的 Kubernetes 集群,提供了托管 Kubernetes 集群服务的云服务商列表可以在 https://kubernetes.io/docs/setup/production-environment/turnkey-solutions/页面查看。

下面介绍如何使用 Amazon Elastic Kubernetes Service(Amazon EKS)创建 Kubernetes 集群。

Amazon EKS 提供了两种方式创建集群,一种是使用 eksctl 命令行工具,另一种是使用 AWS 命令行工具。使用 eksctl 是最快和最简单的方式,下面使用 eksctl 创建一个 Kubernetes 集群。

首先安装 eksctl,macOS 环境可以使用 brew 安装,命令如下:

```
brew tap weaveworks/tap
brew install weaveworks/tap/eksctl
```

如果之前已经安装了 eksctl,则可以升级一下 eksctl,命令如下:

```
brew upgrade eksctl && brew link -- overwrite eksctl
```

查看 eksctl 版本号来验证安装是否成功，命令如下：

```
eksctl version
```

另外需要下载并安装 AWS 命令行工具，安装过程可参考链接 https://docs.aws.amazon.com/cli/latest/userguide/install-cliv2.html。

安装完成后，验证是否成功，命令如下：

```
aws -- version
```

运行命令后输出的信息如下：

```
aws - cli/2.2.0 Python/3.8.8 Darwin/20.1.0 exe/x86_64 prompt/off
```

配置 AWS 命令行工具，使其关联到自己的 AWS 账户，命令如下：

```
aws configure
```

运行命令后输出的信息如下：

```
AWS Access Key ID [ **************** 4KVF]:
```

输入访问密钥 ID 后会输出的信息如下：

```
AWS Secret Access Key [ **************** bGao]:
```

这时输入访问密钥，后续的输出信息如下，可以直接按回车键来忽略：

```
Default region name [None]:
Default output format [None]:
```

AWS 云服务支持的地区称为可用区。接下来的操作都在可用区 us-west-2 进行。

创建密钥对，命令如下：

```
aws ec2 create - key - pair \
-- region us - west - 2 \
-- key - name my - key - pair \
-- query "KeyMaterial"\
-- output text > my - key - pair.pem
```

将文件权限设置为用户只读，命令如下：

```
chmod 400 my-key-pair.pem
```

从 my-key-pair.pem 文件提取公钥，命令如下：

```
ssh-keygen -y -f my-key-pair.pem > my-key-pair.pub
```

使用 eksctl 在可用区 ap-southeast-1 创建集群，命令如下：

```
eksctl create cluster \
--name my-cluster \
--region ap-southeast-1 \
--with-oidc \
--ssh-access \
--ssh-public-key my-key-pair.pub \
--managed
```

运行命令后截取的输出信息如下：

```
2021-05-10 20:52:28 [✓]  saved kubeconfig as "/Users/gsh/.kube/config"
2021-05-10 20:52:28 [□]  no tasks
2021-05-10 20:52:28 [✓]  all EKS cluster resources for "my-cluster" have been created
...
2021-05-10 20:52:31 [□]  kubectl command should work with \
"/Users/gsh/.kube/config", try 'kubectl get nodes'
2021-05-10 20:52:31 [✓]  EKS cluster "my-cluster" in "us-west-2" region is ready
```

输出信息提示把 kubeconfig 写入了本地默认的 kubeconfig 文件中。最后一行输出信息提示集群创建成功。

除了在上述命令中使用的参数，还可以使用--nodes 指定节点数，使用--instance-selector-vcpus 指定 CPU 数量等，查看更多支持的参数以满足对创建的集群的个性化要求，可以使用的命令如下：

```
eksctl create cluster --help
```

查看集群中的节点，命令如下：

```
kubectl get node
```

运行命令后输出的类似信息如下：

```
NAME                                         STATUS   ROLES    AGE   VERSION
ip-192-168-36-195.us-west-2.compute.internal Ready    <none>   14m   v1.19.6-eks-49a6c0
ip-192-168-89-187.us-west-2.compute.internal Ready    <none>   14m   v1.19.6-eks-49a6c0
```

使用 eksctl 删除集群，命令如下：

```
eksctl delete cluster -- name my - cluster -- region us - west - 2
```

运行命令后输出信息中的最后一行如下：

```
2021 - 05 - 10 21:49:19 [✓]  all cluster resources were deleted
```

表示集群删除成功。

7.6 Kubernetes 对象

7.6.1 Kubernetes 对象简介

Kubernetes 对象是持久化的实体，用来表示集群的状态。主要记载以下信息：

哪些容器化应用在运行，哪些资源是可用的，应用运行时采用什么策略（如重启策略、升级策略和容错策略）。

Kubernetes 对象记录了用户的意图或者期望，通过创建 Kubernetes 对象，本质上就是告诉 Kubernetes 集群的期望状态。当创建 Kubernetes 对象后，Kubernetes 会持续工作以保证对象创建成功。

操作 Kubernetes 对象（如创建、更新和删除）需要使用 Kubernetes API。通常使用 kubectl 命令来操作 Kubernetes 对象，kubectl 会代替用户调用 Kubernetes API，也可以在程序中使用 Kubernetes 客户端库来直接调用 Kubernetes API。

Kubernetes 对象几乎包含 Spec 和 Status 字段，Spec 字段设置了对象的期望状态，在创建带有 Spec 字段的对象时，必须指定 Spec。Status 字段用于描述对象当前的状态，Status 字段的内容由 Kubernetes 提供和更新，Kubernetes 控制面会持续地管理对象，使它的实际状态和期望状态相匹配。在不匹配的情况出现时，Kubernetes 会采取相应的操作来修正 Spec 和 Status 的不一致。

查看 Kubernetes API 支持的所有资源，命令如下：

```
kubectl api - resources
```

7.6.2 如何描述 Kubernetes 对象

在创建 Kubernetes 对象前，需要提供对象的基本信息，通常情况下使用 YAML 格式的文件来描述 Kubernetes 对象的基本信息，又称清单文件。下面的示例文件描述了一个 Deployment 类型的 Kubernetes 对象，内容如下：

```yaml
apiVersion: apps/v1
kind: Deployment
metadata:
  name: nginx-deployment
spec:
  selector:
    matchLabels:
      app: nginx
  replicas: 2
  template:
    metadata:
      labels:
        app: nginx
    spec:
      containers:
      - name: nginx
        image: nginx:1.14.2
        ports:
        - containerPort: 80
```

描述 Kubernetes 对象的 YAML 文件中必须包括字段 apiVersion、kind、metadata 和 spec。下面分别介绍。

apiVersion 用于指定创建这个 Kubernetes 对象使用的 Kubernetes API 的版本。

kind 表示这个 Kubernetes 对象的类型。例如 Pod、Deployment 等。

metedata 字段中的内容是这个 Kubernetes 对象的唯一标识，包括 name、UID 及可选的 namespace 字段。

spec 是这个 Kubernetes 对象的期望状态。这个 spec 字段的具体格式由具体的对象类型决定。不同类型的 Kubernetes 对象，spec 字段会有不同的格式。可以参考官方链接 https://kubernetes.io/docs/reference/generated/kubernetes-api/v1.21/查看所有 Kubernetes 对象的 spec 格式。

Kubernetes 集群保存了集群中各个对象的实时配置，通常会保存在 etcd 中。可以通过命令 kubectl get <object-name> -o yaml 来查看对象的实时配置。

7.6.3　如何管理 Kubernetes 对象

使用 kubectl 管理 Kubernetes 对象有 3 种方式，分别是指令式命令、指令式对象配置和声明式对象配置。建议始终采用相同的方式管理，混合使用不同的方式可能会产生未知的结果。

1. 指令式命令

指令式命令的意思是用户指定 kubectl 进行什么操作（如 create、delete、update 等）。按学习难易程度排序，3 种方式中指令式命令最简单，可直接操作实时的 Kubernetes 对象，不需要提供对象配置文件，但是指令式命令只适合在开发环境使用，不适合在生产环境使用。因为这种方式无法将变更加入版本控制，从而不利于变更审核、团队协作和故障修复。

运行命令 kubectl run 会创建一个新的 Pod，并可以此运行容器。例如使用命令式命令来运行 Nginx 容器，命令如下：

```
kubectl run nginx -- image Nginx
```

运行命令后输出的信息如下：

```
pod/nginx created
```

使用指令式命令创建 Kubernetes 对象前可以查看以 kubectl 为对象生成的配置文件，而不是直接发送到 Kubernetes 中去执行，这样可以清楚看到 kubectl 要创建的对象的具体配置。以运行 Nginx 容器为例，命令如下：

```
kubectl run nginx -- image nginx -- dry - run = client - o yaml
```

运行命令后输出的信息如下：

```
apiVersion: v1
kind: Pod
metadata:
  creationTimestamp: null
  labels:
    run: nginx
  name: nginx
spec:
  containers:
  - image: nginx
    name: nginx
    resources: {}
  dnsPolicy: ClusterFirst
  restartPolicy: Always
status: {}
```

使用指令式命令创建一个名为 nginx 的 Deployment，命令如下：

```
kubectl create deployment nginx -- image = nginx
```

使用指令式命令创建 Service 来为名为 nginx 的 Deployment 中的一组 Pod 提供负载均衡，命令如下：

```
kubectl expose deployment nginx -- port 80
```

使用指令式命令水平扩容名为 nginx 的 Deployment 中的 Pod 实例数，假设扩容为 3 个副本，命令如下：

```
kubectl scale deployment nginx --replicas=3
```

指令式命令包含许多子命令，用于修改对象，例如 annotate、label、set、edit、patch 等。其中 edit 命令会将对象的实时配置下载到 YAML 文件中，然后自动通过本地的编辑器打开它，用户编辑配置文件后保存并退出，kubectl 会应用编辑后的配置文件使变更生效。

指令式命令中用于查看对象的子命令包括 get、logs、describe。其中 get 用于查看对象的基本信息，describe 用于查看对象的详细信息，logs 命令用于查看 Pod 中运行的容器所输出到 stdout 和 stderr 的日志。

使用指令式命令删除名为 nginx 的 Deployment，命令如下：

```
kubectl delete deployment nginx
```

2. 指令式对象配置

指令式对象配置的方式需要在 kubectl 命令中指定具体操作（例如 create、delete 等）和至少一个对象配置文件，对象配置文件中包含 YAML 或 JSON 格式的对象定义。

指令式对象配置方式创建对象除了使用本地文件，还支持 URL 指定配置文件。语法如下：

```
kubectl create -f <filename|url>
```

以运行 Nginx 容器为例，需要准备配置文件 nginx.yaml，文件内容如下：

```
apiVersion: v1
kind: Pod
metadata:
  name: nginx
spec:
  containers:
  - name: nginx
    image: nginx
    ports:
    - containerPort: 80
```

创建 Pod 来运行 Nginx 容器，命令如下：

```
kubectl create -f nginx.yaml
```

以指令式对象配置的方式更新对象的命令语法如下：

```
kubectl replace -f <filename|url>
```

以指令式对象配置的方式删除对象的命令语法如下：

```
kubectl delete -f <filename|url>
```

对于没有指定 name 的对象，无法使用上面的命令删除。

以指令式对象配置的方式查看对象的命令语法如下：

```
kubectl get -f <filename|url>
```

3. 声明式对象配置

声明式对象配置的方式和指令式对象配置的方式类似，也需要使用对象配置文件。和指令式对象配置不同的是使用声明式对象配置不需要指定具体操作，kubectl 会根据对象配置文件自动检测针对每个 Kubernetes 对象需要采取的操作（如 create、delete、update 等），并且声明式配置可以把包含对象配置文件的文件夹作为配置文件参数。

声明式对象配置方式需要用到 kubectl apply 命令，kubectl apply 会根据对象配置文件对 Kubernetes 集群做出相应变更。

使用声明式对象配置方式运行 nginx 容器，命令如下：

```
kubectl apply -f nginx.yaml
```

运行命令 kubectl apply 会在对应的每个 Kubernetes 对象中都添加一个 kubectl.kubernetes.io/last-applied-configuration 类型的 annotation。在这个 annotation 中包含了创建对象用到的配置文件内容。

查看上述 nginx.yaml 文件中定义的名为 nginx 的 Pod，命令如下：

```
kubectl get -f nginx.yaml -o yaml
```

运行命令后在输出信息中的 annotations 字段如下：

```
apiVersion: v1
kind: Pod
metadata:
  annotations:
    kubectl.kubernetes.io/last-applied-configuration: |
      {"apiVersion":"v1","kind":"Pod","metadata":{"annotations":{},"name":"nginx",
"namespace":"default"},"spec":{"containers":[{"image":"nginx","name":"nginx","ports":
[{"containerPort":80}]}]}}
  creationTimestamp: "2021-04-16T06:15:45Z"
```

可以看到 kubectl.kubernetes.io/last-applied-configuration 中的内容和 nginx.yaml 是

匹配的,除了有个别字段被设置了默认值。

假设有多个对象配置文件保存在 config 文件夹下,如果要一次性应用文件夹中所有的配置文件,则命令如下:

```
kubectl apply -f config/
```

如果文件夹下嵌套了文件夹,则可以加上参数-R,命令如下:

```
kubectl apply -R -f config/
```

在应用配置文件前,可以先查看 kubectl 会做出哪些变更,命令如下:

```
kubectl diff -f config/
```

其中-f 参数可以是文件、文件夹或 URL。

例如将 nginx.yaml 文件中的 image 字段由 nginx 修改为 nginx:1.14.2,然后运行 kubectl diff 命令查看变化,命令如下:

```
kubectl diff -f nginx.yaml
```

运行命令后输出的类似信息如下:

```
diff -u -N \
/tmp/LIVE-502866992/v1.Pod.default.nginx /tmp/MERGED-236848079/v1.Pod.default.nginx
--- /tmp/LIVE-502866992/v1.Pod.default.nginx 2021-04-16 07:27:40.825976160 +0000
+++ /tmp/MERGED-236848079/v1.Pod.default.nginx \
2021-04-16 07:27:40.833975893 +0000
@@ -80,7 +80,7 @@
    uid: ca9187d5-ee5e-4d16-89f1-934577179d05
 spec:
   containers:
-    - image: nginx
+    - image: nginx:1.14.2
      imagePullPolicy: Always
      name: nginx
      ports:
```

以声明式对象配置方式更新 Kubernetes 对象同样可以使用 kubectl apply 命令,命令如下:

```
kubectl apply -f nginx.yaml
```

运行命令后输出的信息如下:

```
pod/nginx configured
```

查看最新的对象配置，命令如下：

```
kubectl get -f nginx.yaml -o yaml
```

运行命令后在输出信息中 spec 字段的部分信息如下：

```
spec:
    containers:
    - image: nginx:1.14.2
```

表示变更已生效。

同时，在输出信息中的 annotations 字段也发生了变化，annotations 字段如下：

```
metadata:
  annotations:
    kubectl.kubernetes.io/last-applied-configuration: |
{"apiVersion":"v1","kind":"Pod","metadata":{"annotations":{},"name":"nginx","namespace":
"default"},"spec":{"containers":[{"image":"nginx:1.14.2","name":"nginx","ports":
[{"containerPort":80}]}]}}
```

可以看到"image"："nginx：1.14.2"和最新的 nginx.yaml 文件中指定的镜像一致。

在使用 kubectl apply 命令更新对象配置时，会从对象的 last-applied-configuration 注解中读取上次应用的配置文件，如果在上次应用的配置文件中有指定的配置文件中没有出现的字段，则这个字段会被删除。

如果指定的配置文件中的字段值和对象的实时配置中的取值不同，则相应的字段会被添加或者覆盖。

删除 nginx.yaml 文件中定义的对象，命令如下：

```
kubectl delete -f nginx.yaml
```

运行命令后输出的信息如下：

```
pod "nginx" deleted
```

7.7 Node

7.7.1 Node 简介

Kubernetes 中用 Node 来表示虚拟主机或者物理主机，Node 用来运行 Pod。

每个 Node 都必须拥有唯一的名称。Kubernetes 会把具有相同名称的资源看作同一个资源。

Node 支持设置标签，利用 Node 标签和 Pod 定义中的 node selectors 字段可以控制 Pod 调度到哪些 Node 上。在定义了 node selectors 字段的情况下，Pod 只会被调度到根据 node selectors 字段中的标签筛选出来的 Node。

Node 中包含组件 kubelet、容器运行时和 kube-proxy。下面分别介绍。

kubelet 是运行在每个 Node 中的一个代理，它负责保证由 Kubernetes 创建的容器在 Pod 中正常运行。如果容器不是由 Kubernetes 创建的，则 kubelet 不会对它进行管理。

kube-proxy 是 Node 的网络代理，Kubernetes 中的 Service 依赖 kube-proxy 实现。

ContainerRunTime 是负责运行容器的运行时，容器运行时必须实现 Kubernetes CRI 接口。常见的 Kubernetes 容器运行时包括 Docker、Containerd、CRI-O 等。

7.7.2　管理 Node

如果希望停止将新的 Pod 调度到某个 Node 上，例如在需要对某个 Node 进行维护的情况下，则可以将该 Node 标记为不可调度。命令如下：

```
kubectl cardon $NODENAME
```

命令中的 $NODENAME 表示 Node 的名称。运行这个命令不会影响已经在 Node 中运行的 Pod。

查看集群中的所有 Node，命令如下：

```
kubectl get node
```

查看某个 Node 的详细信息，命令如下：

```
kubectl describe node $NODENAME
```

7.7.3　Node 状态

Node 状态信息由 4 部分构成，分别是 Addresses、Conditions、Capacity and Allocatable、Info。

Addresses 表示与 Node 地址相关的信息，包含 HostName、ExternalIP、InternalIP 字段。可使用 kubectl describe node 命令获取的 Addresses 信息，示例如下：

```
Addresses:
  InternalIP:   172.31.24.182
  Hostname:     ip-172-31-24-182
```

HostName 即主机名，通过 Node 所在主机的操作系统内核获取。ExternalIP 是支持从集群外访问的 Node 的 IP 地址。InternalIP 是只可以从集群内访问的 Node 的 IP 地址。

Conditions 用于描述 Node 的具体状态。使用 kubectl describe node 命令获取的 Conditions 信息示例如下（前 3 列）：

```
Conditions:
  Type                 Status    LastHeartbeatTime
  ----                 ------    -----------------
  NetworkUnavailable   False     Wed, 14 Apr 2021 13:15:09 +0800
  MemoryPressure       False     Wed, 14 Apr 2021 14:15:53 +0800
  DiskPressure         False     Wed, 14 Apr 2021 14:15:53 +0800
  PIDPressure          False     Wed, 14 Apr 2021 14:15:53 +0800
  Ready                True      Wed, 14 Apr 2021 14:15:53 +0800
```

第 4 列和第 5 列信息如下：

```
LastTransitionTime                  Reason
------------------                  ------
Wed, 14 Apr 2021 13:15:09 +0800     FlannelIsUp
Wed, 14 Apr 2021 13:14:28 +0800     KubeletHasSufficientMemory
Wed, 14 Apr 2021 13:14:28 +0800     KubeletHasNoDiskPressure
Wed, 14 Apr 2021 13:14:28 +0800     KubeletHasSufficientPID
Wed, 14 Apr 2021 13:15:14 +0800     KubeletReady
```

最后 1 列信息如下：

```
Message
-------
Flannel is running on this node
kubelet has sufficient memory available
kubelet has no disk pressure
kubelet has sufficient PID available
kubelet is posting ready status. AppArmor enabled
```

Capacity and Allocable 表示 Node 可用的资源，包括 CPU、内存和可以调度到此 Node 上的 Pod 的个数。使用 kubectl describe node 命令获取的 Capacity and Allocatable 示例如下：

```
Capacity:
  cpu:                 2
  ephemeral-storage:   8065444Ki
  hugepages-2Mi:       0
  memory:              4028184Ki
  pods:                110
Allocatable:
```

```
cpu:                    2
ephemeral-storage:      7433113179
hugepages-2Mi:          0
memory:                 3925784Ki
pods:                   110
```

Info 用于描述 Node 的基础性信息,例如 Kubernetes 版本、Docker 版本和操作系统名称等,这些信息由 Node 中的 kubelet 搜集而来。使用 kubectl describe node 命令获取的 Info 信息示例如下:

```
System Info:
  Machine ID:                    93c89032a8624d95813da5f3e81f3cf5
  System UUID:                   ec2f919a-064f-95fe-c82e-9339201fa03d
  Boot ID:                       0542fbfc-c57f-4863-af0b-31fd9c80c899
  Kernel Version:                5.4.0-1038-aws
  OS Image:                      Ubuntu 20.04.2 LTS
  Operating System:              Linux
  Architecture:                  amd64
  Container RunTime Version:     docker://20.10.6
  Kubelet Version:               v1.21.0
  Kube-Proxy Version:            v1.21.0
```

7.7.4 Node 控制器

Node 控制器是 Kubernetes 控制面的组件之一。Node 控制器负责管理 Node 和监控 Node 的健康状态。Node 控制器会周期性检查 Node 的状态。如果 Node 的状态未知,例如 Node 宕机时,则 Node 控制器将无法得到 Node 发出的"心跳"信息。如果 Node 持续处于未知状态,则 Node 控制器会使用优雅终止的方式驱逐 Node 上的 Pod。

Node 发出的"心跳"信息有助于确定 Node 的可用性。心跳机制是分布式环境下常见的监测机制。Node"心跳"信息分为 NodeStatus 和 Lease 对象两种。它们都由 Node 上的 kubelet 负责创建和更新。

7.7.5 Node 容量

Node 容量指的是 Node 可用的计算资源容量,例如可用的内存和 CPU。Node 会跟踪它的资源容量变化。通过自注册机制生成的 Node 会在注册期间报告自身的资源容量,如果手动添加了 Node,则需要手动设置 Node 容量。

Kubernetes 调度器会确保 Node 上运行的 Pod 有足够的资源,调度器在调度时会保证 Node 上的所有容器对资源的请求总和不会超过 Node 的容量。计算资源请求总和时只包含由 kubelet 启动的容器所占用的资源,不包括由容器运行时直接启动的容器占用的资源。

7.8 Pod

7.8.1 Pod 简介

Kubernetes 中的 Pod 是一个逻辑意义上的主机，是 Kubernetes 中的原子部署单位。Pod 中可以运行 1 个到多个容器，Pod 中的多个容器往往具有紧密耦合关系，需要相互配合来完成工作。英文 Pod 是豌豆荚的意思，豌豆荚中的豆子紧密排布在一个豌豆荚里。这和一组紧密耦合的容器相似。

例如有 2 个容器，一个容器负责提供 HTTP 服务，另一个容器负责提供 CGI 服务，用于解析 HTTP 访问的动态网页。这 2 个容器就需要部署到一台物理主机上，这样才可以正常工作。类似这样需要紧密配合的容器组就可以用 Pod 的方式部署，因为 Pod 是原子部署单元，Pod 中的容器就不会被分开部署到不同的节点上。在真实场景中，一组容器需要紧密耦合的情况是非常常见的。Pod 解决了对这类容器集合进行部署和调度的难题。保证了它们出现在同一个节点，并且在调度时会把 Pod 中的每个容器所需要的计算资源作为整体考虑在内，以选择可以满足需求的节点进行调度。这就保证了容器组在节点工作时不会遇到计算资源不足的问题。

每个 Pod 都拥有独立的 IP，在同一个 Pod 中运行的容器具有相同的网络命名空间，可以通过 localhost 访问对方。也能通过如 SystemV 信号量或 POSIX 共享内存这类标准的进程间通信方式互相通信。

Pod 是容器运行的环境，所以重启容器并不会导致 Pod 消失。

7.8.2 Pod 使用模式

通常情况下不需要直接创建 Pod 来使用 Pod，而是通过其他 Kubernetes 中的其他资源对象（如 Deployment、StatefulSet 或 Job 等）间接创建 Pod。

Kubernetes 中的 Pod 有 2 种使用方式，一种是 1 个 Pod 中只运行 1 个容器，另一种是 1 个 Pod 中运行多个容器。Kubernetes 通过管理 Pod 间接管理 Pod 中的容器。

单 Pod 单容器是最常见的 Pod 使用模式，这和 Docker 中单容器单进程的模式类似，单 Pod 单容器模式中 Pod 可以看作容器的封装。

单 Pod 多容器模式适用于多个容器需要紧密协作的场景，这种场景下多个容器间往往需要共享某些资源，如文件、网络。Kubernetes 在调度 Pod 时，会保证将 Pod 中的所有容器调度到一个 Node 上。这样就保证了容器间可以互相通信和共享存储。单 Pod 多容器模式相对复杂，非必要的情况下不需要使用这种模式。

7.8.3 Pod 示例

示例 Pod，定义如下：

```yaml
apiVersion: v1
kind: Pod
metadata:
  name: nginx
spec:
  containers:
  - name: nginx
    image: nginx
    ports:
    - containerPort: 80
```

这个 yaml 文件定义了一个运行 nginx 容器的 Pod。

也可以使用指令式命令创建 Pod，命令如下：

```
kubectl run nginx --image nginx
```

7.8.4 Pod 模板

Pod 模板是包含在其他 Kubernetes 对象中用来创建 Pod 的模板，这类对象会根据 Pod 模板来创建和管理相应的 Pod。Kubernetes 中管理 Pod 的对象常见的有 Deployment、StatefulSet、DaemonSet、Job 等。这类对象也称为 workload 资源。

修改 Pod 模板并不会更新已经创建的 Pod，而是根据 Pod 模板创建新的 Pod 替换旧的 Pod。

下面是一个 Deployment 示例，它包含的 templete 字段就是 Pod 模板，这个 Deployment 会按照指定的 Pod 模板创建 Pod 并管理这些 Pod。

```yaml
apiVersion: apps/v1
kind: Deployment
metadata:
  name: nginx-deployment
  labels:
    app: nginx
spec:
  replicas: 3
  selector:
    matchLabels:
      app: nginx
  template:
    metadata:
      labels:
        app: nginx
    spec:
```

```
      containers:
      - name: nginx
        image: nginx:1.14.2
        ports:
        - containerPort: 80
```

这个 Deployment 会运行 3 个 Pod 副本，Pod 中运行 nginx：1.14.2 容器。容器通过暴露 TCP 80 端口来提供服务。

7.8.5 Pod 生命周期

Pod 的生命周期中包含多个阶段，开始阶段称为 Pending，如果 Pod 中至少有 1 个主要容器正常启动，Pod 便进入 Running 阶段，最后进入 Succeeded 或者 Failed 阶段，取决于 Pod 中的容器是否正常退出。如果容器发生错误而退出，则 Pod 会进入 Failed 阶段。如果容器正常退出 Pod，则会进入 Succeeded 阶段。

Pod 阶段是 Pod 在生命周期中所处位置的简单概述，kubectl describe pod 命令的输出信息 status 字段下的 phase 字段就是 Pod 阶段信息。可能的取值有 Pending、Running、Succeeded、Failed 和 Unknown。不同取值的描述如表 7-3 所示。

表 7-3 Pod 阶段不同取值的描述

取　　值	描　　述
Pending(待定)	Pod 已被 Kubernetes 接受，但有一个或者多个容器尚未创建和运行。此阶段包括等待 Pod 被调度的时间和通过网络下载镜像的时间
Running(运行中)	Pod 已经被绑定到了某个节点，Pod 中所有的容器都已被创建。至少有一个容器仍在运行，或者正处于启动或重启状态
Succeeded(成功)	Pod 中的所有容器都已成功终止，并且不会再重启
Failed(失败)	Pod 中的所有容器都已终止，并且至少有一个容器是因为失败而终止。也就是说，容器退出时状态码非 0 或者被系统终止
Unknown(未知)	因为某些原因而无法取得 Pod 的状态，这种情况通常是因为与 Pod 所在主机通信失败

Pod 在整个生命周期中只会被调度一次，也就是指派到 Node 上运行。Pod 会在调度完成后一直在分配的 Node 上运行，直到主动结束运行或者被终止。

Pod 在创建的时候会被分配一个唯一的 ID，也叫 UID，当 Pod 发生异常时，它不会被重新调度，而是可以由 Kubernetes 控制器创建新的几乎相同的 Pod 来替换它。新 Pod 的 UID 和之前的 Pod 是不同的。

进程在终止过程中往往需要时间执行一些清理操作，如果使用 KILL 信号粗暴关闭进程，进程就来不及进行清理操作，这可能会导致一些问题，所以不推荐以这种方式终止进程，建议使用优雅的方式来终止进程。也就是终止进程前，使进程有充足的时间来完成清理操作。Kubernetes 终止 Pod 中的进程同样遵循这个原理。

终止 Pod 时，容器运行时会向 Pod 的每个容器中的主进程发送 TERM 信号（许多容器运行时使用 STOPSIGNAL 信号而不是 TERM），之后容器就进入了优雅关闭的阶段。默认的优雅关闭时间是 30s，kubectl 支持--grace-period=< seconds >参数覆盖默认的优雅关闭时间。当优雅关闭超时后，会发送 KILL 强制终止容器中的进程，然后 Pod 会被从 Kubernetes API 服务器中删除。

7.8.6　Pod 中的容器状态

当调度器把 Pod 调度到某个 Node 之后，kubelet 开始使用容器运行时为 Pod 创建容器。和 Pod 类似，Kubernetes 同样会跟踪 Pod 中容器的状态，并根据不同状态采取相应操作使 Pod 恢复正常。当 Pod 中的容器产生某些错误而退出的时候，kubelet 会根据指定的重启策略来决定是否重启容器。Pod 中容器的状态有 Waiting、Running、Terminated。下面分别介绍。

（1）Waiting 状态表示容器正在执行启动必需的操作，例如下载容器镜像。使用 kubectl 查询包含处于 Waiting 状态的容器的 Pod 时，会看到 Reason 字段。这个字段描述了容器处于 Waiting 状态的原因。

（2）Running 状态表示容器正在正常运行，如果设置了 postStart 回调，则它在容器变为 Running 状态前一定已经成功执行。

（3）Terminated 状态表示容器已经终止，处于 Terminated 状态的容器可能是正常结束也可能是发生异常而终止。可以通过查询 Pod 查看容器退出码和 Reason 来进一步分析。如果设置了 preStop 回调，则它会在容器进入 Terminated 状态前被执行。

7.8.7　Probe

Probe 是 kubelet 定期对容器执行的诊断。借助 Probe，可以自定义容器处于健康状态的条件，以满足复杂的编排需求。执行诊断依赖容器实现相应的处理程序。处理程序有 3 种，分别是 ExecAction、TCPSocketAction、HTTPGetAction。下面分别介绍：

（1）ExecAction 通过在容器中执行自定义命令进行诊断，当执行命令的返回码是 0 时，认为通过了诊断。

（2）TCPSocketAction 通过检查容器中的 TCP 端口来诊断。通过诊断的条件是容器打开了指定的端口。

（3）HTTPGetAction 通过发送 HTTP GET 请求到容器来诊断。如果满足 200≤响应码＜400 就通过了诊断。

执行 Probe 可能会获取 3 种结果：Success、Failure 和 Unknown。Success 表示容器通过了诊断。Failure 表示未通过诊断。Unknown 表示诊断自身失败了，这种情况下 Kubernetes 不会采取任何操作。

Probe 有 3 种类型：livenessProbe、readinessProbe 和 startupProbe。

（1）livenessProbe 指示容器是否正在运行。如果 livenessProbe 探测失败，kubelet 会

终止容器。对于没有指定 livenessProbe 的容器来讲，livenessProbe 默认为成功状态。

适用 livenessProbe 的情况包括：希望容器在 livenessProbe 探测失败后重启。

（2）readinessProbe 指示容器是否就绪，也就是是否可以开始提供服务。如果 readinessProbe 探测失败，则容器所在的 Pod 的 IP 会被端点控制器从匹配到该 Pod 的服务的端点列表中移除，这样容器就不会接收到服务请求了。和 livenessProbe 类似，对于没有指定 readinessProbe 的容器来讲，readinessProbe 默认为成功状态。

如果希望容器在完成启动后才开始接收流量，就可以指定 readinessProbe。这对于启动过程需要执行其他操作的容器比较有用，例如有的容器在启动时需要加载很大的数据或者进行数据迁移等。readinessProbe 和 livenessProbe 可以同时指定，这种情况下，在通过 livenessProbe 探测而没有通过 readinessProbe 探测的时间段，容器不会收到请求流量。

（3）startupProbe 指示容器中的应用是否已经启动。如果指定了 startupProbe，则在 startupProbe 探测成功前，其他所有探测都会被禁用。如果 startupProbe 失败，则 kubelet 将终止容器，和 livenessProbe 类似，对于没有指定 startupProbe 的容器来讲，startupProbe 默认为成功状态。

startupProbe 适用于启动用时较长的容器。这类容器很可能因为启动时间过长而无法通过 livenessProbe 探测。使用 startupProbe 可以允许容器使用超过 livenessProbe 探测间隔的时间来启动。如果设置了 livenessProbe 并且容器启动时间经常超过 initialDelaySeconds + failureThreshold × periodSeconds，就应该指定 startupProbe。

7.8.8 Init 容器

Pod 中除了可以有多个应用容器，也可以有多个 Init 容器，Init 容器是会在 Pod 中应用容器启动前运行的特殊容器，一般用于完成初始化工作等。

Init 容器与普通的应用容器的区别是，普通容器一般需要长期运行，而 Init 容器必须可以成功结束运行，否则 Pod 无法就绪。正因如此，Init 容器不支持设置 livenessProbe、readinessProbe 或者 startupProbe。

如果指定了多个 Init 容器，则它们必须依次运行成功，只有上一个 Init 容器运行成功并结束后，才可以开始运行下一个 Init 容器。

如果 Init 容器运行失败，则 kubelet 会不断重启 Init 容器直到运行成功，而如果 Pod 将 restartPolicy 设置为 Never，则失败的 Init 容器会导致整个 Pod 进入失败状态。

Init 容器通过 Pod 的 initContainers 字段指定，值为容器列表。示例如下：

```
apiVersion: v1
kind: Pod
metadata:
  name: myapp-pod
  labels:
    app: myapp
spec:
```

```
  containers:
  - name: myapp - container
    image: busybox:1.28
    command: ['sh', '-c', 'echo The app is running! && sleep 3600']
  initContainers:
  - name: init - myservice
    image: busybox:1.28
    command: ['sh', '-c', "until nslookup myservice; do echo waiting for myservice; sleep 2; done"]
  - name: init - mydb
    image: busybox:1.28
    command: ['sh', '-c', "until nslookup mydb; do echo waiting for mydb; sleep 2; done"]
```

此文件定义了包含 2 个 Init 容器的 Pod，第 1 个 Init 容器会等待 myservice 就绪，第 2 个 Init 容器会等待 mydb 就绪。只有当 2 个 Init 容器都运行成功后，myapp-container 才会运行。

使用此文件创建 Pod 后，查看 Pod 信息，命令如下：

```
kubectl get pod myapp - pod
```

运行命令后输出的信息如下：

```
NAME          READY   STATUS     RESTARTS   AGE
myapp - pod   0/1     Init: 0/2  0            13s
```

输出显示 myapp-pod 未就绪，并且其中的 2 个 Init 容器都没有成功运行。

查看 init-myservice 容器的日志，命令如下：

```
kubectl logs myapp - pod  - c init - myservice
```

运行命令后输出的信息如下：

```
Server:     10.100.0.10
Address 1: 10.100.0.10 kube - dns.kube - system.svc.cluster.local

waiting for myservice
nslookup: can't resolve 'myservice'
...
```

为了使 Init 容器可以运行成功，需要创建 myservice 和 mydb。

创建文件 services.yaml，编辑如下：

```
---
apiVersion: v1
kind: Service
```

```
metadata:
  name: myservice
spec:
  ports:
   - protocol: TCP
     port: 80
     targetPort: 9376
---
apiVersion: v1
kind: Service
metadata:
  name: mydb
spec:
  ports:
   - protocol: TCP
     port: 80
     targetPort: 9377
```

此文件定义了 2 个 Service，即 myservice 和 mydb。使用此文件创建这两个 Service 之后，再次查看 myapp-pod 的信息。会发现 myapp-pod 已经就绪了，信息如下：

```
NAME        READY  STATUS    RESTARTS  AGE
myapp-pod   1/1    Running   0         5m44s
```

7.9 ReplicaSet

7.9.1 ReplicaSet 简介

在单个 Pod 无法满足高可用和请求压力的场景下，往往需要运行多个相同的 Pod 来解决，例如 Web 前端应用等无状态应用就可以通过运行多个服务器程序实例来提升请求处理能力，实现线性扩展。

手动维护多个 Pod 不仅烦琐而且容易出错，尤其是多个 Pod 分布到不同的节点的情况下手动维护会更复杂。Kubernetes 提供了 ReplicaSet 来维护指定数量的 Pod 副本集合，Pod 副本集合是一组完全相同的 Pod。当节点发生故障时，ReplicaSet 中的 Pod 会自动重新调度到正常的节点，从而保证 Pod 的高可用性。这是手动管理 Pod 难以做到的，所以不推荐直接手动创建和管理 Pod。

ReplicaSet 与 Pod 的关系类似于 supervisor 和进程的关系，当进程意外退出后，supervisor 会重新创建同样的进程，从而保持进程一直处于运行状态。

ReplicaSet 通过定义中的选择器指定的标签匹配 Pod，当集群中满足标签的 Pod 副本数量少于指定数量时，ReplicaSet 根据定义中的 Pod 模板创建新的 Pod，以此补充缺少的

Pod 副本，当集群中 Pod 副本数量多于指定数量时，ReplicaSet 则会删除多余的 Pod。最终使集群中的 Pod 副本数量等于指定的 Pod 副本数量。

由 ReplicaSet 创建的 Pod 会通过 Pod 中的 metadata.ownerReferences 属性和 ReplicaSet 关联。ownerReferences 属性用来表示某个对象所属的对象。如果集群中存在和 ReplicaSet 指定的标签相匹配的 Pod，则这些 Pod 会立即被 ReplicaSet 获取并管理起来。

7.9.2 ReplicaSet 示例

ReplicaSet 定义的示例文件 frontend.yaml 的内容如下：

```
apiVersion: apps/v1
kind: ReplicaSet
metadata:
  name: frontend
  labels:
tier: frontend
spec:
  replicas: 3
  selector:
    matchLabels:
tier: frontend
  template:
    metadata:
      labels:
tier: frontend
    spec:
      containers:
      - name: web
        image: bitmyth/accounts-frontend:v1.0.54
```

下面解释文件中的指令含义。

将 ReplicaSet 的名字定义为 frontend，ReplicaSet 的标签是 tier：frontend，代码如下：

```
metadata:
  name: frontend
  labels:
tier: frontend
```

spec 中的 replicas 字段用于指定 Pod 的副本数量，这里 replicas：3 表示 3 个副本。如果不指定 replicas，会使用默认值 1，代码如下：

```
replicas: 3
```

spec 中的 selector 字段定义了 ReplicaSet 的选择器，选择器包含选择 Pod 使用的标签，代码如下：

```
  selector:
    matchLabels:
tier: frontend
```

spec 中的 template 字段用于定义创建 Pod 时使用的模板。

```
template:
```

使用 frontend.yaml 创建 ReplicaSet，命令如下：

```
kubectl apply -f frontend.yaml
```

运行命令后输出的信息如下：

```
replicaset.apps/frontend created
```

查询 Kubernetes 中的 ReplicaSet，命令如下：

```
kubectl get replicaset
```

也可以使用简写形式，命令如下：

```
kubectl get rs
```

运行命令后输出的类似信息如下：

```
NAME      DESIRED   CURRENT   READY   AGE
frontend  3         3         3       11s
```

查询 ReplicaSet 的详细信息，命令如下：

```
kubectl describe rs frontend
```

运行命令后输出的类似信息如下：

```
Name:          frontend
Namespace:     default
Selector:      tier=frontend
Labels:        tier=frontend
Annotations:   <none>
Replicas:      3 current / 3 desired
Pods Status:   3 Running / 0 Waiting / 0 Succeeded / 0 Failed
Pod Template:
```

```
  Labels:   tier = frontend
  Containers:
   web:
    Image:         bitmyth/accounts - frontend:v1.0.54
    Port:          < none >
    Host Port:     < none >
    Environment:   < none >
    Mounts:        < none >
  Volumes:         < none >
Events:
  Type     Reason             Age   From                    Message
  ----     ------             ----  ----                    --------
  Normal   SuccessfulCreate   13s   replicaset - controller  Created pod: frontend - 4tgr4
  Normal   SuccessfulCreate   13s   replicaset - controller  Created pod: frontend - p6j7j
  Normal   SuccessfulCreate   13s   replicaset - controller  Created pod: frontend - qbqdn
```

根据 Events 字段可以看出 ReplicaSet 创建了 3 个 Pod。

查询 ReplicaSet 中的 Pod，命令如下：

```
kubectl get pod  - l tier = frontend
```

运行命令后输出的类似信息如下：

```
NAME              READY   STATUS    RESTARTS   AGE
frontend - 4tgr4   1/1     Running   0          67s
frontend - p6j7j   1/1     Running   0          67s
frontend - qbqdn   1/1     Running   0          67s
```

查看 ReplicaSet 中 Pod 的 metadata 信息，例如查看 Pod frontend-4tgr4，命令如下：

```
kubectl get pod frontend - 4tgr4  - o yaml
```

运行命令后输出的类似信息如下：

```
apiVersion: v1
kind: Pod
metadata:
  creationTimestamp: "2021 - 04 - 30T03:11:43Z"
  generateName: frontend -
  labels:
    tier: frontend
  name: frontend - 4tgr4
  namespace: default
  ownerReferences:
  - apiVersion: apps/v1
```

```
      blockOwnerDeletion: true
      controller: true
      kind: ReplicaSet
      name: frontend
      uid: 5b7e204d-a39a-43b2-8424-56f8511be6a1
    resourceVersion: "2698"
    uid: 7b1f4e1b-adaf-4fb1-b72d-6f3f1b972a0e
...
```

可以看到 metadata 中包含 ownerReferences 字段，ownerReferences 表示 Pod 所属对象，正式名为 frontend 的 ReplicaSet 对象。

7.9.3 获取模板以外的 Pod

当集群中存在某些并非由 ReplicaSet 创建的 Pod 可以和 ReplicaSet 中的选择器匹配时，这些 Pod 会被 ReplicaSet 获取。

例如有以下 Pod：

```
apiVersion: v1
kind: Pod
metadata:
  name: pod1
  labels:
    tier: frontend
spec:
  containers:
  - name: hello1
    image: nginx
```

当创建这个 Pod 之后，它会被之前创建的 frontend ReplicaSet 获取，命令如下：

```
kubectl get pod -l tier=frontend
```

运行命令后输出的类似信息如下：

```
NAME             READY   STATUS        RESTARTS   AGE
frontend-4tgr4   1/1     Running       0          6m
frontend-p6j7j   1/1     Running       0          6m
frontend-qbqdn   1/1     Running       0          6m
pod1             0/1     Terminating   0          2s
```

可以看到由于 Pod 总数超过了 ReplicaSet 规定的副本数 3，所以 pod1 被终止。说明 pod1 被 ReplicaSet 获取并控制。

为了避免这种意外情况出现，需要在创建 Pod 时确认 Pod 的标签不会匹配到现有的 ReplicaSet 中的选择器。

7.9.4 缩放 ReplicaSet

缩放 ReplicaSet，也就是改变 ReplicaSet 中的副本数，可以通过修改 ReplicaSet 的 .spec.replicas 字段实现，例如将前面的名为 frontend 的 ReplicaSet 的副本数修改为 2，代码如下：

```
spec:
  replicas: 2
```

由于之前 replicas 字段的值是 3，现在的值是 2，ReplicaSet 会删除一个 Pod 使 Pod 数量和指定的 replicas 匹配。

查看 Pod，命令如下：

```
kubectl get pod -l tier=frontend
```

运行命令后输出的类似信息如下：

```
NAME              READY   STATUS       RESTARTS   AGE
frontend-p6j7j    1/1     Running      0          16m
frontend-qbqdn    1/1     Running      0          16m
frontend-4tgr4    0/1     Terminating  0          16m
```

可以看到 ReplicaSet 中的一个 Pod 被终止，现在只运行 2 个 Pod。和 ReplicaSet 中指定的 replicas 值相等。

通常情况下不建议使用 ReplicaSet 管理 Pod，而是使用 ReplicaSet 上层的 Deployment。

7.10 Deployment

7.10.1 Deployment 简介

Deployment 与 ReplicaSet 类似，但是提供了更多高级的功能，例如滚动更新。事实上 Deployment 中包含 ReplicaSet，并通过 ReplicaSet 来维持 Pod 数量的稳定。Deployment 提供了对 Pod 和 ReplicaSet 声明式更新的功能。

Deployment 用于部署无状态应用。

7.10.2 Deployment 示例

Deployment 示例文件 nginx-deployment.yaml 的内容如下：

```
apiVersion: apps/v1
kind: Deployment
metadata:
  name: nginx - deployment
  labels:
    app: nginx
spec:
  replicas: 3
  selector:
    matchLabels:
      app: nginx
  template:
    metadata:
      labels:
        app: nginx
    spec:
      containers:
      - name: nginx
        image: nginx:1.14.2
        ports:
         - containerPort: 80
```

下面解释文件中的指令。

metadata 下的 name：nginx-deployment 表示把 Deployment 命名为 nginx-deployment，代码如下：

```
name: nginx - deployment
```

.spec.replicas 字段将 Pod 副本数量设置为 3。如果不进行设置，则会使用默认值 1，代码如下：

```
replicas: 3
```

.spec.selector 字段用于设置 Pod 选择器，选择器使用的标签是 app＝nginx。.spec.selector.必须和.spec.template.metadata.labels 字段值匹配，代码如下：

```
selector:
  matchLabels:
    app: nginx
```

.spec.template 用于定义 Deployment 创建 Pod 时使用的模板。

```
template:
```

. spec. template. metadata 字段用于将 Pod 标签定义为 app=nginx。

```
metadata:
  labels:
    app: nginx
```

. spec 字段用于指定 Pod 中运行容器的规约，代码如下：

```
spec:
```

. spec. containers 字段用于定义 Pod 中运行 nginx：1. 14. 2 容器，容器命名为 nginx，代码如下：

```
containers:
    - name: nginx
      image: nginx: 1.14.2
```

容器暴露 TCP 80 端口来提供服务，代码如下：

```
ports:
    - containerPort: 80
```

使用 nginx-deployment. yaml 文件创建 Deployment，命令如下：

```
kubectl apply - f nginx - deployment.yaml -- record
```

命令中的--record 参数会把当前运行的 kubectl 命令添加为 Deployment 的注解。

运行命令后输出的信息如下：

```
deployment.apps/nginx - deployment created
```

获取 Deployment 创建的 Pod，使用标签 app=nginx 筛选，命令如下：

```
kubectl get pod - l app = nginx
```

运行命令后输出的类似信息如下：

```
NAME                                READY   STATUS    RESTARTS   AGE
nginx - deployment - 66b6c48dd5 - krjs8    1/1     Running   0          9m
nginx - deployment - 66b6c48dd5 - nrcxw    1/1     Running   0          9m
nginx - deployment - 66b6c48dd5 - ts48w    1/1     Running   0          9m
```

可以看到 Deployment 创建了 3 个 Pod。

以 nginx-deployment-66b6c48dd5-krjs8 为例，查看 Pod 的详细信息，命令如下：

```
kubectl get pod nginx-deployment-66b6c48dd5-krjs8 -o yaml
```

运行命令后输出的类似信息如下：

```
apiVersion: v1
kind: Pod
metadata:
  creationTimestamp: "2021-05-01T09:37:01Z"
  generateName: nginx-deployment-66b6c48dd5-
  labels:
    app: nginx
    pod-template-hash: 66b6c48dd5
  name: nginx-deployment-66b6c48dd5-krjs8
  namespace: default
  ownerReferences:
  - apiVersion: apps/v1
    blockOwnerDeletion: true
    controller: true
    kind: ReplicaSet
    name: nginx-deployment-66b6c48dd5
    uid: b940768a-f9b1-4aae-bfac-d242b42c31ea
  resourceVersion: "11500"
  uid: 3df2ffb1-ca3f-4fb5-93d8-4bd5fd9b783c
...
```

可以看到 nginx-deployment-66b6c48dd5-krjs8 这个 Pod 的 ownerReferences 是名为 nginx-deployment-66b6c48dd5 的 ReplicaSet。由于 Deployment 通过 ReplicaSet 实现 Pod 副本数量的稳定，所以 Deployment 中的 Pod 并不是直接由 Deployment 创建的，而是由 ReplicaSet 创建的。

查看这个 ReplicaSet 的详细信息，命令如下：

```
kubectl get rs nginx-deployment-66b6c48dd5 -o yaml
```

运行命令后输出的类似信息如下：

```
apiVersion: apps/v1
kind: ReplicaSet
metadata:
  annotations:
    deployment.kubernetes.io/desired-replicas: "3"
    deployment.kubernetes.io/max-replicas: "4"
    deployment.kubernetes.io/revision: "1"
  creationTimestamp: "2021-05-01T09:37:01Z"
  generation: 1
```

```
    labels:
      app: nginx
      pod-template-hash: 66b6c48dd5
    name: nginx-deployment-66b6c48dd5
    namespace: default
    ownerReferences:
    - apiVersion: apps/v1
      blockOwnerDeletion: true
      controller: true
      kind: Deployment
      name: nginx-deployment
      uid: 903f73d0-07cd-4686-ad13-185ae1b5f639
    resourceVersion: "11512"
    uid: b940768a-f9b1-4aae-bfac-d242b42c31ea
...
```

输出信息显示 ReplicaSet 对象 nginx-deployment-66b6c48dd5 所属对象是名为 nginx-deployment 的 Deployment。说明这个 ReplicaSet 对象是由 Deployment 创建的。

这个 ReplicaSet 对象名 nginx-deployment-66b6c48dd5 中的后缀 66b6c48dd5 是它的 pod-template-hash 的标签值，信息如下：

```
pod-template-hash: 66b6c48dd5
```

这个标签是 ReplicaSet 的 Pod 模板的哈希值，由 Deployment 为 ReplicaSet 添加。

7.10.3 更新 Deployment

当 Deployment 的 Pod 模板变更后，会触发 Deployment 进行 rollout(发布)。也就是更新 Deployment 中的 Pod。

例如把上述 nginx-deployment 中的 Pod 使用的镜像从 nginx：1.14.2 升级到 nginx：1.16.1，命令如下：

```
kubectl set image deploymentnginx-deployment nginx=nginx:1.16.1 --record
```

命令中的--record 参数会把当前运行的 kubectl 命令添加为 Deployment 的注解，注解如下：

```
kubernetes.io/change-cause: kubectl set image deploymentnginx-deployment nginx=nginx:1.16.1
        --record=true
```

运行命令后输出的信息如下：

```
deployment.apps/nginx-deployment image updated
```

也可以直接编辑集群中的 Deployment 配置文件修改镜像，命令如下：

```
kubectl edit deploymentnginx - deployment
```

还可以修改本地的 nginx-deployment.yaml 文件后，使用 kubectl apply 命令实现更新 Deployment，这是最推荐的方式，因为这样可以把变更加入版本控制中。

查看这次 rollout 的状态，命令如下：

```
kubectl rollout status deployment nginx - deployment
```

运行命令后输出的类似信息如下：

```
Waiting for deployment "nginx - deployment" rollout to finish: 2 out of 3 new replicas have been updated...
Waiting for deployment "nginx - deployment" rollout to finish: 2 out of 3 new replicas have been updated...
Waiting for deployment "nginx - deployment" rollout to finish: 2 out of 3 new replicas have been updated...
Waiting for deployment "nginx - deployment" rollout to finish: 1 old replicas are pending termination...
Waiting for deployment "nginx - deployment" rollout to finish: 1 old replicas are pending termination...
deployment "nginx - deployment" successfully rolled out
```

这时 Deployment 更新成功，查看 nginx-deployment，命令如下：

```
kubectl get deployment
```

运行命令后输出的信息如下：

```
NAME               READY   UP - TO - DATE   AVAILABLE   AGE
nginx - deployment  3/3     3                3           31m
```

再次查看 ReplicaSet，命令如下：

```
kubectl get rs
```

运行命令后输出的类似信息如下：

```
NAME                              DESIRED   CURRENT   READY    AGE
nginx - deployment - 559d658b74   3         3         3        31m
nginx - deployment - 66b6c48dd5   0         0         0        31m
```

显示有两个 ReplicaSet，使用旧镜像 nginx：1.14.2 的 ReplicaSet 从 3 个 Pod 缩容为

1个。使用新镜像 nginx：1.16.1 的 ReplicaSet 扩容为 3 个。

可以在修改镜像之前在另一个终端窗口使用带有--watch 参数的查询命令观察这个过程，--watch 参数可以持续监视对象的变化，可简写为-w。命令如下：

```
kubectl get rs -- watch
```

运行命令后输出的类似信息如下：

```
NAME                          DESIRED   CURRENT   READY    AGE
nginx - deployment - 559d658b74   0         0         0        4h17m
nginx - deployment - 66b6c48dd5   3         3         3        4h17m
nginx - deployment - 559d658b74   0         0         0        4h17m
nginx - deployment - 559d658b74   1         0         0        4h17m
nginx - deployment - 559d658b74   1         0         0        4h17m
nginx - deployment - 559d658b74   1         1         0        4h17m
nginx - deployment - 559d658b74   1         1         1        4h17m
nginx - deployment - 66b6c48dd5   2         3         3        4h18m
nginx - deployment - 559d658b74   2         1         1        4h17m
nginx - deployment - 66b6c48dd5   2         3         3        4h18m
nginx - deployment - 559d658b74   2         1         1        4h17m
nginx - deployment - 66b6c48dd5   2         2         2        4h18m
nginx - deployment - 559d658b74   2         2         1        4h17m
nginx - deployment - 559d658b74   2         2         2        4h17m
nginx - deployment - 66b6c48dd5   1         2         2        4h18m
nginx - deployment - 66b6c48dd5   1         2         2        4h18m
nginx - deployment - 559d658b74   3         2         2        4h17m
nginx - deployment - 66b6c48dd5   1         1         1        4h18m
nginx - deployment - 559d658b74   3         2         2        4h17m
nginx - deployment - 559d658b74   3         3         2        4h17m
nginx - deployment - 559d658b74   3         3         3        4h17m
nginx - deployment - 66b6c48dd5   0         1         1        4h18m
nginx - deployment - 66b6c48dd5   0         1         1        4h18m
nginx - deployment - 66b6c48dd5   0         0         0        4h18m
nginx - deployment - 66b6c48dd5   0         0         0        39m
```

输出信息显示 2 个 ReplicaSet 交替进行缩放，最后完成了 Deployment 的更新与发布。保证了发布过程中服务不间断运行。如果直接终止所有运行旧镜像的 Pod，然后创建新 Pod，则会导致一定时间内服务不可用。

Deployment 会确保更新 Pod 的过程中一直有运行状态的 Pod。任意时刻只有一部分 Pod 会被终止，默认不可用 Pod 的最大比例为 25%。同时也确保创建的 Pod 不会超过期望 Pod 数量的 25%，这种更新策略称为滚动更新。更新策略通过.spec.strategy 字段设置，默认策略为 RollingUpdate，RollingUpdate 策略定义示例如下：

```
spec:
  strategy:
    rollingUpdate:
      maxSurge: 25%
      maxUnavailable: 25%
    type: RollingUpdate
```

maxUnavailable 表示在更新期间不可用 Pod 的最大数值。它的值可以是整数(例如 5),也可以是百分数。如果是百分数形式(例如 20%),不可用 Pod 的绝对数值由百分数计算且向下取整。向下取整可以尽可能减少不可用的 Pod 的数量。在更新 Deployment 的整个过程中不可用的 Pod 的数量会始终保持不超过 maxUnavailable 指定的值。

maxSurge 表示在更新期间新创建的 Pod 数量超过期望数的最大数值。它的值可以是整数(例如 5),也可以是百分数(例如 20%)。如果采用百分数形式,新创建 Pod 的绝对数值由百分数计算且向上取整。在更新 Deployment 的整个过程中超过期望副本数的 Pod 的数量会始终保持不超过 maxSurge 指定的值。

如果将 maxSurge 指定为 0,则 maxUnavailable 不可以设置为 0。否则 Deployment 无法完成更新。因为删除旧 Pod 无法满足 maxUnavailable 的限制,创建新 Pod 无法满足 maxSurge 的限制,陷入进退两难的境地。

假设 maxSurge 和 maxUnavailable 都设置为 0,则在创建 Deployment 时会失败,提示如下:

```
The Deployment "nginx-deployment" is invalid: \
spec.strategy.rollingUpdate.maxUnavailable: Invalid value: \
intstr.IntOrString{Type:0, IntVal:0, StrVal:""}: may not be 0 when 'maxSurge' is 0
```

另外一种策略是 Recreate。采用 Recreate 策略的 Deployment 在更新时会先把现有的 Pod 全部终止,然后创建新的 Pod。Recreate 策略定义示例如下:

```
spec:
  strategy:
    type: Recreate
```

查询 nginx-deployment 的详细信息,命令如下:

```
kubectl describe deployment nginx-deployment
```

输出信息中包含 Deployment 更新策略,信息如下:

```
StrategyType:           RollingUpdate
RollingUpdateStrategy:  25% max unavailable, 25% max surge
```

在更新镜像前,可以通过监视 Pod 观察 Deployment 的升级过程,命令如下:

```
kubectl get pod -w
```

运行命令后输出的信息如下:

```
NAME                                  READY STATUS   RESTARTS AGE
nginx-deployment-66b6c48dd5-g6cdb 1/1   Running 0   3m42s
nginx-deployment-66b6c48dd5-kn9jt   1/1   Running  0 3m44s
nginx-deployment-66b6c48dd5-mfllc   1/1   Running  0 3m41s
nginx-deployment-559d658b74-cb8gk 0/1   Pending   0   0s
nginx-deployment-559d658b74-cb8gk 0/1   Pending   0   0s
nginx-deployment-559d658b74-cb8gk 0/1   ContainerCreating 0       0s
nginx-deployment-559d658b74-cb8gk   1/1   Running              0          2s
nginx-deployment-66b6c48dd5-mfllc   1/1  Terminating         0          3m49s
nginx-deployment-559d658b74-xqds4   0/1   Pending              0         0s
nginx-deployment-559d658b74-xqds4   0/1   Pending             0         0s
nginx-deployment-559d658b74-xqds4   0/1   ContainerCreating  0   0s
nginx-deployment-559d658b74-xqds4   1/1  Running             0         1s
nginx-deployment-66b6c48dd5-mfllc   0/1   Terminating          0         3m50s
nginx-deployment-66b6c48dd5-g6cdb   1/1   Terminating         0         3m51s
nginx-deployment-559d658b74-nj7hw   0/1   Pending              0         0s
nginx-deployment-559d658b74-nj7hw   0/1   Pending              0         0s
nginx-deployment-559d658b74-nj7hw   0/1   ContainerCreating  0        0s
nginx-deployment-66b6c48dd5-g6cdb   0/1   Terminating          0         3m52s
nginx-deployment-559d658b74-nj7hw   1/1   Running             0         1s
nginx-deployment-66b6c48dd5-kn9jt   1/1   Terminating          0         3m54s
nginx-deployment-66b6c48dd5-mfllc   0/1   Terminating          0         3m51s
nginx-deployment-66b6c48dd5-mfllc   0/1   Terminating          0          3m51s
nginx-deployment-66b6c48dd5-kn9jt 0/1   Terminating          0          3m55s
nginx-deployment-66b6c48dd5-kn9jt   0/1   Terminating          0              4m1s
nginx-deployment-66b6c48dd5-kn9jt   0/1   Terminating          0              4m1s
nginx-deployment-66b6c48dd5-g6cdb   0/1   Terminating          0              3m59s
nginx-deployment-66b6c48dd5-g6cdb   0/1   Terminating          0              3m59s
```

输出信息显示新旧 Pod 交替进行创建和终止,最终所有旧 Pod 升级为新 Pod。

仔细观察这个过程可以发现 Deployment 首先创建了一个新的 Pod 而不是终止一个旧 Pod。这是由于计算出的最大不可用 Pod 的数量为期望副本数乘以 maxUnavailable(3×25%=0.75),向下取整得 0,Pod 最大增加数为期望副本数乘以 maxSurge(3×0.25%=0.75),向上取整得 1。如果先终止一个旧 Pod,则不可用 Pod 的数量变成 1,大于最大不可用 Pod 的数量,而先创建一个新的 Pod,超过期望副本数的 Pod 的数量是 1,满足 maxSurge 限制,所以只能优先创建新的 Pod。

Deployment(nginx-deployment)的详细信息中的 Events 字段也记录了 ReplicaSet 的变化过程,信息如下:

```
Events:
  Type    Reason             Age   From                   Message
  ----    ------             ----  ----                   -------
  Normal  ScalingReplicaSet  4m3s  deployment-controller  Scaled up replica set nginx-deployment-66b6c48dd5 to 3
  Normal  ScalingReplicaSet  20s   deployment-controller  Scaled up replica set nginx-deployment-559d658b74 to 1
  Normal  ScalingReplicaSet  19s   deployment-controller  Scaled down replica set nginx-deployment-66b6c48dd5 to 2
  Normal  ScalingReplicaSet  19s   deployment-controller  Scaled up replica set nginx-deployment-559d658b74 to 2
  Normal  ScalingReplicaSet  17s   deployment-controller  Scaled down replica set nginx-deployment-66b6c48dd5 to 1
  Normal  ScalingReplicaSet  17s   deployment-controller  Scaled up replica set nginx-deployment-559d658b74 to 3
  Normal  ScalingReplicaSet  16s   deployment-controller  Scaled down replica set nginx-deployment-66b6c48dd5 to 0
```

Events 中清楚地显示了如何对新旧 ReplicaSet 交替进行缩放的过程。

下面介绍 Deployment 如何同时处理多个更新的情况。

并发更新 Deployment 称为 rollover，也就是在上一次更新还未完成的时刻再次进行更新，此时 Deployment 控制器会马上创建新的 ReplicaSet 并开始扩容新的 ReplicaSet 中的副本数。同时立即对旧的 ReplicaSet 进行缩容，终止其中的 Pod。

例如有一个指定 5 个 Pod 副本的 Deployment 正在发布，发布过程会创建 5 个 nginx：1.14.2 镜像的 Pod，在发布未完成的时刻，假设已经创建了 3 个运行 nginx：1.14.2 镜像的 Pod，这时更新 Deployment 的 Pod 模板，改为使用 nginx：1.16.1 镜像。这种情况下，Deployment 不会等待之前的 5 个 nginx：1.14.2 Pod 创建完成，而是立刻开始终止旧的 nginx：1.14.2 Pod，并开始创建 5 个 nginx：1.16.1 Pod。

7.10.4 回滚 Deployment

在发布 Deployment 的过程中如果出现错误，则可以进行回滚。假设把 nginx-deployment 的镜像修改为错误的 nginx：1.161，这个镜像并不存在，所以发布会失败。修改镜像命令如下：

```
kubectl set image deployment.v1.apps/nginx-deployment nginx=nginx:1.161 --record=true
```

运行命令后输出的内容如下：

```
deployment.apps/nginx-deployment image updated
```

这次发布会卡住，查看此次发布状态来验证，命令如下：

```
kubectl rollout status deployment nginx-deployment
```

运行命令后输出的内容如下：

```
Waiting for deployment "nginx-deployment" rollout to finish: 1 out of 3 new replicas have been
updated...
```

查看 Pod 状态，命令如下：

```
kubectl get pod
```

运行命令后输出的内容如下：

```
AME                                READY   STATUS             RESTARTS   AGE
nginx-deployment-66b6c48dd5-6kdwb   1/1     Running            0          20m
nginx-deployment-66b6c48dd5-dzj62   1/1     Running            0          20m
nginx-deployment-66b6c48dd5-xkzn8   1/1     Running            0          20m
nginx-deployment-66bc5d6c8-mlzr2    0/1     ImagePullBackOff   0          6m38s
```

输出信息显示 Pod nginx-deployment-66bc5d6c8-mlzr2 发生错误 ImagePullBackOff。可以看到在发生错误后，Deployment 控制器自动中止了这次发布，没有继续创建新的 Pod。这是由 Deployment 滚动更新策略中的 maxUnavailable 值决定的，默认为 25%。

回滚之前查看 Deployment 的发布历史，命令如下：

```
kubectl rollout history deployment nginx-deployment
```

运行命令后输出的内容如下：

```
deployment.apps/nginx-deployment
REVISION   CHANGE-CAUSE
1          kubectl apply \
 --filename=/Users/gsh/k8s/examples/controllers/nginx-deployment.yaml --record=true
2          kubectl set image deployment/nginx-deployment nginx=nginx:1.16.1 --record=true
3          kubectl set image deployment/nginx-deployment nginx=nginx:1.161 --record=true
```

查看具体某次发布或修订的具体信息，命令如下：

```
kubectl rollout history deployment nginx-deployment --revision=2
```

运行命令后输出的内容如下：

```
deployment.apps/nginx-deployment with revision #2
Pod Template:
  Labels:   app=nginx
```

```
    pod-template-hash=559d658b74
  Annotations:  kubernetes.io/change-cause: kubectl set image deployment/nginx-deployment
nginx=nginx:1.16.1 --record=true
  Containers:
   nginx:
    Image:  nginx:1.16.1
    Port:  80/TCP
    Host Port:  0/TCP
    Environment:  <none>
    Mounts:  <none>
  Volumes:  <none>
```

版本 2 是当前 nginx-deployment 的上一个版本，现在回滚到版本 2，命令如下：

```
kubectl rollout undo deployment nginx-deployment
```

可以使用参数--to-revision 指定回滚到的具体版本，例如--to-revision=2。

运行命令后输出的内容如下：

```
deployment.apps/nginx-deployment rolled back
```

再次查看发布历史，命令如下：

```
kubectl rollout history deployment nginx-deployment
```

运行命令后输出的信息如下：

```
deployment.apps/nginx-deployment
REVISION  CHANGE-CAUSE
1         kubectl apply \
--filename=/Users/gsh/k8s/examples/controllers/nginx-deployment.yaml --record=true
3         kubectl set image deployment/nginx-deployment nginx=nginx:1.161 --record=true
4         kubectl set image deployment/nginx-deployment nginx=nginx:1.16.1 --record=true
```

在回滚之前在另一个终端窗口监视 Pod 的变化，命令如下：

```
kubectl get pod --watch
```

在回滚过程中输出的内容如下：

```
NAME                               READY   STATUS      RESTARTS   AGE
nginx-deployment-559d658b74-4v62q   1/1     Running     0          22m
nginx-deployment-559d658b74-bdf5v   1/1     Running     0          22m
nginx-deployment-559d658b74-lfmnh   1/1     Running     0          22m
```

```
nginx - deployment - 66bc5d6c8 - dw92c     0/1     ImagePullBackOff     0          10m
nginx - deployment - 66bc5d6c8 - dw92c     0/1      Terminating         0          10m
nginx - deployment - 66bc5d6c8 - dw92c     0/1      Terminating         0          11m
nginx - deployment - 66bc5d6c8 - dw92c     0/1      Terminating         0          11m
```

显示 Deployment 终止了出现 ImagePullBackOff 的 Pod。

为了验证回滚是否成功，可查看 nginx-deployment 的信息，命令如下：

```
kubectl get deployment nginx - deployment
```

运行命令后输出的信息如下：

```
NAME                 READY    UP - TO - DATE   AVAILABLE    AGE
nginx - deployment   3/3       3 3             53m
```

显示一共存在 3 个 Pod，有 3 个处于运行状态。

查看 nginx-deployment 的详细信息，命令如下：

```
kubectl describe deployment nginx - deployment
```

运行命令后输出的截取信息如下：

```
Name:                      nginx - deployment
Namespace:                 default
CreationTimestamp:        Wed, 05 May 2021 10:45:54  + 0800
Labels:                    app = nginx
Annotations:               deployment.kubernetes.io/revision: 4
                            kubernetes.io/change - cause: kubectl set image deployment/nginx -
deployment nginx = nginx:1.16.1  -- record = true
Selector:                  app = nginx
Replicas:                  3 desired | 3 updated | 3 total | 3 available | 0 unavailable
StrategyType:              RollingUpdate
MinReadySeconds:           0
RollingUpdateStrategy:    25 % max unavailable, 25 % max surge
Pod Template:
  Labels:   app = nginx
  Containers:
   nginx:
    Image:          nginx:1.16.1
    Port:          80/TCP
    Host Port:     0/TCP
    Environment:   < none >
    Mounts:        < none >
  Volumes:          < none >
```

在输出信息的 PodTemplate 字段中显示现在 nginx-deployment 使用的镜像回滚到了之前的版本，

```
Image: Nginx: 1.16.1
```

证明这次回滚成功。

Deployment 的修订历史存储在它控制的 ReplicaSet 中。当旧的 ReplicaSet 被删除，就会失去回滚到旧 ReplicaSet 对应版本的能力。默认会保留 10 个旧 ReplicaSet。这个数值通过 Deployment 的. spec. revisionHistoryLimit 字段设置，默认值为 10。如果把这个字段设置为 0，则副本数为 0 的旧 ReplicaSet 都会被清除，Deployment 从而会失去回滚的能力。

7.10.5 缩放 Deployment

缩放 Deployment，也就是改变 Deployment 中的 Pod 副本数，例如把 nginx-deployment 缩放为 10 个副本，命令如下：

```
kubectl scale deployment nginx - deployment  -- replicas = 10
```

运行命令后输出的信息如下：

```
deployment.apps/nginx - deployment scaled
```

查看缩放后的 nginx-deployment 的信息，命令如下：

```
kubectl get deployment nginx - deployment
```

运行命令后输出的信息如下：

```
NAME                 READY   UP - TO - DATE   AVAILABLE   AGE
nginx - deployment   10/10   10               10          23h
```

显示已经有 10 个 Pod 在运行。

7.10.6 暂停和恢复 Deployment

Deployment 支持暂停和恢复发布，在暂停期间可以对 Deployment 进行多次更新，这些更新都不会生效，也不会记录到发布历史(Rollout History)中。在恢复之后，暂停期间所做的更新会合并起来一起发布，产生一次发布记录。利用这个功能可以将 Deployment 的多次更新统一发布，而不会产生多个不必要的发布记录。

下面删除之前创建的 nginx-deployment，重新使用 nginx-deployment. yaml 文件新创建 Deployment，先后进行两次更新，第一次修改 Pod 模板中的镜像，第二次修改资源限制。在更新前暂停 Deployment，在完成两次更新后，恢复 Deployment，然后观察发布历史。

查看新创建的 Deployment 的详细信息，命令如下：

```
kubectl describe deployment nginx-deployment
```

运行命令后输出的截取信息如下：

```
Name:                    nginx-deployment
Namespace:               default
CreationTimestamp:       Thu, 06 May 2021 17:57:18 +0800
Labels:                  app=nginx
Annotations:             deployment.kubernetes.io/revision: 1
Selector:                app=nginx
Replicas:                  3 desired | 3 updated | 3 total | 3 available | 0 unavailable
StrategyType:              RollingUpdate
MinReadySeconds:           0
RollingUpdateStrategy:     25% max unavailable, 25% max surge
Pod Template:
  Labels:   app=nginx
  Containers:
   nginx:
    Image:          nginx:1.14.2
    Port:           80/TCP
    Host Port:      0/TCP
    Environment:    <none>
    Mounts:         <none>
  Volumes:          <none>
Conditions:
  Type           Status    Reason
  ----           ------    ------
  Available      True      MinimumReplicasAvailable
  Progressing    True      NewReplicaSetAvailable
OldReplicaSets:  <none>
NewReplicaSet:   nginx-deployment-66b6c48dd5 (3/3 replicas created)
```

更新前首先暂停 Deployment，命令如下：

```
kubectl rollout pause deployment nginx-deployment
```

运行命令后输出的信息如下：

```
deployment.apps/nginx-deployment paused
```

查询 nginx-deployment 的信息，以 yaml 格式输出，命令如下：

```
kubectl get deployments.apps -o yaml
```

运行命令后输出信息中的 spec 字段如下：

```
spec:
    paused: true
```

输出信息中 conditions 字段最后一个元素的信息如下：

```
- lastTransitionTime: "2021-05-06T10:52:49Z"
    lastUpdateTime: "2021-05-06T10:52:49Z"
    message: Deployment is paused
    reason: DeploymentPaused
    status: Unknown
    type: Progressing
```

这些信息可以验证 Deployment 处于暂停状态。

现在对 nginx-deployment 进行第一次更新，把 nginx-deployment 使用的镜像从 nginx：1.14.2 修改为 nginx：1.16.1，命令如下：

```
kubectl set image deployment nginx-deployment nginx=nginx:1.16.1
```

运行命令后输出的信息如下：

```
deployment.apps/nginx-deployment image updated
```

查看发布历史，命令如下：

```
kubectl rollout history deployment nginx-deployment
```

运行命令后输出的信息如下：

```
deployment.apps/nginx-deployment
REVISION   CHANGE-CAUSE
1          <none>
```

说明这次更新启动产生了新的发布。

这时再次查看 nginx-deployment 信息，命令如下：

```
kubectl describe deployment nginx-deployment
```

运行命令后输出的截取信息如下：

```
Name:                   nginx-deployment
Namespace:              default
CreationTimestamp:      Thu, 06 May 2021 17:57:18 +0800
```

```
Labels:                    app = nginx
Annotations:               deployment.kubernetes.io/revision: 1
Selector:                  app = nginx
Replicas:                  3 desired | 0 updated | 3 total | 3 available | 0 unavailable
StrategyType:              RollingUpdate
MinReadySeconds:           0
RollingUpdateStrategy:     25 % max unavailable, 25 % max surge
Pod Template:
  Labels:  app = nginx
  Containers:
   nginx:
    Image:          nginx:1.16.1
    Port:           80/TCP
    Host Port:      0/TCP
    Environment:    < none >
    Mounts:         < none >
  Volumes:          < none >
Conditions:
  Type           Status    Reason
  ----           ------    ------
  Available      True      MinimumReplicasAvailable
  Progressing    Unknown   DeploymentPaused
OldReplicaSets:  nginx - deployment - 66b6c48dd5 (3/3 replicas created)
NewReplicaSet:   < none >
```

显示 Deployment 的 Pod 模板中的镜像已经更新为 nginx：1.16.1，但是 NewReplicaSet 为 none，表示没有创建新的 ReplicaSet。

现在修改资源限制，命令如下：

```
kubectl set resources deployment nginx - deployment  - c = nginx \
 -- limits = cpu = 200m, memory = 512Mi
```

运行命令后输出的信息如下：

```
deployment.apps/nginx - deployment resource requirements updated
```

这时再次查看发布历史，命令如下：

```
kubectl rollout history deployment nginx - deployment
```

运行命令后输出的信息如下：

```
deployment.apps/nginx - deployment
REVISION  CHANGE - CAUSE
1         < none >
```

表示这次更新同样没有启动新的发布。

再次查看 nginx-deployment 信息，命令如下：

```
kubectl describe deployment nginx-deployment
```

运行命令后输出的截取信息如下：

```
Pod Template:
  Labels:   app=nginx
  Containers:
   nginx:
    Image:        nginx:1.16.1
    Port:         80/TCP
    Host Port:    0/TCP
    Limits:
      cpu:         200m
      memory:      512Mi
    Environment:  <none>
    Mounts:       <none>
  Volumes:        <none>
Conditions:
  Type           Status    Reason
  ----           ------    ------
  Available      True      MinimumReplicasAvailable
  Progressing    Unknown   DeploymentPaused
OldReplicaSets:  nginx-deployment-66b6c48dd5 (3/3 replicas created)
NewReplicaSet:   <none>
```

输出信息显示 nginx-deployment 的 Pod 模板发生了变化，增加的字段如下：

```
Limits:
  cpu:        200m
  memory:     512Mi
```

完成两次更新后，恢复 Deployment，命令如下：

```
kubectl rollout resume deployment nginx-deployment
```

运行命令后输出的信息如下：

```
deployment.apps/nginx-deployment resumed
```

再次查看 nginx-deployment，截取的输出信息如下：

```
Pod Template:
  Labels:   app = nginx
  Containers:
   nginx:
    Image:       nginx:1.16.1
    Port:        80/TCP
    Host Port:   0/TCP
    Limits:
      cpu:        200m
      memory:     512Mi
    Environment:  <none>
    Mounts:       <none>
  Volumes:        <none>
Conditions:
  Type           Status    Reason
  ----           ------    ------
  Available      True      MinimumReplicasAvailable
  Progressing    True      NewReplicaSetAvailable
OldReplicaSets:  <none>
NewReplicaSet:   nginx-deployment-84864d5954 (3/3 replicas created)
```

输出信息显示 Deployment 创建了新的 ReplicaSet(nginx-deployment-84864d5954)。

查看发布历史，输出的信息如下：

```
deployment.apps/nginx-deployment
REVISION  CHANGE-CAUSE
1         <none>
2         <none>
```

查看编号为 2 的发布版本信息，命令如下：

```
kubectl rollout history deployment nginx-deployment  --revision=2
```

运行命令后输出的信息如下：

```
deployment.apps/nginx-deployment with revision #2
Pod Template:
  Labels:   app=nginx
    pod-template-hash=84864d5954
  Containers:
   nginx:
    Image:   nginx:1.16.1
    Port:   80/TCP
    Host Port:   0/TCP
    Limits:
```

```
        cpu:  200m
        memory:  512Mi
      Environment:  <none>
      Mounts:  <none>
    Volumes:  <none>
```

输出信息显示两次更新的内容合并到了一个发布版本中。这个版本既修改了镜像，又设置了资源限制。

查看 ReplicaSet，命令如下：

```
kubectl get rs
```

运行命令后输出的信息如下：

```
NAME                              DESIRED   CURRENT   READY   AGE
nginx-deployment-66b6c48dd5       0         0         0       52m
nginx-deployment-84864d5954       3         3         3       6m45s
```

新的 ReplicaSet 中有 3 个 Pod 在运行，证明发布成功。

查看这次发布的状态，命令如下：

```
kubectl rollout status deployment nginx-deployment
```

运行命令后输出的信息如下：

```
deployment "nginx-deployment" successfully rolled out
```

显示发布成功。

7.11 StatefulSet

7.11.1 StatefulSet 简介

StatefulSet 用于运行有状态的应用。与 Deployment 类似，StatefulSet 可以对一组 Pod 的集合进行部署和缩放。与 Deployment 不同的是，StatefulSet 可以保证集合中 Pod 的顺序和唯一性。

对于有 N 个 Pod 副本的 ReplicaSet，它的每个 Pod 都会被分配一个整数序号，序号范围是 $0 \sim N-1$。每个序号在 ReplicaSet 中是唯一的。

Deployment 创建的 Pod 会被分配随机的主机名，而 ReplicaSet 每次创建的 Pod 都会分配一个有黏性的唯一标识作为主机名。即使 Pod 重新调度也会保持之前的唯一标识。

Pod 主机名由 StatefulSet 的名字和 Pod 的序号生成，格式为 \$(statefulset 名字)-\$(序号)。

例如名为 web 且 Pod 副本数为 3 的 ReplicaSet 创建的 3 个 Pod 名字分别是 web-0、web-1、web-2。

7.11.2 StatefulSet 示例

下面是 StatefulSet 示例的配置文件 statefulset.yaml：

```yaml
apiVersion: v1
kind: Service
metadata:
  name: nginx
  labels:
    app: nginx
spec:
  ports:
  - port: 80
    name: web
  clusterIP: None
  selector:
    app: nginx
---
apiVersion: apps/v1
kind: StatefulSet
metadata:
  name: web
spec:
  selector:
    matchLabels:
      app: nginx
  serviceName: "nginx"
  replicas: 3
  template:
    metadata:
      labels:
        app: nginx
    spec:
      terminationGracePeriodSeconds: 10
      containers:
      - name: nginx
        image: k8s.gcr.io/nginx-slim:0.8
        ports:
        - containerPort: 80
          name: web
        volumeMounts:
        - name: www
          mountPath: /usr/share/nginx/html
```

```
    volumeClaimTemplates:
    - metadata:
        name: www
      spec:
        accessModes: [ "ReadWriteOnce" ]
        storageClassName: "standard"
        resources:
          requests:
            storage: 1Gi
```

下面解释文件中的指令。

文件中被 YAML 分隔符(---)分为两部分,第一部分定义了一个无头 Service。无头 Service 用来设置 StatefulSet 中 Pod 的网络域。第二部分定义了一个 ReplicaSet。

把 ReplicaSet 命名为 web,ReplicaSet 名字必须是合法的 DNS 子域名,指令如下:

```
name: web
```

定义 ReplicaSet 的 Pod 选择器,选择器使用标签 app=nginx。

.spec.selector.matchLabels 字段必须和.spec.template.metadata.labels 匹配。否则创建时会校验失败,指令如下:

```
selector:
  matchLabels:
    app: nginx
```

将 ReplicaSet 所属的服务名定义为 nginx,指令如下:

```
serviceName: "nginx"
```

将 ReplicaSet 中 Pod 的副本数指定为 3 个,指令如下:

```
replicas: 3
```

volumeClaimTemplates 用于指定 Pod 使用的 PVC 的模板。模板指定了 PVC 申请的存储大小是 1Gi,存储类型名为 standard。存储的访问模式是 ReadWriteOnce(可读可写且同时只能一个节点挂载),Kubernetes 会根据这个模板为 ReplicaSet 中的每个 Pod 创建 PVC 和 PV,指令如下:

```
volumeClaimTemplates:
- metadata:
    name: www
  spec:
```

```
    accessModes: [ "ReadWriteOnce" ]
    storageClassName: "standard"
    resources:
      requests:
        storage: 1Gi
```

现在使用 statefulset.yaml 文件创建 StatefulSet，命令如下：

```
kubectl apply -f statefulset.yaml
```

运行命令前可以在另一个终端窗口运行命令来观察 Pod 的创建过程，命令如下：

```
kubectl get pod -w
```

运行命令后输出的类似信息如下：

```
NAME    READY    STATUS              RESTARTS   AGE
web-0   0/1      Pending             0          7m48s
web-0   0/1      Pending             0          7m58s
web-0   0/1      ContainerCreating   0          7m58s
web-0   1/1      Running             0          8m15s
web-1   0/1      Pending             0          0s
web-1   0/1      Pending             0          0s
web-1   0/1      Pending             0          9s
web-1   0/1      ContainerCreating   0          9s
web-1   1/1      Running             0          20s
web-2   0/1      Pending             0          0s
web-2   0/1      Pending             0          0s
web-2   0/1      Pending             0          9s
web-2   0/1      ContainerCreating   0          9s
web-2   1/1      Running             0          20s
```

输出显示 3 个 Pod 按 Pod 序号从 0 到 2 依次创建。这个方式是由 ReplicaSet 的 Pod 管理策略 OrderedReady 决定的。

OrderedReady 是 ReplicaSet 默认的 Pod 管理策略。在这种策略下，3 个 Pod 中按照 web-0、web-1、web-2 的顺序启动，web-1 在 web-0 就绪前不会创建，同样地 web-2 在 web-1 就绪前也不会创建。

如果 web-2 创建之前，web-0 发生故障而退出运行，则 web-2 会停止创建，直到 web-0 重新创建成功并进入就绪状态，web-2 才会继续创建。

OrderedReady 策略下的 Pod 在终止时也会顺序执行，只不过顺序与创建时相反，即从 $N-1$ 到 0。

除 OrderedReady 之外还支持 Parallel 管理策略。Parallel 的意思是并行，在这个策略下 StatefulSet 中的所有 Pod 在创建和终止时都会并发执行，不再按顺序依次执行。

StatefulSet 的 Pod 管理策略可以通过 StatefulSet 的 podManagementPolicy 字段设置，示例如下：

```
spec:
  podManagementPolicy: OrderedReady
```

查看 StatefulSet 的信息，命令如下：

```
kubectl get statefulset -o wide
```

运行命令后输出的类似信息如下：

```
NAME   READY   AGE   CONTAINERS   IMAGES
web    3/3     23m   nginx        k8s.gcr.io/nginx-slim:0.8
```

输出显示 3 个 Pod 已经全部运行成功。

7.11.3 稳定的网络标识

上述 StatefulSet 中的 Pod(web-0、web-1、web-2)均会获得独立的 DNS 子域，域名为 StatefulSet 中的 Pod 提供了稳定的网络标识。

域名格式如下：

```
$(podname).$(service name).$(namespace).svc.cluster.local
```

其中 cluster.local 是 Kubernetes 集群默认的域名，$(namespace)是 StatefulSet 的命名空间，$(service name)是 ReplicaSet 使用的无头 Service 的名称，$(podname)是 StatefulSet 中 Pod 的名称。

以 web-0 为例，它的域名是 web-0.nginx.default.svc.cluster.local。

可以在集群中运行 ping 命令验证，命令如下：

```
kubectl run --rm -it alpine --image alpine -- \
ping web-0.nginx.default.svc.cluster.local
```

运行命令后输出的类似信息如下：

```
If you don't see a command prompt, try pressing enter.
64 Bytes from 192.168.82.69: seq=3 ttl=254 time=0.059 ms
```

输出信息表示 web-0 的域名被解析到 IP 192.168.82.69。

```
kubectl get pod web-0 --template={{.status.podIP}}
```

命令输出如下：

```
192.168.82.69
```

和域名 web-0. nginx. default. svc. cluster. local 解析到的 IP 地址一致。

7.11.4　稳定的存储

在 7.11.2 节中的 ReplicaSet 创建成功后，Kubernetes 会为每个 Pod 创建 PVC。

以 web-0 为例，查看它的 volumes 信息，命令如下：

```
kubectl get pod web-0 -o yaml
```

输出信息中的. spec. volumes 字段如下：

```
volumes:
- name: www
  persistentVolumeClaim:
    claimName: www-web-0
- name: default-token-cphzk
  secret:
    defaultMode: 420
    secretName: default-token-cphzk
```

输出显示 web-0 拥有名为 www-web-0 的 PVC。

继续查看 web-1 的 volumes 信息，命令如下：

```
kubectl get podweb-1 \
-o jsonpath='{.spec.volumes[*].persistentVolumeClaim}'
```

运行命令后输出的信息如下：

```
{"claimName":"www-web-1"}
```

继续查看 web-2 的 volumes 信息，命令如下：

```
kubectl get pod web-2\
-o jsonpath='{.spec.volumes[*].persistentVolumeClaim}'
```

运行命令后输出的信息如下：

```
{"claimName":"www-web-2"}
```

可以看出 Pod 对应的 PVC 的命名格式是 .spec.volumeClaimTemplates.metadata.name 字段和 Pod 名称并用短横线(-)连接。

即使 ReplicaSet 中的 Pod 调度到了其他节点，Kubernetes 仍然可以找到之前与 Pod 关联的 PVC 并把 PVC 对应的 PV 挂载到 Pod 中。这样就实现了稳定的存储。

查看 Kubernetes 中的 PVC 信息，命令如下：

```
kubectl get pvc
```

运行命令后输出的类似信息如下(前 4 列)：

```
NAME          STATUS    VOLUME                                       CAPACITY
www-web-0     Bound     pvc-b9bb7c04-9b32-499b-b1f8-eaed3d4e9dd1     1Gi
www-web-1     Bound     pvc-fce97969-91bb-4586-8a45-936a860a93ef     1Gi
www-web-2     Bound     pvc-f7348d28-c96a-4023-93fb-92fa53f018ec     1Gi
```

后 3 列信息如下：

```
ACCESS MODES   STORAGECLASS   AGE
RWO             standard       21m
RWO             standard       12m
RWO             standard       12m
```

输出显示 Kubernetes 为 ReplicaSet 创建了 3 个 PVC。它们被分别关联到 3 个 Pod。

查看 PV 信息，命令如下：

```
kubectl get pv
```

运行命令后输出的类似信息如下(前 3 列)：

```
NAME                                         CAPACITY   ACCESS MODES
pvc-b9bb7c04-9b32-499b-b1f8-eaed3d4e9dd1     1Gi          RWO
pvc-f7348d28-c96a-4023-93fb-92fa53f018ec     1Gi          RWO
pvc-fce97969-91bb-4586-8a45-936a860a93ef     1Gi          RWO
```

后 6 列信息如下：

```
RECLAIM POLICY   STATUS   CLAIM                 STORAGECLASS   REASON   AGE
Retain            Bound     default/www-web-0    standard                13m
Retain            Bound     default/www-web-2    standard                12m
Retain            Bound     default/www-web-1    standard                13m
```

输出显示 PV 都已经处于绑定状态(Bound)。

7.12 DaemonSet

7.12.1 DaemonSet 简介

DaemonSet 用于部署需要在集群中每个节点都运行一个相同实例的程序。DaemonSet 可以保证任意时刻每个节点都运行一个相同的 Pod，当有新的节点加入集群时，DaemonSet 会自动在新节点创建一个 Pod。当节点被删除后，DaemonSet 在节点上的 Pod 也会被自动回收。

常见的需要部署到集群中的每个节点的程序包括日志收集程序、监控程序等。这些都可以采用 DaemonSet 进行部署。

7.12.2 DaemonSet 示例

创建 DaemonSet 示例配置文件 fluentbit.yaml，编辑内容如下：

```yaml
apiVersion: apps/v1
kind: DaemonSet
metadata:
  name: fluent-bit
spec:
  selector:
      matchLabels:
        name: fluent-bit #Label selector that determines which Pods belong to the DaemonSet
  template:
    metadata:
      labels:
        name: fluent-bit #Pod template's label selector
    spec:
      containers:
      - name: fluent-bit
        image: gcr.io/cloud-solutions-images/fluent-bit:1.6
        resources:
          limits:
            memory: 200Mi
          requests:
            cpu: 100m
            memory: 200Mi
```

使用此文件创建 DaemonSet，命令如下：

```
kubectl apply -f fluentbit.yaml
```

运行命令后输出的信息如下：

```
daemonset.apps/fluent-bit created
```

查看 DaemonSet 中的 Pod 信息，命令如下：

```
kubectl get pod -l name=fluent-bit
```

运行命令后输出的类似信息如下：

```
NAME               READY   STATUS    RESTARTS   AGE
fluent-bit-bmcqm   1/1     Running   0          7m6s
fluent-bit-mwprd   1/1     Running   0          7m6s
```

由于集群中有 2 个工作节点，所以 DaemonSet 共创建了 2 个 Pod，分别运行在不同的节点上。

7.12.3 DaemonSet 扩缩容

当集群扩容到 3 个工作节点时，观察 DaemonSet 自动在新创建的节点上创建 Pod 的过程，命令如下：

```
kubectl get pod -l name=fluent-bit -w
```

运行命令后输出的类似信息如下：

```
NAME               READY   STATUS              RESTARTS   AGE
fluent-bit-bmcqm   1/1     Running             0          22h
fluent-bit-mwprd   1/1     Running             0          22h
fluent-bit-djrjx   0/1     Pending             0          0s
fluent-bit-djrjx   0/1     Pending             0          0s
fluent-bit-djrjx   0/1     ContainerCreating   0          0s
fluent-bit-djrjx   1/1     Running             0          9s
```

输出显示 DaemonSet 在新的节点上自动创建了 Pod fluent-bit-djrjx。

当集群缩容到 2 个工作节点时，观察 DaemonSet 自动在新创建的节点上创建 Pod 的过程，命令如下：

```
kubectl get pod -l name=fluent-bit -w
```

运行命令后输出的类似信息如下：

```
NAME               READY   STATUS    RESTARTS   AGE
fluent-bit-bmcqm   1/1     Running   0          24h
fluent-bit-djrjx   1/1     Running   0          125m
fluent-bit-mwprd   1/1     Running   0          24h
```

```
fluent-bit-bmcqm   1/1   Running       0   24h
fluent-bit-bmcqm   1/1   Terminating   0   24h
fluent-bit-bmcqm   1/1   Terminating   0   24h
```

输出显示 DaemonSet 中的 1 个 Pod fluent-bit-bmcqm 自动被终止。

7.13 Job

7.13.1 Job 简介

Job 通常用于执行任务。通过创建一个或多个 Pod,在 Pod 中执行任务。当任务成功执行后,Job 便进入完成状态。

Job 支持并行和非并行的任务。对于非并行的任务,Job 只创建一个 Pod 执行任务。

7.13.2 Job 示例

下面是 Job 定义的示例文件 job.yaml,这个 Job 会将 π 计算到小数点后 200 位,并将结果打印出来。这是一个非并行的任务,所以只会创建一个 Pod。文件内容如下:

```
apiVersion: batch/v1
kind: Job
metadata:
  name: pi
spec:
  template:
    spec:
      containers:
      - name: pi
        image: perl
        command: ["perl",  "-Mbignum=bpi", "-wle", "print bpi(200)"]
      restartPolicy: Never
  backoffLimit: 4
```

下面解释文件中的指令。

spec.template 用于指定 Job 创建 Pod 用到的 Pod 模板,指令如下:

```
spec:
  template:
```

当 Job 中的 Pod 失败时,不重新创建 Pod。默认 Job 中的 Pod 失败后会重新创建 Pod 来重试任务,指令如下:

```
restartPolicy: Never
```

Job 在失败以后重试的次数限制为 4。spec. backoffLimit 用于设置 Job 的失败回退策略，当失败重试的次数到达 backoffLimit 后，Job 会被视为失败。backoffLimit 的默认值为 6。Job 中失败的 Pod 会被 Job 控制器重建，然后重试，重试的间隔时间呈指数增长（从 10s 到 20s 再到 40s），最长时间是 6min。

```
backoffLimit: 4
```

使用 job. yaml 创建 Job，命令如下：

```
kubectl apply -f job.yaml
```

运行命令后输出的信息如下：

```
job.batch/pi created
```

查询 Job 信息，命令如下：

```
kubectl get job pi
```

运行命令后输出的类似信息如下：

```
NAME   COMPLETIONS   DURATION   AGE
pi     1/1           39s        104s
```

COMPLETIONS 列显示 1/1，表示一共运行了 1 个 Pod，其中 1 个 Pod 运行结束。DURATION 列表示 Pod 运行的时间为 39s。

查询 Job 的详细信息，命令如下：

```
kubectl describe job pi
```

运行命令后输出的类似信息如下：

```
Name:            pi
Namespace:       default
Selector:        controller-uid=501cb9d3-7704-4608-8750-7d926fe90ac6
Labels:          controller-uid=501cb9d3-7704-4608-8750-7d926fe90ac6
                  job-name=pi
Annotations:     <none>
Parallelism:     1
Completions:     1
Start Time:     Sat, 01 May 2021 15:37:34 +0800
Completed At:   Sat, 01 May 2021 15:38:13 +0800
Duration:        39s
Pods Statuses:   0 Running / 1 Succeeded / 0 Failed
```

```
Pod Template:
  Labels:   controller - uid = 501cb9d3 - 7704 - 4608 - 8750 - 7d926fe90ac6
            job - name = pi
  Containers:
   pi:
    Image:       perl
    Port:        < none >
    Host Port:   < none >
    Command:
      perl
      - Mbignum = bpi
      - wle
      print bpi(2000)
    Environment:   < none >
    Mounts:        < none >
  Volumes:         < none >
Events:
  Type     Reason              Age     From              Message
  ----     -------             ----    ----              --------
  Normal   SuccessfulCreate    119s    job - controller  Created pod: pi - nwc9j
  Normal   Completed           80s     job - controller  Job completed
```

查看 Job 创建的 Pod,命令如下：

```
kubectl get pod  -- selector = job - name = pi
```

运行命令后输出的类似信息如下：

```
NAME       READY  STATUS     RESTARTS  AGE
pi-nwc9j   0/1    Completed  0         9m52s
```

为了查看任务执行的结果,需要查看 pi-nwc9j 这个 Pod 的日志,命令如下：

```
kubectl logs pi - nwc9j
```

运行命令后输出的信息如下：

```
3.14159265358979323846264338327950288419716939937510582097494459230781640628620899862803482534211706798214808651328230664709384460955058223172535940812848111745028410270193852110555964462294895493038200
```

输出信息是 π 到小数点后 200 位。

7.13.3　Job 清理

Job 执行完成后不会删除 Job 中的 Pod。可以查看保留的 Pod 中的日志输出。Job 本身也会被保留下来。可以使用 kubectl 命令手动删除 Job。删除 Job 会将 Job 中的 Pod 一起删除。

删除之前使用 job.yaml 创建的 Job，命令如下：

```
kubeclt delete -f job.yaml
```

运行命令后输出的信息如下：

```
job.batch "pi" deleted
```

7.14 ConfigMap

7.14.1 ConfigMap 简介

应用通常会用到一些非机密的配置信息，例如配置监听端口、日志存储路径等。Kubernetes 提供了 ConfigMap，用于把非机密的数据以键值对的形式保存起来。ConfigMap 不会对数据加密，如果要保存机密信息，则可以使用 Secret。

使用 ConfigMap 可以实现配置和容器解耦，提高容器的可移植性。当需要在生产环境、开发环境等不同环境中运行应用时，只需使用不同的 ConfigMap 提供不同的配置，而不需要修改应用容器。

ConfigMap 的存储容量有限，不能用于存储大量的数据。ConfigMap 不可以存储超过 1MB 的数据。如果数据超过这个限制，则需要考虑使用卷、数据库或者文件系统来存储。

当 ConfigMap 通过 Volume 挂载到容器中使用时，ConfigMap 的变更会最终同步到 Volume 中。同步变更是由 kubelet 完成的，kubelet 会周期性地对 ConfigMap 的变更进行同步。由于 kubelet 会在本地缓存 ConfigMap 中的数据，并利用缓存获取当前 ConfigMap 中的数据，所以 Pod 中数据同步的实际延时等于 kubelet 同步周期加上缓存传播的延时。

当以环境变量的形式使用 ConfigMap 中的数据时，ConfigMap 变更之后环境变量不会自动变更，这种情况下需要重启 Pod 来更新环境变量以获取最新的 ConfigMap 数据。

如果确定 ConfigMap 中的数据不会变更，则可以在该 ConfigMap 中加入指令 immutable:true 来创建 Immutable(不可变更的)ConfigMap。kubelet 不会监视和周期性同步不可变更的 ConfigMap，在成百上千的 Pod 引用了这个相同的 ConfigMap 的情况下，使用不可变更的 ConfigMap 可以显著降低 Kubernetes API server 的压力，另一方面由于不可变更的 ConfigMap 无法进行修改，避免了对 ConfigMap 的意外改动，这样会提高应用的稳定性。

对于不可变更的 ConfigMap，只能通过删除后重建来改变它的数据。

7.14.2 ConfigMap 示例

ConfigMap 命名必须是合法的 DNS 子域名。ConfigMap 定义中没有 Kubernetes 对象常见的 spec 字段，而是包含了 data 和 binaryData 字段，这两个字段都是可选的，值形式为键值对。其中 data 字段用于存储 UTF 格式的字节序列，binaryData 字段用于存储二进制数据并使用 base64 编码。data 和 binaryData 中的键名只能由字母、数字、-、_和.构成。

下面是一个 ConfigMap 定义的示例文件 configmap.yaml，内容如下：

```
apiVersion: v1
kind: ConfigMap
metadata:
  name: env-config
  namespace: default
data:
  log_level: INFO
  ui.properties: |
    color.good=purple
    color.bad=yellow
    allow.textmode=true
```

定义文件中的 data 字段包含 2 个键值对，其中 ui.properties 对应的值为多行文本，所以使用 YAML 语法中规定的竖线"|"开头。

使用 configmap.yaml 文件创建 ConfigMap，命令如下：

```
kubectl apply -f configmap.yaml
```

运行命令后输出的信息如下：

```
configmap/env-config configured
```

查询 env-config 的详细信息，命令如下：

```
kubectl describe configmaps env-config
```

运行命令后输出的信息如下：

```
Name:         env-config
Namespace:    default
Labels:       <none>
Annotations:  <none>

Data
====
log_level:
```

```
----
INFO
ui.properties:
----
color.good = purple
color.bad = yellow
Events:  <none>
```

输出信息显示的是明文的配置数据。

7.14.3 Pod 使用 ConfigMap

Pod 可以通过 4 种方式使用 ConfigMap 来读取配置内容。

Pod 通过环境变量读取 ConfigMap，示例文件 configmap-env.yaml 的内容如下：

```
apiVersion: v1
kind: Pod
metadata:
  name: configmap-env
spec:
  containers:
    - name: demo
      image: alpine
      command: ["sleep", "3600"]
      env:
        - name: LOG_LEVEL
          valueFrom:
            configMapKeyRef:
              name: env-config
              key: log_level
```

创建 Pod，命令如下：

```
kubectl apply -f configmap-env.yaml
```

查看环境变量是否生效，在创建的 Pod 中执行命令 env 打印环境变量，命令如下：

```
kubectl exec configmap-env -- env
```

运行命令后输出的类似信息如下：

```
PATH = /usr/local/sbin:/usr/local/bin:/usr/sbin:/usr/bin:/sbin:/bin
HOSTNAME = configmap-env
TERM = xterm
LOG_LEVEL = INFO
```

```
KUBERNETES_SERVICE_HOST = 10.96.0.1
KUBERNETES_SERVICE_PORT = 443
KUBERNETES_SERVICE_PORT_HTTPS = 443
KUBERNETES_PORT = tcp://10.96.0.1:443
KUBERNETES_PORT_443_TCP = tcp://10.96.0.1:443
KUBERNETES_PORT_443_TCP_PROTO = tcp
KUBERNETES_PORT_443_TCP_PORT = 443
KUBERNETES_PORT_443_TCP_ADDR = 10.96.0.1
HOME = /root
```

在输出信息中第4行可以看到环境变量LOG_LEVEL,信息如下:

```
LOG_LEVEL = INFO
```

说明LOG_LEVEL设置成功,和前面名为env-config的ConfigMap中log_level键的值一致。

通过Volume使用ConfigMap示例文件configmap-volume.yaml的内容如下:

```
apiVersion: v1
kind: Pod
metadata:
  name: configmap - volume
spec:
  containers:
    - name: demo
      image: alpine
      command: ["sleep", "3600"]
      volumeMounts:
      - name: config
        mountPath: "/config"
        readOnly: true
  volumes:
    - name: config
      configMap:
        name: env - config
```

文件中的指令解释如下。

首先引用名为config的volume,并挂载到容器demo中的/config文件夹下。将文件的模式设置为只读。ConfigMap中的每个键都会变成/config下的一个文件名。这样容器中的应用就可以读取文件以便获取ConfigMap中的数据,指令如下:

```
volumeMounts:
- name: config
  mountPath: "/config"
  readOnly: true
```

定义一个 volume，命名为 config。volume 中的数据来自名为 env-config 的 ConfigMap，指令如下：

```
volumes:
  - name: config
    configMap:
      name: env-config
```

创建 Pod，命令如下：

```
kubectl apply -f configmap-volume.yaml
```

查看容器中的/config 文件夹的内容，验证 volume 是否挂载成功，命令如下：

```
kubectl exec configmap-volume -- ls /config
```

运行命令后输出的信息如下：

```
log_level
ui.properties
```

可以看到/config 文件夹下的文件名和 ConfigMap 中的键名一致。ConfigMap 中的所有键值对以文件形式挂载到了容器中的指定文件夹/config 下。

/config 文件夹中的文件 log_level 和 ui.properties 是软链，而不是普通的文件。查看文件的详细信息，命令如下：

```
kubectl exec configmap-volume -- ls -la /config/
```

运行命令后输出的类似信息如下：

```
total 12
drwxrwxrwx    3 root    root         4096 May  2 08:52 .
drwxr-xr-x    1 root    root         4096 May  2 11:52 ..
drwxr-xr-x    2 root    root         4096 May  2\
 08:52 ..2021_05_02_08_52_18.603026737
lrwxrwxrwx    1 root    root         31 May  2 08:52 ..data\
 -> ..2021_05_02_08_52_18.603026737
lrwxrwxrwx    1 root    root         20 May  2 08:52\
 log_level -> ..data/log_level
lrwxrwxrwx    1 root    root         13 May  2 08:52 ui.env\
 -> ..data/ui.properties
```

跟随软链可以找到真实文件的位置在..2021_05_02_08_52_18.603026737 文件夹下，列出这个文件夹的内容，命令如下：

```
kubectl exec configmap-volume -- ls -la\
 /config/..2021_05_02_08_52_18.603026737/
```

运行命令后输出的类似信息如下：

```
total 16
drwxr-xr-x          2 root      root            4096 May  2 08:52 .
drwxrwxrwx          3 root      root            4096 May  2 08:52 ..
-rw-r--r--          1 root      root            4 May     2 08:52 log_level.txt
-rw-r--r--          1 root      root            55 May    2 08:52 ui.env
```

查看容器中文件 log_level 的内容，命令如下：

```
kubectl exec configmap-volume -- cat /config/log_level
```

运行命令后输出的信息如下：

```
INFO
```

输出内容和 ConfigMap 中的键 log_level 的值一致。

查看容器中文件 ui.properties 的内容，命令如下：

```
kubectl exec configmap-volume -- cat /config/ui.properties
```

运行命令后输出的信息如下：

```
color.good = purple
color.bad = yellow
allow.textmode = true
```

输出内容和 ConfigMap 中的键 ui.properties 的值一致。

如果希望将 ConfigMap 中的键映射到 Pod 中的文件名，则可以为每个键单独指定映射的文件名。修改 configmap-volume.yaml 文件中的.spec.volumes 字段，修改后的命令如下：

```
volumes:
  - name: config
    configMap:
      name: env-config
      items:
      - key: "ui.properties"
        path: "ui.env"
      - key: log_level
        path: log_level.txt
```

现在删除之前创建的 Pod，重新创建 Pod。

然后查询 Pod 中/config 文件夹下的内容，命令如下：

```
kubectl exec configmap-volume -- ls /config
```

运行命令后输出的信息如下：

```
log_level.txt
ui.env
```

可以看到使用新配置创建的 ConfigMap 中的 log_level 在 Pod 中的文件名会变为 log_level.txt，ui.properties 在 Pod 中的文件名为 ui.env。

7.15 Secret

7.15.1 Secret 简介

对于密码、密钥、凭证等机密信息，使用 ConfigMap 明文存储有安全风险，可以使用 Secret 来存储机密信息。

通常数据保存在 Secret 中的 data 或者 stringData 字段，其中 data 字段中的所有键对应的值必须进行 base64 编码，而 stringData 中的值则可以保存任意数据而不需要编码。stringData 中的所有键值对会合并到 data 字段中，如果 data 和 stringData 字段中出现了相同的键，则 stringData 中的键对应的值会有更高优先级。

Secret 有不同的类型，类型保存在 Secret 的 type 字段。Kubernetes 内置了许多 Secret 类型，例如 Opaque、kubernetes.io/service-account-token 等，如果 Secret 的 type 字段值为空，则会被当作 Opaque 类型。除了内置的 Secret 类型，还可以自定义 Secret 类型。

下面演示 Opaque 类型的 Secret。

创建 Opaque 类型的 Secret，命令如下：

```
kubectl create secret generic empty-secret
```

查看 empty-secret 这个 Secret 的信息，命令如下：

```
kubectl get secrets empty-secret
```

运行命令后输出的信息如下：

```
NAME           TYPE     DATA   AGE
empty-secret   Opaque   0      29s
```

7.15.2 创建 Secret

假设创建一个用于保存用户名和密码的 Secret，用户名和密码分别保存在文件 username.txt 和 password.txt 中。

创建这两个文件，命令如下：

```
echo -n 'admin'> ./username.txt
echo -n '123'> ./password.txt
```

命令中的-n 参数是为了防止 echo 输出换行符。

创建 Secret，Secret 的数据从这两个文件中读取，命令如下：

```
kubectl create secret generic db-user-pass \
  --from-file=./username.txt \
  --from-file=./password.txt
```

生成的 Secret 中的数据键名默认为文件名，也可以使用--from-file=[key=]source 的方式自定义键名，示例如下：

```
kubectl create secret generic db-user-pass \
  --from-file=username=./username.txt \
  --from-file=password=./password.txt
```

除了文件外还可以通过--from-literal=< key >=< value >参数直接提供 Secret 数据的来源，命令如下：

```
kubectl create secret generic db-user-pass \
  --from-literal=username=devuser \
  --from-literal=password='S!B\*d$zDsb='
```

查看创建的 Secret 信息，命令如下：

```
kubectl get secrets
```

运行命令后输出的信息如下：

```
NAME            TYPE                                  DATA      AGE
db-user-pass    Opaque                                  2       10s
```

查看 db-user-pass 的详细信息，命令如下：

```
kubectl describe secrets/db-user-pass
```

运行命令后输出的信息如下：

```
Name:          db-user-pass
Namespace:     default
Labels:        <none>
Annotations:   <none>

Type:  Opaque

Data
====
password.txt:   3 Bytes
username.txt:   5 Bytes
```

可以看到数据并没有明文显示。这样可以防止机密信息被意外泄露，例如被保存在终端日志中。

7.15.3 查看 Secret 数据

查看 Secret 数据，命令如下：

```
kubectl get secret db-user-pass -o jsonpath='{.data}'
```

运行命令后输出的信息如下：

```
{"password.txt":"MTIz","username.txt":"YWRtaW4="}
```

使用 base64 解码，命令如下：

```
echo MTIz| base64 --decode
```

运行命令后输出的信息如下：

```
123
```

7.16 Kubernetes 存储

7.16.1 Volume

volume（卷）是 Kubernetes 中对存储的抽象概念。可以类比为计算机中的磁盘。volume 的核心是一个文件夹，可供 Pod 中的容器访问。Kubernetes 支持许多类型的 volume，Pod 可以同时使用多种类型的 volume。volume 解决了存储的问题和 Pod 中容器间共享文件数据的问题。

对于任意类型的 volume，volume 和 Pod 中的容器的生命周期无关，Pod 中的容器重启不会导致 volume 中的数据丢失。

volume 按生命周期长短可以分为短暂 volume 和持久 volume。短暂 volume 和 Pod 拥有同样的生命周期，在 Pod 终止退出后，短暂 volume 会被清除。持久 volume 不受 Pod 生命周期的影响，在 Pod 退出后，持久 volume 仍会被继续保留。

7.16.2 PersistentVolume

PersistentVolume(持久卷)简称 PV，表示 Kubernetes 集群中的一块存储。PV 是存储的具体实现，PV 对象保存了底层存储实现的所有细节。PV 可以由集群管理员手动创建，也可以由 Kubernetes 动态供应。

手动创建 PV 也称为静态供应，管理员提前创建需要用到的 PV，Pod 直接引用创建好的 PV 来存储数据。

动态供应是指在创建 Pod 时，动态创建需要用到的 PV。动态供应的优势是免去了手动创建的烦琐步骤。

PV 有许多内置的实现类型，以插件的形式提供。

下面是 hostPath 类型的 PV 示例定义的文件 pv-volume.yaml，hostPath 类型的 PV 可以把宿主机的文件或文件夹挂载到 Pod 中，这种 PV 只适用于开发和测试环境。文件内容如下：

```
apiVersion: v1
kind: PersistentVolume
metadata:
  name: pv - volume
  labels:
    type: local
spec:
  storageClassName: manual
  capacity:
    storage: 10Gi
  volumeMode: Filesystem
  accessModes:
    - ReadWriteOnce
  hostPath:
    path: "/mnt/data"
```

文件中的指令介绍如下。

将 PV 名称指定为 pv-volume，指令如下：

```
name: pv - volume
```

将 storageClassName 指定为 manual,PVC 和 PV 必须具有相同的 storageClass 才可以进行绑定,指令如下:

```
storageClassName: manual
```

. spec. capacity 用于指定 PV 的容量,这里将容量的大小指定为 10Gi。

```
capacity:
  storage: 10Gi
```

. spec. volumeMode 用于指定 PV 的卷模式,Kubernetes 支持两种卷模式,即 Filesystem 和 Block。volumeMode 为可选项,默认值是 Filesystem。

如果 volumeMode 为 Filesystem 的 PV,则会被挂载到 Pod 中的一个文件夹中。如果 PV 的底层存储设备是空的块设备,Kubernetes 在挂载前会在存储设备上创建文件系统。

如果 volumeMode 为 Block 的 PV,则可以作为原始块设备使用,原始块设备上没有任何文件系统。这种模式可以为 Pod 提供极快的速度访问 PV,指令如下:

```
volumeMode: Filesystem
```

. spec. accessModes 表示 PV 被访问的模式,一个 PV 可以支持多种访问模式,但是同一时刻只能使用一种模式挂载。PV 支持的访问模式由底层的存储提供者的能力决定。所有的访问模式包括 ReadWriteOnce、ReadOnlyMany 和 ReadWriteMany。

ReadWriteOnce 表示可以被一个节点以读写方式挂载。CLI 中缩写为 RWO。

ReadOnlyMany 表示可以被多个节点以只读方式挂载。CLI 中缩写为 ROX。

ReadWritetMany 表示可以被多个节点以读写方式挂载。CLI 中缩写为 RWX。

```
accessModes:
  - ReadWriteOnce
```

把宿主机的/mnt/data 文件夹作为挂载源。Pod 中读写这个 PV 的结果会保存到宿主机/mnt/data 文件夹中,指令如下:

```
hostPath:
  path: "/mnt/data"
```

使用 pv-volume. yaml 创建 PV,命令如下:

```
kubectl apply - f pv - volume. yaml
```

运行命令后输出的信息如下:

```
persistentvolume/pv-volume created
```

在宿主机创建文件夹/mnt/data，命令如下：

```
sudo mkdir /mnt/data
```

在/mnt/data 文件夹下创建文件 index. html，命令如下：

```
sudo sh -c "echo 'Hello from Kubernetes storage'> /mnt/data/index.html"
```

命令会在/mnt/data/index. html 文件中写入字符串 Hello from Kubernetes storage。可以使用 cat 命令验证，命令如下：

```
cat /mnt/data/index.html
```

查看 test-pv-volume 信息，命令如下：

```
kubectl get pv pv-volume
```

运行命令后输出的信息如下（前 5 列）：

```
NAME          CAPACITY   ACCESS MODES   RECLAIM POLICY   STATUS
pv-volume     2Gi        RWO            Retain           Available
```

后 4 列信息如下：

```
CLAIM   STORAGECLASS   REASON   AGE
        manual                  79m
```

其中 Status 列的值为 Available，即可用状态，表示这个 PV 还没有绑定到 PVC。

7.16.3　PersistentVolumeClaim

PersistentVolumeClaim（持久卷申领）简称 PVC，表示用户对存储的需求。PVC 对用户屏蔽了存储的细节，用户只需通过 PVC 指定存储需求（大小、读写权限等）而不必关心底层存储是如何供应的。实现了存储使用和存储供应的解耦。PVC 与 PV 的更新类似于面向对象编程语言中接口与实现的关系。

PVC 与 PV 的绑定是一对一的双向映射关系，已经绑定到 PVC 的 PV 无法绑定到其他 PVC。

在创建 PVC 之后，Kubernetes 会寻找匹配的 PV，然后把匹配的 PV 绑定到 PVC。如果没有合适的 PV，则 PVC 会一直处于未绑定状态直到出现匹配的 PV。例如集群中只存在多个大小为 50Gi 的 PV，申请 100Gi 的 PVC 就会无法绑定。

正在被 Pod 使用中的 PVC 和已经绑定到 PVC 的 PV 不会被系统删除，以保证数据不会丢失。当删除一个正在被 Pod 使用的 PVC 时，PVC 不会被立即删除，而是推迟到它不再被任何 Pod 使用的时候。当删除一个已经被绑定的 PV 时，PV 的删除会被推迟到不再绑定到 PVC。这个删除保护功能称为 Storage Object in Use Protection。

下面是 PVC 示例定义的文件 pv-claim.yaml，文件内容如下：

```
apiVersion: v1
kind: PersistentVolumeClaim
metadata:
  name: pv - claim
spec:
  storageClassName: manual
  accessModes:
    - ReadWriteOnce
  resources:
    requests:
      storage: 2Gi
```

使用这个文件创建 PVC，命令如下：

```
kubectl apply - f pv - claim.yaml
```

运行命令后输出的信息如下：

```
persistentvolumeclaim/pv - claim created
```

查看 test-pv-claim 信息，命令如下：

```
kubectl get pvc pv - claim
```

运行命令后输出的信息如下(前 5 列)：

```
NAME          STATUS   VOLUME          CAPACITY   ACCESS MODES
pv - claim     Bound     pv - volume     2Gi         RWO
```

后 2 列信息如下：

```
STORAGECLASS   AGE
manual         71s
```

输出信息显示 STATUS 列为 Bound，即已绑定状态。说明 Kubernetes 已经找到了满足 pv-claim 要求的 PV，VOLUME 列为 pv-volume，表示绑定到了之前创建的 pv-volume 这个 PV 上。

再次查看 pv-volume，命令如下：

```
kubectl get pvpv - volume
```

运行命令后输出的信息如下(前 5 列)：

```
NAME           CAPACITY   ACCESS MODES   RECLAIM POLICY   STATUS
pv-volume      2Gi        RWO             Retain           Bound
```

后 4 列信息如下：

```
CLAIM   STORAGECLASS   REASON     AGE
default/pv-claim manual           129m
```

输出信息显示 test-pv-volume 为 Bound，即已绑定状态，CLAIM 列为 default/pv-claim，表示 PV 绑定到了 pv-claim 这个 PVC 上。

7.16.4　Pod 使用 PersistentVolumeClaim

Pod 通过 PVC 来使用存储，对 Pod 来讲 PVC 可以视为卷。

下面使用了 PVC 的 Pod 示例配置的文件 pv-pod.yaml，文件内容如下：

```
apiVersion: v1
kind: Pod
metadata:
  name: pv - pod
spec:
  volumes:
    - name: pv - storage
      persistentVolumeClaim:
        claimName: pv - claim
  containers:
    - name: pv - container
      image: nginx
      ports:
        - containerPort: 80
          name: "http - server"
      volumeMounts:
        - mountPath: "/usr/share/nginx/html"
          name: pv - storage
```

下面解释文件中指令的含义。

定义一个 volume，volume 名字命名为 pv-storage。这个 volume 引用的是名为 pv-claim 的 PVC。pv-claim 是 7.13.4 节创建好的。从 Pod 角度来看，这个 PVC 就是 volume，指令

如下：

```
volumes:
  - name: pv - storage
    persistentVolumeClaim:
      claimName: pv - claim
```

Pod 中的容器使用 nginx 镜像，容器命名为 pv-container，指令如下：

```
containers:
  - name: pv - container
    image: nginx
```

容器暴露 80 端口，端口命名为 http-server，指令如下：

```
ports:
  - containerPort: 80
    name: "http - server"
```

将 pv-storage 挂载到容器中的/usr/share/nginx/html 文件夹下，这个文件夹是 nginx 默认的 Web 服务根目录。HTTP 请求的文件会在这个文件夹下寻找，指令如下：

```
volumeMounts:
  - mountPath: "/usr/share/nginx/html"
    name: pv-storage
```

使用这个文件创建 Pod，命令如下：

```
kubectl apply  - f pv - pod.yaml
```

运行命令后输出的信息如下：

```
pod/pv - pod created
```

查看新创建的 Pod 的信息，命令如下：

```
kubectl get pod pv/pod
```

运行命令后输出的信息如下：

```
NAME      READY   STATUS    RESTARTS    AGE
pv - pod  1/1     Running   0           41s
```

表示 Pod 运行成功。

查看 Pod 中/usr/share/nginx/html 文件夹的内容，验证挂载是否成功，命令如下：

```
kubectl exec pv-pod -- ls /usr/share/nginx/html
```

运行命令后输出的信息如下：

```
index.html
```

在 Pod 中发送 http 请求，命令如下：

```
kubectl exec pv-pod -- curl -s localhost
```

运行命令后输出的信息如下：

```
Hello from Kubernetes storage
```

请求的地址是 localhost，在没有指定路径的情况下，nginx 会输出 web 根目录下的 index.html 文件内容。这个 index.html 文件就是挂载的 pv-storage 中的 index.html。

7.16.5 StorageClass

StorageClass（存储类型）用来描述 PV 的类型，不同类型对应不同的服务质量等级或备份策略等，或者集群管理员指定的任意策略。StorageClass 实现了对 PV 类型的抽象，是实现存储动态供应的关键。StorageClass 描述了对动态供应存储的具体要求，相当于存储的模板。

当有新的指定了 StorageClass 的 PVC 创建时，Kubernetes 会根据 StorageClass 中的配置动态地创建新的 PV 以绑定到 PVC。如果 PVC 没有指定 StorageClass，并且 Kubernetes 中指定了默认的 StorageClass，则 Kubernetes 会使用默认的 StorageClass 完成存储的动态供应。

StorageClass 的名字十分重要，PVC 通过指定 StorageClass 的名字来获得动态存储供应。一旦创建成功，StorageClass 就无法进行更新。

设置默认 StorageClass 的方法是给对应的 StorageClass 对象添加如下注解：

```
storageclass.kubernetes.io/is-default-class: "true"
```

只能把一个 StorageClass 设置为默认，如果同时设置多个默认 StorageClass，则系统会拒绝创建没有指定 storageClassName 字段的 PVC。

下面是 StorageClass 示例配置的文件 storageclass-ebs.yaml，文件内容如下：

```
apiVersion: storage.k8s.io/v1
kind: StorageClass
```

```
metadata:
  name: standard
  annotations:
    storageclass.kubernetes.io/is-default-class: "true"
provisioner: kubernetes.io/aws-ebs
parameters:
  type: gp2
  fsType: ext4
reclaimPolicy: Retain
allowVolumeExpansion: true
mountOptions:
  - debug
volumeBindingMode: Immediate
```

下面解释文件中的指令。

指定存储供应者(provisioner)是 kubernetes.io/aws-ebs，指令如下：

```
provisioner: kubernetes.io/aws-ebs
```

指定存储回收策略是 Retain(保留)，即删除 PVC 对象后，保留与其绑定的 PV 和外部的数据卷。用户可以手动进行删除。即使删除 PV，与之关联的外部存储(例如 AWSEB、GCEPD 等)依旧会保留。除了 Retain 还有 Delete 和 Recycle(已废弃)回收策略。

对于支持 Delete 回收策略的存储插件来讲，采用 Delete 回收策略会使 PVC 对象删除后对应的 PV 及外部的数据卷也会被删除。

```
reclaimPolicy: Retain
```

使用文件 storageclass-ebs.yaml 创建 StorageClass，命令如下：

```
kubectl apply -f storageclass-ebs.yaml
```

运行命令后输出的信息如下：

```
storageclass.storage.k8s.io/standard created
```

查询 StorageClass 信息，命令如下：

```
kubectl get storageclass standard
```

运行命令后输出的信息如下(前 4 列)：

```
NAME                PROVISIONER             RECLAIMPOLICY   VOLUMEBINDINGMODE
standard (default)  kubernetes.io/aws-ebs   Retain          Immediate
```

后 2 列信息如下：

```
ALLOWVOLUMEEXPANSION    AGE
true                    3m28s
```

动态卷供应的实现依赖于 StorageClass 对象，集群中可以定义多个 StorageClass，每个 StorageClass 都必须指定 provisioner 字段来描述使用的存储卷插件和在供应卷时 provisioner 需要用到的参数。

7.16.6 动态卷供应

动态卷供应即在用户需要存储时自动创建和供应存储卷。在使用存储前手动创建存储卷的方式比较低效，而动态卷供应可以使存储供应自动化。在需要用到存储的时候才供应存储卷，而不需要提前准备存储卷。

动态卷供应的实现依赖于 StorageClass 对象，集群中可以定义多个 StorageClass，每个 StorageClass 都必须指定 provisioner 字段来描述使用的存储卷插件和在供应卷时 provisioner 需要用到的参数。常见的 provisioner 如表 7-4 所示。

表 7-4 常见的 provisioner

卷插件	是否由 Kubernetes 内部支持	卷插件	是否由 Kubernetes 内部支持
AWSElasticBlockStore	是	iSCSI	否
AzureFile	是	QuoByte	是
AzureDisk	是	NFS	否
CephFS	否	RBD	是
Cinder	是	VsphereVolume	是
FC	否	PortworxVolume	是
FlexVolume	否	ScaleIO	是
Flocker	是	StorageOS	是
GCEPersistentDisk	是	Local	否
Glusterfs	是		

由 Kubernetes 内部支持的 provisioner 称为内部 provisioner，内部 provisioner 的名字开头是 kubernetes.io 前缀，并且和 Kubernetes 一起发布。例如 AWSElasticBlockStore 卷插件对应的 provisioner 名字是 kubernetes.io/aws-ebs，GCEPersistentDisk 对应的 provisioner 名字是 kubernetes.io/gce-pd。

不由 Kubernetes 提供的 provisioner 称为外部 provisioner，例如 NFS。

开启动态卷供应功能需要提前创建一个或多个 StorageClass 对象，在创建了 StorageClass 后，创建 PVC 可以指定 storageClassName 来引用某个已经存在的 StorageClass 对象，storageClassName 必须和某个 StorageClass 的 name 匹配。系统将根据 StorageClass 自动供应 PV 和用户的 PVC 绑定。这种设计避免了将供应存储的细节暴露给用户，降低了使用存储的难度。

在系统中已经存在 StorageClass(standard)的情况下(名为 standard 的 StorageClass 在 7.13.5 节中已经创建),新建一个 storageClassName 为 standard 的 PVC,观察存储动态供应的过程。

创建 PVC 前查看是否有 PV,命令如下:

```
kubectl get pv
```

运行命令后输出的信息如下:

```
No resources found
```

输出显示没有 PV 资源。

下面开始创建 PVC。

创建 PVC 配置文件 pvc-100.yaml,内容如下:

```
apiVersion: v1
kind: PersistentVolumeClaim
metadata:
  name: pvc - 100
spec:
  accessModes:
    - ReadWriteOnce
  storageClassName: standard
  resources:
    requests:
      storage: 100Gi
```

PVC 的 storageClassName 值为 standard。系统会根据 standard StorageClass 为这个 PVC 供应存储。

等待数秒,然后查看 PVC 信息,命令如下:

```
kubectl get pvc
```

运行命令后输出的信息如下:

```
NAME      STATUS   VOLUME                                     CAPACITY
pvc-100   Bound    pvc-ada75236-61b2-4a0c-bca0-778b665ad9a7   100Gi
```

后 3 列信息如下:

```
ACCESS MODES   STORAGECLASS   AGE
RWO            standard       8m42s
```

输出显示 pvc-100 绑定了自动供应的卷(pvc-ada75236-61b2-4a0c-bca0-778b665ad9a7)。

查看 pv 信息,命令如下:

```
kubectl get pv
```

运行命令后输出的信息如下(前 3 列):

```
NAME                                      CAPACITY   ACCESS MODES
pvc-ada75236-61b2-4a0c-bca0-778b665ad9a7   100Gi      RWO
```

后 6 列信息如下:

```
RECLAIM POLICY   STATUS   CLAIM             STORAGECLASS   REASON   AGE
Retain           Bound    default/pvc-100   standard                4m47s
```

输出显示系统中自动创建了 100Gi 大小的 PV,存储类型为 standard。已经和 pvc-100 这个 PVC 绑定。

查看自动创建的 PV 的详细信息,命令如下:

```
kubectl get pv pvc-ada75236-61b2-4a0c-bca0-778b665ad9a7 -o yaml
```

运行命令后输出的类似信息如下:

```
apiVersion: v1
kind: PersistentVolume
metadata:
  annotations:
    kubernetes.io/createdby: aws-ebs-dynamic-provisioner
    pv.kubernetes.io/bound-by-controller: "yes"
    pv.kubernetes.io/provisioned-by: kubernetes.io/aws-ebs
  creationTimestamp: "2021-05-13T08:48:39Z"
  finalizers:
  - kubernetes.io/pv-protection
  labels:
    failure-domain.beta.kubernetes.io/region: us-west-2
    failure-domain.beta.kubernetes.io/zone: us-west-2a
  name: pvc-ada75236-61b2-4a0c-bca0-778b665ad9a7
  resourceVersion: "59973"
  selfLink: /api/v1/persistentvolumes/pvc-ada75236-61b2-4a0c-bca0-778b665ad9a7
  uid: 713fe2a3-cdd8-4632-81aa-8de7cd2c6838
spec:
  accessModes:
  - ReadWriteOnce
  awsElasticBlockStore:
    fsType: ext4
```

```
      volumeID: aws://us-west-2a/vol-02b717d90026fcc7a
    capacity:
      storage: 100Gi
    claimRef:
      apiVersion: v1
      kind: PersistentVolumeClaim
      name: pvc-100
      namespace: default
      resourceVersion: "59956"
      uid: ada75236-61b2-4a0c-bca0-778b665ad9a7
    mountOptions:
    - debug
    nodeAffinity:
      required:
        nodeSelectorTerms:
        - matchExpressions:
          - key: failure-domain.beta.kubernetes.io/zone
            operator: In
            values:
            - us-west-2a
          - key: failure-domain.beta.kubernetes.io/region
            operator: In
            values:
            - us-west-2
    persistentVolumeReclaimPolicy: Delete
    storageClassName: standard
    volumeMode: Filesystem
  status:
    phase: Bound
```

输出信息显示 PV 被自动设置了 nodeAffinity(节点亲和性)字段。节点亲和性可以限制 PV 被哪些节点访问。使用了这个 PV 的 Pod 会被自动调度到根据 PV 节点亲和性规则选中的节点上。对于大部分卷类型,不需要手动设置这个字段。它会根据卷类型自动设置,例如这个示例使用的卷类型 AWSEBS,以及 GCE PD、Azure Disk 等都支持这个功能。

7.16.7 AWS EBS 使用示例

使用命令行工具 aws 创建 EBS,命令如下:

```
aws ec2 create-volume \
--availability-zone=us-west-2a\
--size=10 \
--volume-type=gp2 \
--region=us-west-2
```

命令中将可用区指定为 us-west-2a，将区域指定为 us-west-2，并将存储空间指定为 10Gi。此外，将卷类型指定为 gp2。不同的卷类型对应不同的可用性。卷类型的详细信息可参考官方文档链接 https://docs.aws.amazon.com/AWSEC2/latest/UserGuide/ebs-volume-types.html。

运行命令后输出的信息如下：

```
{
"AvailabilityZone": "us-west-2a",
"CreateTime": "2021-05-11T02:08:14+00:00",
"Encrypted": false,
"Size": 10,
"SnapshotId": "",
"State": "creating",
"VolumeId": "vol-030b3c4742d647098",
"Iops": 100,
"Tags": [],
"VolumeType": "gp2",
"MultiAttachEnabled": false
}
```

创建成功后，登录 AWS 后台也可以看到对应的条目。

下面定义 Pod 文件 ebs-pod.yaml，Pod 中使用了刚刚创建的 EBS 卷，文件如下：

```
apiVersion: v1
kind: Pod
metadata:
  name: test-ebs
spec:
  containers:
  - image: nginx
    name: test-container
    volumeMounts:
    - mountPath: /test-ebs
      name: test-volume
  volumes:
  - name: test-volume
    awsElasticBlockStore:
      volumeID: "vol-030b3c4742d647098"
      fsType: ext4
```

文件中的 volumeID 字段的值是之前创建的 EBS 卷的 volumeId 值。

使用文件 ebs-pod.yaml 创建 Pod，命令如下：

```
kubectl apply -f ebs-pod.yaml
```

运行命令后输出的信息如下：

```
pod/test - ebs created
```

查看 Pod 信息，命令如下：

```
kubectl get pod
```

运行命令后输出的信息如下：

```
NAME        READY   STATUS     RESTARTS   AGE
test-ebs    1/1     Running    0          104s
```

表示 Pod 运行成功。

接下来在 Pod 中的/test-ebs 文件夹下新建文件 ebs. txt 并写入字符串，命令如下：

```
kubectl exec test - ebs  --  sh  - c 'echo hello ebs > /test - ebs/ebs.txt'
```

查看文件内容，命令如下：

```
kubectl exec test - ebs  --  cat /test - ebs/ebs.txt
```

运行命令后输出的信息如下：

```
hello ebs
```

删除 test-ebs，然后重新创建，会发现 Pod 中/test-ebs/ebs. txt 文件仍然保持不变。说明 EBS 中的数据不受 Pod 生命周期影响，保证了数据在 Pod 终止后不会丢失。

7.17 Kubernetes Service

7.17.1 Service 简介

Service 用于把 Pod 中的应用暴露为集群外为可访问的服务。没有 ServicePod 中的应用只能通过集群内访问。另外 Service 还提供对一组 Pod 的负载均衡。

虽然 Kubernetes 会为每个 Pod 提供 IP，但是通过 IP 访问 Pod 中的服务不是很方便，并且在 Kubernetes 设计理念中，Pod 是临时的对象，会被创建或销毁，以此保持集群一直处于预期的状态。Pod 失败后重新创建的 Pod 会分配到不同的 IP。这样就无法使用之前的 IP 访问 Pod 中的服务。

使用服务发现机制是解决这个问题的有效方式。Service 实现了服务发现的功能。通过 Service 名称可以解析到一组 Pod。调用方可以不关心服务背后的 Pod 的地址变化，使用固定的 Service 名称就可以访问服务。完成了调用方和服务方的地址解耦。

Service 通过选择器来决定选择哪些 Pod 提供服务，定义了选择器的 Service 会自动创建一个与 Service 同名的 Endpoint 对象。Endpoint 对象中保存了与 Pod 地址相关的信息。如果没有为 Service 定义选择器，则可以手动创建 Endpoint 对象来控制 Service 的流量去向。

Kubernetes 中的控制器会持续扫描与 Service 选择器匹配的 Pod，匹配成功后会更新与 Service 同名的 Endpoint 对象。

Service 对后端 Pod 起到流量代理的作用，代理的实现由 kube-proxy 组件完成，在 Kubernetes 中的每个节点都会启动一个 kube-proxy。

7.17.2　Service 示例

下面是示例配置的文件 nginx-service. yaml，内容如下：

```
apiVersion: v1
kind: Service
metadata:
  name: nginx - service
spec:
  ports:
   - port: 8000
 targetPort: 80
  protocol: TCP
  selector:
    app: nginx
```

下面解释文件中指令的含义。

将对象类型定义为 Service，指令如下：

```
kind: Service
```

将 Service 名称指定为 nginx-service，指令如下：

```
name: nginx-service
```

将访问服务的端口指定为 TCP 8000。将 Service 转发到 Pod 的 TCP 80 端口。Service 会把 8000 端口的流量导向 Pod 上的 80 端口。port 和 targetPort 可以任意设置，一般为了方便会把它们设置为同样的值。TCP 是 Service 默认使用的协议，可以按需使用其他支持的协议，如 UDP、SCTP、HTTP 等，指令如下：

```
ports:
- port: 8000
```

```
targetPort: 80
protocol: TCP
```

指定选择 Pod 使用的标签。所有具有标签 app=nginx 的 Pod 都会被这个 Service 选中，以此提供服务。如果不定义 selector 字段，流量不会转发到 Pod，这对于需要将流量转发到除了 Pod 以外目标的场景很有用。例如转发到集群外部的数据库，或者转发到处于其他命名空间的 Service 及转发到还未部署到 Kubernetes 中的其他系统等，指令如下：

```
selector:
  app: nginx
```

另外需要注意对于暴露多个端口的 Service，需要为所有的端口命名，例如一个暴露了 80 和 443 端口的 Service 需要定义如下：

```
apiVersion: v1
kind: Service
metadata:
  name: my-service
spec:
  selector:
    app: MyApp
  ports:
    - name: http
      protocol: TCP
      port: 80
      targetPort: 9376
    - name: https
      protocol: TCP
      port: 443
      targetPort: 9377
```

这里把 80 端口命名为 http，把 443 端口命名为 https。端口名称遵循 Kubernetes 名称的命名规范，只能使用小写英文字符、数字和中画线，并且不能以中画线开头。例如-web 是错误的命名。

使用 nginx-service. yaml 创建 Service，命令如下：

```
kubectl apply -f nginx-service.yaml
```

运行命令后输出的信息如下：

```
service/nginx-service created
```

查看 nginx-service，命令如下：

```
kubectl get svc nginx-service
```

运行命令后输出的信息如下：

```
NAME            TYPE        CLUSTER-IP       EXTERNAL-IP   PORT(S)     AGE
nginx-service   ClusterIP   10.100.164.22    <none>        8000/TCP    49s
```

TYPE 列为 ClusterIP，表示 Service 的类型是 ClusterIP，ClusterIP 为 Service 的默认类型。CLUSTER-IP 列显示的 10.100.164.22 是 Kubernetes 分配给 Service 的一个集群内部的 IP。这个 IP 只可以从集群内访问。

Service 对象的 spec.clusterIP 字段保存了这个 IP，查询命令如下：

```
kubectl get svc nginx-service -o jsonpath={.spec.clusterIP}
```

因为 nginx-service 指定了 selector，所以 Kubernetes 会自动创建名为 nginx-service 的 Endpoint 对象。

查看自动创建的 Endpoint 的信息，命令如下：

```
kubectl get endpoints
```

运行命令后输出的信息如下：

```
NAME            ENDPOINTS                                    AGE
kubernetes      192.168.155.145:443,192.168.164.55:443      12h
nginx-service   <none>    3s
```

因为集群内没有 Pod 具有 app=nginx 标签，所以 ENDPOINTS 列的信息为空。

创建一个包含 3 个 Pod 副本的 Deployment，Deployment 配置文件 nginx-deployment 的内容如下：

```
apiVersion: apps/v1
kind: Deployment
metadata:
  name: nginx-deployment
  labels:
    app: nginx
spec:
  replicas: 3
  selector:
    matchLabels:
      app: nginx
  template:
    metadata:
```

```
      labels:
        app: nginx
    spec:
      containers:
      - name: nginx
        image: nginx:1.14.2
        ports:
        - containerPort: 80
```

使用这个文件创建 Deployment，命令如下：

```
kubectl apply -f nginx-deployment.yaml
```

再次查看 nginx-service Endpoint 信息，命令如下：

```
kubectl get endpoints nginx-service
```

运行命令后输出的类似信息如下：

```
NAME            ENDPOINTS                                                   AGE
nginx-service   192.168.29.27:80,192.168.5.10:80,192.168.88.239:80   34m
```

ENDPOINTS 列显示了 3 个地址，它们是 nginx-deployment 中的 3 个 Pod 的 IP。

查看 nginx-deployment 中的 Pod 的 IP 信息，命令如下：

```
kubectl get pod -l app=nginx \
-o custom-columns=NAME:.metadata.name,IP:.status.podIP
```

运行命令后输出的类似信息如下：

```
NAME                                IP
nginx-deployment-66b6c48dd5-cdc6f   192.168.5.10
nginx-deployment-66b6c48dd5-nlcqg   192.168.88.239
nginx-deployment-66b6c48dd5-tgpdk   192.168.29.27
```

IP 这一列显示的 3 个 IP 和 Endpoint(nginx-service)中的 3 个 IP 一致。

查看 nginx-service Endpoint 的详细信息，命令如下：

```
kubectl get endpoints nginx-service -o yaml
```

运行命令后输出的类似信息如下：

```
apiVersion: v1
kind: Endpoints
```

```
metadata:
  annotations:
    endpoints.kubernetes.io/last-change-trigger-time: "2021-05-14T02:57:54Z"
  creationTimestamp: "2021-05-14T02:31:05Z"
  name: nginx-service
  namespace: default
  resourceVersion: "9634"
  selfLink: /api/v1/namespaces/default/endpoints/nginx-service
  uid: 2d2d34fd-4fa6-48b9-a7b2-5e23ab5cf976
subsets:
- addresses:
  - ip: 192.168.29.27
    nodeName: ip-192-168-8-34.us-west-2.compute.internal
    targetRef:
      kind: Pod
      name: nginx-deployment-66b6c48dd5-tgpdk
      namespace: default
      resourceVersion: "9623"
      uid: b415ca5f-0063-4196-a926-67a708d72ac9
  - ip: 192.168.5.10
    nodeName: ip-192-168-8-34.us-west-2.compute.internal
    targetRef:
      kind: Pod
      name: nginx-deployment-66b6c48dd5-cdc6f
      namespace: default
      resourceVersion: "9626"
      uid: a24077c5-effc-4a79-9d34-183791eb0dad
  - ip: 192.168.88.239
    nodeName: ip-192-168-74-110.us-west-2.compute.internal
    targetRef:
      kind: Pod
      name: nginx-deployment-66b6c48dd5-nlcqg
      namespace: default
      resourceVersion: "9632"
      uid: 61675f3b-0708-4142-b80d-a57cafb9b7ea
  ports:
  - port: 80
    protocol: TCP
```

输出信息中的 subsets.addresses 字段包含了 3 个与 Pod 地址相关的信息。

为了验证服务是否可用，可在集群内的一个 Pod 中请求服务地址，命令如下：

```
kubectl run --rm -it test --image nginx -- bash
```

运行命令并进入容器后看到的提示符如下：

```
root@test:/#
```

在容器内运行命令如下：

```
curlnginx-service:8000
```

输出信息截取如下：

```
<!DOCTYPE html>
<html>
<head>
<title>Welcome to nginx!</title>
...
```

输出内容是 nginx 的默认首页，说明 nginx-service 运行正常。

为了观察负载均衡，修改 3 个 Pod 的默认 index.html 文件的内容，把 Pod 主机名（也就是 Pod 名称）写入 index.html 文件，命令如下：

```
kubectl exec nginx-deployment-66b6c48dd5-nlcqg -- sh -c 'echo 'hostname'> /usr/share/nginx/html/index.html'
kubectl exec nginx-deployment-66b6c48dd5-tgpdk -- sh -c 'echo 'hostname'> /usr/share/nginx/html/index.html'
kubectl exec nginx-deployment-66b6c48dd5-nlcqg -- sh -c 'echo 'hostname'> /usr/share/nginx/html/index.html'
```

接着在之前创建的名为 test 的 Pod 中请求 nginx-service，以便观察响应内容，验证 nginx-service 是否对请求进行负载均衡，命令如下：

```
for i in {1..10}; do curl nginx-service:8000; done
```

运行命令会将 10 次请求发送到 nginx-service，输出的信息如下：

```
nginx-deployment-66b6c48dd5-nlcqg
nginx-deployment-66b6c48dd5-nlcqg
nginx-deployment-66b6c48dd5-cdc6f
nginx-deployment-66b6c48dd5-tgpdk
nginx-deployment-66b6c48dd5-tgpdk
nginx-deployment-66b6c48dd5-nlcqg
nginx-deployment-66b6c48dd5-cdc6f
nginx-deployment-66b6c48dd5-tgpdk
nginx-deployment-66b6c48dd5-nlcqg
nginx-deployment-66b6c48dd5-cdc6f
```

输出信息显示 nginx-service 会把请求负载均衡到 3 个 Pod 中。

7.17.3　代理模式

Kubernetes 中的每个节点都运行着 kube-proxy 示例。kube-proxy 为非 ExternalName 类型的 Service 提供了虚拟 IP。

kube-proxy 代理的模式包括用户空间代理模式、iptables 和 IPVS。

如果希望某个客户端每次发送到 Service 的流量都转发到相同的后端 Pod,实现黏性会话,则可以将 Service 的. spec. sessionAffinity 字段设置为 ClientIP。同时可以通过 Service 的. spec. sessionAffinityConfig. clientIP. timeoutSeconds 字段控制黏性会话的超时时间,单位为秒,默认值为 10800s,也就是 3h。黏性会话示例配置如下:

```
spec:
  sessionAffinity: ClientIP
  sessionAffinityConfig:
    clientIP:
      timeoutSeconds: 10800
```

1. 用户空间代理模式

在用户空间代理模式下,kube-proxy 会监视 Service 与 Endpoint 对象的创建和删除操作。当发现新的 Service 时,kube-proxy 会在本地节点上打开一个随机选择的端口。后续当 kube-proxy 捕获到 Service 的 clusterIP 和端口上的流量后会转发到这个端口。这个端口接收的请求会被转发到对应 Service 的某个后端 Pod。

在用户空间代理模式下,kube-proxy 使用 round-robin 负载均衡算法选择后端 Pod。在遇到 Service 的后端 Pod 没有响应的情况下,会自动重试其他 Pod。

2. iptables 代理模式

在 iptables 代理模式下 kube-proxy 同样会监视 Service 与 Endpoint 对象的创建和删除操作。kube-proxy 为每个 Service 设置 iptables 规则以捕获 Service 的流量,然后转发到 Service 的后端 Pod。这种模式下 kube-proxy 会随机选取后端 Pod。

使用 iptables 转发流量带来的系统开销较小也更加稳定,因为流量由 Linux 内核的 nettfilter 处理而无须在用户空间和内核空间之间切换。

以 iptables 模式工作的 kube-proxy 在遇到 Service 的后端 Pod 没有响应的情况下,不会自动重试其他 Pod,所以会导致这次连接失败。为了避免 kube-proxy 将流量转发到失败的 Pod,可以为 Pod 配置 readiness probe(就绪探测)。

3. IPVS 代理模式

在 IPVS 代理模式下 kube-proxy 同样会监视 Service 与 Endpoint 对象的创建和删除操作。kube-proxy 通过 netlink 接口创建 IPVS 规则并与 Service 和 Endpoint 周期性地进行同步。IPVS 将捕获 Service 的流量并转发到 Service 后端的 Pod。

与 iptables 模式类似,IPVS 模式下的 kube-proxy 同样工作在内核空间,但是 IPVS 使用哈希表作为底层的数据结构,所以转发流量的延时比 iptables 模式更低,性能更好。与其他代理模式相比,IPVS 支持更高的网络流量吞吐。

IPVS 支持更多的负载均衡算法,包括 rr(round-bobin)、lc(least connection)、dh(destination hash)、sh(source hashing)、sed(shortest expected delay)、nq(never queue)。

使用 IPVS 代理模式前,需要确保节点内核的 IPVS 模块可用。如果没有检测到可用的 IPVS 模块,kube-proxy 会回退到 iptables 代理模式。

7.17.4 服务发现

访问网站时,用户通常使用域名而不是网站 IP。与域名相比 IP 记忆起来比较困难,所以产生了将域名解析为 IP 的 DNS 系统。同样在访问服务时也存在类似的问题,直接访问服务的 IP 地址不仅容易出错而且维护困难,当服务地址改动时,调用方也需要跟着改动。使用服务发现可以解决这个问题。通过服务名称解析服务地址称为服务发现。

Kubernetes 支持两种服务发现模式,即环境变量模式和 DNS 模式。

在创建 Pod 时,kubelet 组件会在 Pod 中添加与当前 Kubernetes 中的 Service 对应的一组环境变量。Service 名称中小写字母会转换成大写,中画线会转换成下画线。

以前面创建的 nginx-service 为例,新创建的 Pod 中会添加与之对应的环境变量,环境变量如下:

```
NGINX_SERVICE_PORT = tcp://10.100.16.163:8000
NGINX_SERVICE_PORT_8000_TCP = tcp://10.100.16.163:8000
NGINX_SERVICE_SERVICE_PORT = 8000
NGINX_SERVICE_PORT_8000_TCP_PROTO = tcp
NGINX_SERVICE_PORT_8000_TCP_PORT = 8000
NGINX_SERVICE_PORT_8000_TCP_ADDR = 10.100.16.163
NGINX_SERVICE_SERVICE_HOST = 10.100.16.163
NGINX_VERSION = 1.14.2 - 1~stretch
```

当新创建的 Pod 需要使用环境变量的方式访问 Service 时,对应的 Service 必须在 Pod 之前创建。否则 Pod 中不会添加与这个 Service 相关的环境变量。

另一种服务发现模式是 DNS 模式,使用前需要在 Kubernetes 中启用 DNS 服务,如 CoreDNS。

启用 DNS 服务后,DNS 服务会监视 Kubernetes 中的 Service,当有新的 Service 创建时,DNS 服务会为之创建相关的 DNS 记录。这样 Kubernetes 中所有的 Pod 就可以通过 DNS 域名解析到对应的 Service 了。

DNS 查询会受到发出查询的 Pod 所在的命名空间影响,没有指定命名空间的 DNS 查询只会返回 Pod 所在命名空间内的结果。访问命名空间与 Pod 不同的 Service,必须在 DNS 查询中指定 Service 命名空间。

例如有一个名为 my-service 的服务,并且命名空间为 my-ns,DNS 服务会为之创建一个 my-service.my-ns 的 DNS 记录。这个记录会解析到服务的 clusterIP。和这个服务处于相同命名空间的 Pod 可以通过 my-service 域名访问这个服务，而处于与 my-ns 不同命名空间的 Pod 可以通过 my-service. my-ns 域名访问该服务。例如在命名空间 prod 中的 Pod 访问 my-service 不会返回任何结果。

不仅会为 Service 创建 DNS 记录，还会为 Service 中的命名端口创建 DNS 记录。例如上述 my-service 服务定义了一个名为 http 的 TCP 端口，那么通过_http. _tcp. my-service. my-ns 就可以解析到这个端口号了。

7.17.5 Service 类型

有些 Service 需要从 Kubernetes 集群外部访问，如前端项目，这需要对 Service 的类型进行设置，或者使用 Ingress 来暴露 Service。

Service 支持设置不同的类型，对应不同的方式暴露服务。类型包括 ClusterIP、NodePort、LoadBalancer 和 ExternalName，通过 Service 的. spec. type 字段指定。

ClusterIP 通过集群内部的 IP 暴露服务，以这种方式暴露的服务只能从集群内部访问。ClusterIP 是 Service 默认的类型。

NodePort 通过在集群中的每个节点上的某个固定端口暴露服务，在集群外部通过地址 <NodeIP>: <NodePort>可以访问 Service。创建 NodePort 类型的服务会自动创建对应的 ClusterIP 类型的服务。

LoadBalancer 通过云服务商的负载均衡器暴露服务。创建 LoadBalancer 类型的服务会自动创建对应的 LoadBalancer 和 ClusterIP 类型的服务。

ExternalName 通过创建 CNAME 类型的 DNS 记录来暴露服务，访问 ExternalName 类型的 Service 不需要 Kubernetes 创建任何类型的代理。

7.17.6 ClusterIP 类型

对于没有指定. spec. type 字段的 Service，默认类型为 ClusterIP。以 7.14.2 节中的 nginx-service 为例，这个 Service 没有指定 type，查看它的 Service 类型，命令如下：

```
kubectl get svc nginx - service
```

运行命令后输出的信息如下：

```
NAME             TYPE         CLUSTER - IP      EXTERNAL - IP   PORT(S)     AGE
nginx - service  ClusterIP    10.100.43.200    <none>          8000/TCP    13m
```

TYPE 列将 nginx-service 的类型显示为 ClusterIP。

也可以使用 jsonpath 直接获取. spec. type 字段，命令如下：

```
kubectl get svc nginx - service  - o jsonpath = {.spec.type}
```

运行命令后输出的信息如下：

```
ClusterIP
```

7.17.7 NodePort 类型

Kubernetes 会为 NodePort 类型的 Service 分配一个端口号，分配端口号的范围默认为 30000～32767。可以使用 kube-proxy 的启动参数--service-node-port-range 设置端口范围。集群中的所有节点都会监听这个端口。端口号会保存到 Service 的. spec. ports[*]. nodePort 字段。

如果希望自定义端口号，则可以手动设置 Service 的. spec. ports[*]. nodePort 字段，命令如下：

```
spec:
  ports
  - nodePort: 31122
    port: 8000
    protocol: TCP
    targetPort: 80
```

这样配置的 Service 就会使用节点上的 31122 端口暴露服务。手动指定端口需要注意避免端口冲突，并且端口不能超出--service-node-port-range 参数设置的端口范围。

如果需要自定义端口工作的 IP，需要设置 kube-proxy 的启动参数--nodeport-addresses 或者 kube-proxy 配置文件中的 nodePortAddresses 字段来指定 IP 段。参数格式是逗号分隔的 IP 地址段，例如(10.0.0.0/8,192.2.3.0/25)。--nodeport-addresses 参数的默认值是空列表，意味着 kube-proxy 会把节点主机上的所有网卡的 IP 地址都用来代理 nodePort 端口上的流量。

修改 nginx-service. yaml，将. spec. type 字段修改为 NodePort，命令如下：

```
apiVersion: v1
kind: Service
metadata:
  name: nginx-service
spec:
  type: NodePort
  ports:
  - port: 8000
    targetPort: 80
    protocol: TCP
  selector:
    app: nginx
```

查看 nginx-service 信息，命令如下：

```
kubectl get svc nginx-service
```

运行命令后输出的信息如下：

```
NAME          TYPE      CLUSTER-IP     EXTERNAL-IP  PORT(S)         AGE
nginx-service NodePort  10.100.43.200  <none>       8000:31169/TCP  5h8m
```

TYPE 列将 nginx-service 的类型变为 NodePort。分配的 NodePort 为 31169。

此时登录到 Kubernetes 集群中的任意 worker 节点，查看节点主机的网卡列表，假设有以下网卡：

```
1: lo: <LOOPBACK,UP,LOWER_UP> mtu 65536 qdisc noqueue state UNKNOWN group default qlen 1000
    link/loopback 00:00:00:00:00:00 brd 00:00:00:00:00:00
    inet 127.0.0.1/8 scope host lo
       valid_lft forever preferred_lft forever
    inet6 ::1/128 scope host
       valid_lft forever preferred_lft forever
2: eth0: <BROADCAST,MULTICAST,UP,LOWER_UP> mtu 9001 qdisc mq state UP group default qlen 1000
    link/ether 0a:58:c3:e1:f8:9d brd ff:ff:ff:ff:ff:ff
    inet 192.168.68.100/19 brd 192.168.95.255 scope global dynamic eth0
       valid_lft 3061sec preferred_lft 3061sec
    inet6 fe80::858:c3ff:fee1:f89d/64 scope link
       valid_lft forever preferred_lft forever
3: eth1: <BROADCAST,MULTICAST,UP,LOWER_UP> mtu 9001 qdisc mq state UP group default qlen 1000
    link/ether 0a:14:1d:71:7e:21 brd ff:ff:ff:ff:ff:ff
    inet 192.168.88.94/19 brd 192.168.95.255 scope global eth1
       valid_lft forever preferred_lft forever
    inet6 fe80::814:1dff:fe71:7e21/64 scope link
       valid_lft forever preferred_lft forever
```

3 个网卡的 IP 地址是 127.0.0.1、192.168.68.100、192.168.88.94。分别使用这 3 个 IP 和 NodePort 访问 nginx-service。命令如下：

```
curl localhost:31169
curl 192.168.68.100:31169
curl 192.168.68.100:31169
```

3 次请求都可以访问 nginx-service。

7.17.8 LoadBalancer 类型

LoadBalancer 类型的 Service 会在支持外部负载均衡器的集群环境中获得云服务商创

建的一个负载均衡器。创建过程是异步的,负载均衡器相关信息会记录到 Service 对象的 .status.loadBalancer 字段。

外部的负载均衡器会把流量转发到 Service 的后端 Pod,负载均衡算法由云服务商决定。

修改 nginx-service.yaml,将.spec.type 字段设置为 LoadBalancer。代码如下:

```
apiVersion: v1
kind: Service
metadata:
  name: nginx-service
spec:
  type: LoadBalancer
  ports:
  - port: 8000
    targetPort: 80
    protocol: TCP
  selector:
    app: nginx
```

使更新生效,命令如下:

```
kubectl apply -f nginx-service.yaml
```

查看 nginx-service 信息,命令如下:

```
kubectl get svc nginx-service -o yaml
```

运行命令后输出的信息如下(前 3 列):

```
NAME            TYPE           CLUSTER-IP
nginx-service   LoadBalancer   10.100.43.200
```

后 3 列信息如下:

```
EXTERNAL-IP                                                PORT(S)          AGE
9b7ff38d61f-2010923911.us-west-2.elb.amazonaws.com   8000:30820/TCP   8h
```

EXTERNAL-IP 列作了截取,EXTERNAL-IP 列的值为负载均衡器的地址。

查看 nginx-service 的详细信息,命令如下:

```
kubectl get svc nginx-service -o yaml
```

运行命令后输出的信息如下:

```
apiVersion: v1
kind: Service
metadata:
  annotations:
    kubectl.kubernetes.io/last - applied - configuration: |
{"apiVersion":"v1","kind":"Service","metadata":{"annotations":{},"name":"nginx - service",
"namespace":"default"},"spec":{"ports":[{"port":8000,"protocol":"TCP","targetPort":
80}],"selector":{"app":"nginx"},"type":"LoadBalancer"}}
  creationTimestamp: "2021 - 05 - 17T02:16:19Z"
  finalizers:
  - service.kubernetes.io/load - balancer - cleanup
  name: nginx - service
  namespace: default
  resourceVersion: "248127"
  selfLink: /api/v1/namespaces/default/services/nginx - service
  uid: 0e4123b7 - 0b48 - 4901 - 8c51 - 9b7ff38d61f6
spec:
  clusterIP: 10.100.43.200
  externalTrafficPolicy: Cluster
  ports:
  - nodePort: 30820
    port: 8000
    protocol: TCP
    targetPort: 80
  selector:
    app: nginx
  sessionAffinity: None
  type: LoadBalancer
status:
  loadBalancer:
    ingress:
    - hostname: \
    a0e4123b70b4849018c519b7ff38d61f - 2010923911.us - west - 2.elb.amazonaws.com
```

输出信息中的.spec.ports[*].nodePort字段表示Kubernetes为LoadBalancer类型的nginx-service同样分配了节点端口。如果想禁止为LoadBalancer类型的Service分配节点端口，可以将Service的.spec.allocateLoadBalancerNodePort字段值设置为false，这个功能需要Kubernetes版本不低于v1.20。

.status.loadBalancer字段可以看到云服务商创建的负载均衡器信息，这里使用的是AWS托管的集群，会为LoadBalancer类型的Serviceh创建ELB。ELB的主机名比较长。通过负载均衡器访问nginx-service，命令如下：

```
curl \
a0e4123b70b4849018c519b7ff38d61f - 1c19c3acfa09fec7.elb.us - west - 2.amazonaws.com:8000
```

运行命令后输出的信息如下：

```
nginx - deployment - 66b6c48dd5 - 5v2sf
```

7.17.9 ExternalName 类型

ExternalName 类型的 Service 为集群外部 DNS 名称提供内部别名。客户端使用内部 DNS 名称发出请求，请求会被重定向到外部名称。

下面是 ExternalName 类型 Service 示例定义的文件：

```
apiVersion: v1
kind: Service
metadata:
  name: my - service
  namespace: prod
spec:
  type: ExternalName
  externalName: my.database.example.com
```

文件中定义的 Service 的 DNS 名称是 my-service. prod. svc. cluster. local。当内部客户端向 my-service. prod. svc. cluster. local 发出请求时，Kubernetes 中的 DNS 服务会返回一个 CNAME 记录，值为 my. database. example. com，这样请求就会被重定向到 my. database. example. com。

使用 ExternalName 类型的 Service 需要 kube-dns 的版本不低于 1.7 或者 CoreDNS 的版本不低于 0.0.8。

7.17.10 headless Service

如果不需要为 Service 分配集群内部 IP 及 Service 的负载均衡功能，可以使用 headless Service。由于 headless Service 没有 clusterIP，DNS 会根据服务名直接解析出后端一组 Pod 的 IP。与其他具备 clusterIP 的 Service 的流量路径相比少了 clusterIP 这一步，所以称为 headless（无头）服务。由于 headless Service 无法使用 Kubernetes 内置的负载均衡功能，所以用户可以自由选择外部的负载均衡组件。

创建 headless Service，需要把 Service 的. spec. clusterIP 字段设置为 None。

kube-proxy 不会处理 headless Service 的流量，不进行负载均衡和网络代理。

根据 headless Service 是否定义了选择器（. spec. selector 字段），DNS 的处理分为下面 2 种情况：

（1）定义了选择器的 headless Service，Kubernetes 会创建相关的 Endpoint 对象，Service 的 DNS 记录会解析到后端的一组 Pod 的 IP。

（2）没有对应选择器的 headless Service，Kubernetes 不会创建 Endpoint 对象。如果

Service 类型是 ExternalName，DNS 会配置一条 CNAME 记录，对于其他 Service 类型，DNS 会根据与 Service 同名的 Endpoint 配置一条记录。

下面是 headless Service 示例配置文件 headless-service.yaml，内容如下：

```
apiVersion: v1
kind: Service
metadata:
  name: headless
spec:
  clusterIP: None
  ports:
  - port: 80
    name: web
  selector:
    name: busybox
```

使用 headless-service.yaml 创建 Service，命令如下：

```
kubectl apply -f headless-service.yaml
```

运行命令后输出的信息如下：

```
service/headless created
```

由于定义了 selector，会创建相关的 Endpoint 对象，查看 Endpoint 信息，命令如下：

```
kubectl get endpoints
```

运行命令后输出的信息如下：

```
NAME         ENDPOINTS                                    AGE
headless     <none>                                       2m24s
kubernetes   192.168.103.219:443,192.168.131.212:443      2d5h
```

创建后端 Pod，新建文件 headless-pod.yaml，内容如下：

```
apiVersion: v1
kind: Pod
metadata:
  name: busybox1
  labels:
    name: busybox
spec:
  hostname: busybox-1
  subdomain: default-subdomain
```

```
  containers:
  - image: busybox:1.28
    command:
      - sleep
      - "3600"
    name: busybox
---
apiVersion: v1
kind: Pod
metadata:
  name: busybox2
  labels:
    name: busybox
spec:
  hostname: busybox-2
  subdomain: default-subdomain
  containers:
  - image: busybox:1.28
    command:
      - sleep
      - "3600"
    name: busybox
```

文件定义了 2 个使用相同镜像并且具有相同标签的 Pod。Pod 标签与 headless Service 的选择器匹配。接下来需要运行的命令如下：

```
kubectl apply -f headless-pod.yaml
```

运行命令后输出的信息如下：

```
pod/busybox1 created
pod/busybox2 created
```

查看 Endpoint 信息，命令如下：

```
kubectl get endpoints headless
```

运行命令后输出的信息如下：

```
NAME       ENDPOINTS                              AGE
headless   192.168.27.200:80,192.168.77.95:80   9m41s
```

输出信息 ENDPOINTS 列显示有 2 个后端 Pod 地址。说明 headless Service 匹配到了新创建的 2 个 Pod。

启动一个 Pod，在 Pod 中使用 nslookup 命令查看 DNS 解析 headless Service，命令如下：

```
kubectl run --rm -it test --image alpine -- sh
```

运行命令后输出的信息如下：

```
If you don't see a command prompt, try pressing enter.
/ #
```

进入 Pod 后，查询 DNS 域名 headless，命令如下：

```
nslookup headless
```

运行命令后输出的信息如下：

```
Server:   10.100.0.10
Address:   10.100.0.10:53

Name:   headless.default.svc.cluster.local
Address: 192.168.77.95
Name:   headless.default.svc.cluster.local
Address: 192.168.27.200

** server can't find headless.svc.cluster.local: NXDOMAIN

** server can't find headless.cluster.local: NXDOMAIN

*** Can't find headless.default.svc.cluster.local: No answer

** server can't find headless.svc.cluster.local: NXDOMAIN

** server can't find headless.cluster.local: NXDOMAIN

** server can't find headless.us-west-2.compute.internal: NXDOMAIN

** server can't find headless.us-west-2.compute.internal: NXDOMAIN
```

输出显示域名 headless.default.svc.cluster.local 解析的结果是 2 个 IP：192.168.77.95、192.168.27.200。这 2 个 IP 就是后端 2 个 Pod 的 IP 地址。

查看 Pod 信息验证，命令如下：

```
kubectl get pod -l name=busybox -o wide
```

截取运行命令后输出信息的前 6 列，输出信息如下：

```
NAME        READY   STATUS     RESTARTS   AGE    IP
busybox1    1/1     Running    2          127m   192.168.27.200
busybox2    1/1     Running    2          127m   192.168.77.95
```

输出信息 IP 列显示的 2 个 IP 和上述域名 headless. default. svc. cluster. local 的解析结果一致。

7.18 Kubernetes DNS

Kubernetes 会为 Service 和 Pod 创建 DNS 记录。Kubernetes 提供了内置的 DNS 服务 kube-dns 负责分配 DNS 域名。

7.18.1 DNS 服务

查看 Kubernetes 中的 DNS 服务,命令如下:

```
kubectl get svc kube-dns  -n kube-system
```

运行命令后输出的类似信息如下:

```
NAME        TYPE        CLUSTER-IP    EXTERNAL-IP   PORT(S)          AGE
kube-dns    ClusterIP   10.100.0.10   <none>        53/UDP,53/TCP    130m
```

输出信息显示 kube-dns 的类型为 ClusterIP。暴露了 TCP 53 和 UDP 53 两个端口。

查看 kube-dns 的详细信息,命令如下:

```
kubectl describe svc kube-dns  -n kube-system
```

运行命令后输出的类似信息如下:

```
Name:              kube-dns
Namespace:         kube-system
Labels:            eks.amazonaws.com/component=kube-dns
                   k8s-app=kube-dns
                   kubernetes.io/cluster-service=true
                   kubernetes.io/name=CoreDNS
Annotations:       prometheus.io/port: 9153
                   prometheus.io/scrape: true
Selector:          k8s-app=kube-dns
Type:              ClusterIP
IP Families:       <none>
IP:                10.100.0.10
IPs:               <none>
Port:              dns  53/UDP
```

```
    TargetPort:          53/UDP
    Endpoints:           192.168.40.32:53,192.168.94.154:53
    Port:                dns-tcp  53/TCP
    TargetPort:          53/TCP
    Endpoints:           192.168.40.32:53,192.168.94.154:53
    Session Affinity:    None
    Events:              <none>
```

输出信息显示 kube-dns 使用标签 k8s-app=kube-dns 选择后端 Pod。

查看 kube-dns 的后端 Pod，命令如下：

```
kubectl get pod -n kube-system -l k8s-app=kube-dns
```

运行命令后输出的类似信息如下：

```
NAME                        READY   STATUS    RESTARTS   AGE
coredns-6548845887-c4chb    1/1     Running   0          135m
coredns-6548845887-hv4hj    1/1     Running   0          135m
```

集群中每个节点上都运行了 coredns Pod。

7.18.2 Service DNS

Service 会被分配一条 DNSA 记录或者 AAAA 记录。记录中的域名格式如下：

```
$(service name).$(namespace).svc.cluster.local
```

其中 cluster.local 是 Kubernetes 集群的默认域名，$(namespace)是 Service 的命名空间，$(service name)是 Service 的名称。

例如之前创建的 nginx-service 的域名如下：

```
nginx-service.default.svc.cluster.local
```

对于非 headless 的普通 Service，它的 DNS 记录解析结果是 Service 的 clusterIP，而 headless Service 的 DNS 记录会返回解析到该 Service 后端一组 Pod 的 IP 集合。

7.18.3 PodDNS

为 Pod 分配的 DNS 域名格式如下：

```
$(pod ip).$(namespace).pod.cluster.local
```

其中 cluster.local 是 Kubernetes 集群的默认域名，$(namespace)是 Pod 的命名空间，$(pod ip)是 Pod 的 IP 地址。

例如 IP 为 192.168.28.228 的 Pod，可以通过 192-168-28-228.default.pod.cluster.local 解析到。

DNS 查询会用到 Pod 中的/etc/resolv.conf 文件进行扩展。Kubelet 会为每个 Pod 设置这个文件。

创建一个运行 alpine 镜像的 Pod 并进入 Pod，命令如下：

```
kubectl run --rm -it  test --image alpine  -- sh
```

进入 Pod 后，查看 Pod 中/etc/resolv.conf 的文件内容，命令如下：

```
cat /etc/resolv.conf
```

运行命令后输出的信息如下：

```
nameserver 10.100.0.10
search default.svc.cluster.local svc.cluster.local \
   cluster.local us-west-2.compute.internal
options ndots:5
```

如果需要自定义 Pod 中的 DNS 配置，可以设置 Pod 的 dnsConfig 字段，dnsConfig 字段为可选字段，可以与任何 dnsPolicy 搭配使用。当 Pod 的 dnsPolicy 字段值为 None 时，必须指定 dnsConfig。

dnsConfig 字段支持指定以下属性。

(1) nameservers：指定 Pod 的 DNS 服务器 IP 地址列表。最多可以指定 3 个 IP 地址。当 Pod 的 dnsPolicy 设置为 None 时，列表必须包含至少一个 IP 地址。服务器地址将合并到从指定的 DNS 策略生成的基本域名服务器，并删除重复的地址。

(2) searches：指定在 Pod 中用于查找主机名的 DNS 搜索域列表，此属性是可选的。当指定此属性时，所提供的列表将合并到根据所选 DNS 策略生成的基本搜索域名中。重复的域名将被删除。Kubernetes 最多允许指定 6 个搜索域。

(3) options：可选的对象列表，其中每个对象可能具有必需的 name 属性和可选的 value 属性。此属性中的内容将合并到从指定的 DNS 策略生成的选项，重复的条目将被删除。

下面是指定了 dnsConfig 的 Pod 示例配置文件 custom-dns.yaml 的文件内容：

```
apiVersion: v1
kind: Pod
metadata:
  namespace: default
  name: dns-example
spec:
  containers:
```

```
    - name: test
      image: nginx
  dnsPolicy: "None"
  dnsConfig:
    nameservers:
      - 1.2.3.4
    searches:
      - ns1.svc.cluster-domain.example
      - my.dns.search.suffix
    options:
      - name: ndots
        value: "2"
      - name: edns0
```

使用 custom-dns.yaml 文件创建 Pod,命令如下：

```
kubectl apply -f custom-dns.yaml
```

运行命令后输出的信息如下：

```
pod/dns-example created
```

查看 Pod 中/etc/resolv.conf 的文件内容,命令如下：

```
kubectl exec dns-example -- cat /etc/resolv.conf
```

运行命令后输出的信息如下：

```
nameserver 1.2.3.4
search ns1.svc.cluster-domain.example my.dns.search.suffix
options ndots:2 edns0
```

看到/etc/resolv.conf 文件内容和 dnsConfig 字段值相匹配。说明 Pod 自定义 DNS 设置已生效。

7.19 Kubernetes Ingress

7.19.1 Ingress 简介

使用 HTTP 和 HTTPS 协议访问的应用,可以使用 Ingress 将标准端口的 Service 暴露到集群外部。Ingress 支持设置路由规则来将请求转发到不同的后端 Service。另外 Ingress 还支持负载均衡、SSL 终结和虚拟主机。

Ingress 不支持暴露使用非 HTTP 和 HTTPS 标准端口的 Service,可以使用 NodePort

或者 LoadBalancer 类型控制器暴露此类 Service。

使用 Ingress 前，需要在集群中安装 Ingress 控制器，否则 Ingress 不会生效。Ingress 控制器可以选择 Kubernetes 官方维护的 ingress-nginx 或者第三方提供的其他 Ingress 控制器。这些 Ingress 控制器的功能大致相同。

7.19.2 Ingress 示例

下面是 Ingress 示例配置文件 ingress-demo.yaml 的内容：

```
apiVersion: networking.k8s.io/v1
kind: Ingress
metadata:
  name: ingress-demo
  annotations:
    nginx.ingress.kubernetes.io/rewrite-target: /
spec:
  rules:
  - http:
      paths:
      - path: /testpath
        pathType: Prefix
        backend:
          service:
            name: nginx-service
            port:
              number: 8000
```

下面解释文件中指令的含义。

把 Ingress 命名为 ingress-demo，Ingress 名称必须是合法的 DNS 子域名，指令如下：

```
name: ingress-demo
```

通过注解为 Ingress 中的 nginx 设置参数。注解内容取决于集群中安装的 Ingress 控制器。不同的控制器支持不同的配置参数，从而需要不同的注解。这里假设使用的 Ingress 控制器为 ingress-nginx，指令如下：

```
annotations:
  nginx.ingress.kubernetes.io/rewrite-target: /
```

.spec.rules 字段用于指定 Ingress 规则。流量会根据 Ingress 规则转发到相应的后端 Service。这里指定了一条规则，当 HTTP 请求的路径前缀是/testpath 时，它会被转发到名为 nginx-service 的 Service，Service 暴露的端口号为 8000。

这条规则没有指定 host，意味着适用于所有与路径匹配的 HTTP 请求，当指定 host

后，请求的主机需要和 Ingress 规则的 host 匹配才会被 Ingress 处理，指令如下：

```
spec:
  rules:
  - http:
      paths:
      - path: /testpath
        pathType: Prefix
        backend:
          service:
            name: nginx-service
            port:
              number: 8000
```

7.19.3　Ingress 规则

Ingress 规则可以通过指定 host 字段匹配主机名，例如 app.com。如果指定了 host，则规则只对指定的 host 生效。如果没有指定 host，则规则对所有 HTTP 请求的流量生效。

主机名可以精确匹配，例如 app.com，也可以使用通配符（*）匹配，例如 *.app.com 会匹配 foo.app.com 或者 bar.app.com。

使用通配符的主机名不可以匹配层级不同的域名。例如 *.app.com 不会匹配 foo.bar.app.com 或者 app.com。

Ingress 规则还支持匹配请求路径，例如/app。每个路径都要关联一个后端服务，通过 service.name 字段指定后端服务名称，通过 service.port 字段指定服务的端口信息。

当 Ingress 规则指定的 host 和路径与 HTTP 请求匹配时，HTTP 请求会被负载均衡器转发到 Ingress 规则对应的后端服务。

当 Ingress 规则指定了路径时，必须将 pathType 字段设置为路径指定类型，没有设置 pathType 的路径会校验失败。路径类型包括 ImplementationSpecific、Exact 和 Prefix。

ImplementationSpecific：由 IngressClass 决定匹配策略。

Exact：精确匹配请求 URL 路径并且区分大小写。

Prefix：根据请求 URL 路径的前缀进行匹配。匹配过程是先把路径使用斜杠分隔符分隔，然后依次匹配分隔后的路径元素。注意：当规则中的路径最后一个元素是请求 URL 路径的最后一个元素的子字符串时，匹配失败，例如假设有一个 Prefix 类型的路径为/a/b，请求路径是/a/bb，由于 b 是 bb 的子串，所以匹配失败。当请求路径是/a/b、/a/b/或/a/b/c 时都会匹配成功。

7.19.4　Ingress 控制器

为了使 Ingress 正常工作需要在 Kubernetes 集群中安装 Ingress 控制器。Ingress 控制器不是 Kubernetes 内置的控制器，不会自动启动，需要用户自行选择合适的 Ingress 控制器并部署到集群中。Kubernetes 支持部署多个 Ingress 控制器，在创建 Ingress 的时候建议添加 ingress.class 注解以指定要使用的 Ingress 控制器。

Ingress 控制器有许多实现，分别由 Kubernetes 和第三方维护，由 Kubernetes 维护的 Ingress 控制器有 AWS、GCE 和 nginx。

第三方维护的 Ingress 控制器有 Ambassador、Contour、HAPorxy Ingress、Istio Ingress、Traefik Ingress、Voyager 等。

7.19.5 默认后端

如果希望 HTTP 请求在没有匹配到任何 Ingress 规则的时候有一个默认的后端来处理请求，则可以为 Ingress 设置默认后端(Default Backend)。类似于编程语言中的 switch 分支语句中的 default 分支。

没有设置规则的 Ingress 会把流量转发到默认后端。

如果 Ingress 设置了规则并且 HTTP 请求的主机和路径没有匹配到任何规则，请求则会被 DefaultBackend 处理。

默认后端示例配置如下：

```
apiVersion: networking.k8s.io/v1
kind: Ingress
metadata:
  name: ingress-default
  annotations:
    nginx.ingress.kubernetes.io/rewrite-target: /
spec:
  defaultBackend:
    service:
      name: nginx-service
      port:
        number: 8000
```

这个 Ingress 没有定义任何规则，所以会把流量都转发到 nignx-service：8000。

7.19.6 资源后端

Ingress 支持把 Kubernetes 中的资源对象设置为后端，称为资源后端(Resource Backend)，字段名为 resource。设置了 resource 就不可以设置 service，它们是互斥的，否则会导致校验失败。资源后端通常用来服务静态资源文件。

下面是资源后端示例配置文件的内容：

```
apiVersion: networking.k8s.io/v1
kind: Ingress
metadata:
```

```
  name: ingress - resource - backend
spec:
  defaultBackend:
    resource:
      apiGroup: k8s.example.com
      kind: StorageBucket
      name: static - assets
  rules:
    - http:
        paths:
          - path: /icons
            pathType: ImplementationSpecific
            backend:
              resource:
                apiGroup: k8s.example.com
                kind: StorageBucket
                name: icon - assets
```

7.19.7　fanout 示例

fanout 类型的 Ingress 会根据请求的 URL 把单个 IP 地址接收到的流量转发到多个后端 Service。多个 Service 共用一个 Ingress 的负载均衡器。

创建 fanout 类型 Ingress 的配置文件 simple-fanout-example.yaml，文件内容如下：

```
apiVersion: networking.k8s.io/v1
kind: Ingress
metadata:
  name: simple - fanout - example
spec:
  rules:
  - host: foo.bar.com
    http:
      paths:
      - path: /foo
        pathType: Prefix
        backend:
          service:
            name: service1
            port:
              number: 4200
      - path: /bar
        pathType: Prefix
        backend:
          service:
```

```
            name: service2
            port:
              number: 8080
```

这个 Ingress 会把/foo 请求转发到 service1：4200，把/bar 请求转发到 service2：8000。

7.19.8 虚拟主机示例

虚拟主机支持将同一 IP 地址接收的针对多个主机名的 HTTP 流量路由到相应后端。Ingress 根据请求头的 Host 字段选择路由。

创建虚拟主机配置文件 name-virtual-host-ingress.yaml，文件内容如下：

```
apiVersion: networking.k8s.io/v1
kind: Ingress
metadata:
  name: name-virtual-host-ingress
spec:
  rules:
  - host: foo.bar.com
    http:
      paths:
      - pathType: Prefix
        path: "/"
        backend:
          service:
            name: service1
            port:
              number: 80
  - host: bar.foo.com
    http:
      paths:
      - pathType: Prefix
        path: "/"
        backend:
          service:
            name: service2
            port:
              number: 80
```

这个 Ingress 会把访问 foo.bar.com 主机的请求转发到 service1：80，把访问 bar.foo.com 主机的请求转发到 service2：80。

7.19.9 TLS 示例

Ingress 支持 TLS 加密和 TLS 终止，Ingress 只支持 443 作为 TLS 端口。TLS 终止的

意思是只加密客户端和 Ingress 之间的流量，流量会被解密为明文后转发到 Service。

使用 TLS 需要用到一个包含 TLS 私钥和证书的 Secret。Secret 的键名必须是 tls.crt 和 tls.key，tls.crt 用于保存证书，tls.key 用于保存私钥。

Secret 示例配置如下：

```
apiVersion: v1
kind: Secret
metadata:
  name: testsecret-tls
  namespace: default
data:
  tls.crt: base64 encoded cert
  tls.key: base64 encoded key
type: kubernetes.io/tls
```

下面使用 openssl 命令生成密钥和证书。

生成密钥和证书，命令如下：

```
openssl req -x509 -nodes -days 365 -newkey rsa:2048 -keyout /tmp/tls.key -out /tmp/tls.
crt -subj "/CN=my-tls/O=my-tls"
```

把证书转换为 base64 格式，命令如下：

```
cat /tmp/tls.crt | base64
```

把输出信息复制并粘贴到 tls-secret.yaml，替换 tls.crt 键的值。

同样把密钥转换为 base64 格式，命令如下：

```
cat /tmp/tls.key | base64
```

把输出信息复制并粘贴到 tls-secret.yaml，替换 tls.key 键的值。

然后使用文件 tls-secret.yaml 创建 Ingress，命令如下：

```
kubectl apply -f tls-secret.yaml
```

也可以使用 kubectl create 命令创建 Secret，命令如下：

```
kubectl create secret tls tls-secret --key /tmp/tls.key --cert /tmp/tls.crt
```

运行命令后输出的信息如下：

```
secret/tls-secret created
```

创建使用 TLS 的 Ingress 示例配置文件 tls-example-ingress. yaml，内容如下：

```
apiVersion: networking.k8s.io/v1
kind: Ingress
metadata:
  name: tls-example-ingress
spec:
  tls:
  - hosts:
      - https-example.com
    secretName: testsecret-tls
  rules:
  - host: https-example.com
    http:
      paths:
      - path: /
        pathType: Prefix
        backend:
          service:
            name: nginx-service
            port:
              number: 8000
```

使用文件 tls-example-ingress. yaml 创建 Ingress，命令如下：

```
kubectl apply -f tls-example-ingress.yaml
```

运行命令后输出的信息如下：

```
ingress.networking.k8s.io/tls-example-ingress created
```

查看 tls-example-ingress 信息，命令如下：

```
kubectl get ingress.networking.k8s.io tls-example-ingress
```

运行命令后输出的信息如下：

```
NAME                  CLASS    HOSTS              ADDRESS    PORTS     AGE
tls-example-ingress   <none>   https-example.com  ...        80, 443   26m
```

通过 443 端口访问 Ingress，验证 TLS 是否正常工作，命令如下：

```
curl -k \
https://aedc1e2bd620e492e8ee7787dfc82536-0404fc7994ffd793.elb.us-west-2.amazonaws.
com:443
```

运行命令后输出的信息如下：

```
<!DOCTYPE html>
<html>
<head>
<title>Welcome to nginx!</title>
...
```

请求成功，说明 TLS 配置正确。

7.20 Kubernetes 身份认证

7.20.1 Kubernetes 用户

Kubernetes 中有两类用户，一类是一般意义上的正常用户，表示人的身份；另一类是表示进程身份的服务账号(Service Account)。

不过 Kubernetes 中并没有表示正常用户的对象，所以不存在用于添加正常用户的 API。正常用户可以通过 Kubernetes 中的 Certificate Authority(CA)签名的有效证书来通过认证。Kubernetes 会把证书中 subject 部分的 Common Name 字段值(例如/CN=bob)作为用户名。认证结束后 Kubernetes 中的基于角色的访问控制系统会判断用户是否有权限执行某操作。

与正常用户不同的是，Service Account 由 Kubernetes 进行管理，支持通过调用 API 创建，也可以由 API 自动创建。

请求 Kubernetes API 服务器必须通过认证，没有关联正常用户或 Service Account 的请求会被 Kubernetes 视为匿名请求。

7.20.2 认证策略

Kubernetes 通过客户端证书、令牌、认证代理或者 HTTP basic auth 等身份认证插件完成 API 请求的身份认证。

身份认证插件会把用户名、用户 ID、用户组及附加字段关联到请求。这些关联到请求的属性会被用于鉴权。

集群中至少需要启用一个身份认证插件，用于认证 Service Account，启动一到多个插件用于认证正常用户。当集群中启用了多个身份认证插件后，Kubernetes 会逐个运行并进行认证尝试，第 1 个成功完成身份认证的插件会结束身份认证流程，类似于电路中的短路效果。多个插件的使用顺序并不固定。

通过身份认证的用户或者 Service Account 会被加入 system：authenticated 组中。

7.20.3 证书认证方式

下面演示如何使用证书来认证用户。假设要认证的用户的名称是 blue，blue 所在的用户组是 developer。

首先为用户生成私钥，保存到 blue.key 文件，命令如下：

```
openssl genrsa -out blue.key 2048
```

运行命令后输出的信息如下：

```
Generating RSA private key, 2048 bit long modulus
...............................................................+++
.....+++
e is 65537 (0x10001)
```

接着生成 CSR 文件，命令如下：

```
openssl req -new -key blue.key -out blue.csr -subj "/CN=blue/O=developer"
```

CSR 的 CN 和 O 属性非常重要，Kubernetes 会读取 CN 的值作为用户名，读取 O 的值作为用户组名。

下面创建 CertificateSigningRequest 配置文件 csr.yaml，内容如下：

```yaml
apiVersion: certificates.k8s.io/v1
kind: CertificateSigningRequest
metadata:
  name: myuser
spec:
  groups:
   - system:authenticated
  request: LS0tLS1CRUdJTiBDRVJUSUZJQ0FURSBSRVFVRVNULS0tLS0KTUlJQ2FEQ0NBVkFDQVFBd0l6RU5NQXN
HQTFVRUF3d0VZbXgxWlRFU01CQUdBMVVFQ2d3SlpHVjJaV3h2Y0dWeQpNSUlCSWpBTkJna3Foa2lHOXcwQkFRRUZBQ
U9DQVE4QU1JSUJDZ0tDQVFFQXUvOFlCTndFZjdaaXE3VmtmSnVkCllPVGFldlR0M0lpVFlqbmlTaWl5ek04ZTFob0p
aRlNwaU1DMXFlcUY2NnZGcmVZOGNsNU9TWkx0MDBvclBMTHgKSjVXWXpEcDRxdzBtcmtPS1pzdldDQlJzY3VkZ2VvV
FdPZkVGcHpHU1NKWWxwaXFuMEtiRzdZZ2lZdk44UzRhSQp1dG9IUkVraTZqU3VkTEQ4RnFQOXB1elRjWlR3dEF3VVZ
aUzFJZnV1NU1GN0diZ2plMnlHRTFYRXkzc2pLYlVyCkpMaU1WRHcrbWQyb0NrdUtuTXh4SEdodGRxNTJhKy9ja2Zqc
HROeE1hVll4OHRUUm9ZT2xlSjZqSUxqUU9Na0sKQzZwc1BiWHgrbGNvWUw2MnpmVStic0JaY2dCRGh4eTZ5dkltOE9
TVjZGZVBmZWZiQVZPN0l3bjZYWW51dWk0UQpNd0lEQVFBQm9BQXdEUVlKS29aSWh2Y05BUUVMQlFBRGdnRUJBQ0hTV
UkvS00zV2xUTnRkcDNIQjlTOUVhQzE3CnJIeS9Md0N2cE9HMVNOd0NndjE5QUhvQThtbmtCc0U1SE13Y0dzUWJUSUl
tU0ZTbTErTW1sTXZpdjBlR3h3K1AKNXpyRTJYaEtUNUh0SE8zdlpMVFM4VE9zczdyRllZeVlNUllMY29oWWpSUWhVc
E9zSTRaUFE3YmkyNUlGTEduUAoycUttWndUTWZxWWo1UHpWVXFkdVlOOHdKaXVlU0J1SEpJUENOdS9XSDRGRUxGdG9
ZUHZ5L2lFSGFEUDZkVkZDCjlrZFVETGNabjVFK09VN0RkQlUvWk8yQ240cXFIWEo4U0tTSHJKSk9IQldFOUhkTzI2a
2ptRXNobHM2c3Z6U3oKNEdJeHVnVFIzVVQrZlZRUUtQenRLM3lNNW5UTHlya1FHUzRWQ0VnZUVJVjBEVG04Lzl5aUV
iQjNvb009Ci0tLS0tRU5EIENFUlRJRklDQVRFIFJFUVVFU1QtLS0tLQo=
```

```
  signerName: kubernetes.io/kube-apiserver-client
  usages:
  - client auth
```

文件中.spec.usages 字段的值必须是 client auth。

文件中.spec.request 字段的值是 CSR 文件通过 base64 编码得到的，命令如下：

```
cat blue.csr | base64 | tr -d "\n"
```

使用此文件生成 CertificateSigningRequest，命令如下：

```
kubectl apply -f csr.yaml
```

运行命令后输出的信息如下：

```
certificatesigningrequest.certificates.k8s.io/blue created
```

查看 CSR 信息，命令如下：

```
kubectl get csr blue
```

运行命令后输出的信息如下：

```
NAMEAGE SIGNERNAME                                REQUESTOR          CONDITION
blue 72s kubernetes.io/kube-apiserver-client   kubernetes-admin   Pending
```

输出显示状态为 Pending，这是因为此 CSR 还没有被批准。

批准此 CSR，命令如下：

```
kubectl certificate approve blue
```

运行命令后输出的信息如下：

```
certificatesigningrequest.certificates.k8s.io/blue approved
```

再次查看此 CSR 的信息，会看到以下输出信息：

```
NAME AGE SIGNERNAME                                  REQUESTOR           CONDITION
blue4m31s kubernetes.io/kube-apiserver-client   kubernetes-admin   Approved,Issued
```

输出信息中 CONDITION 列的值变成了 Approved 和 Issued。表示 CSR 已获批准，并且签发了证书。

查看证书，命令如下：

```
kubectl get csr blue - o yaml
```

运行命令后输出的信息如下：

```
apiVersion: certificates.k8s.io/v1
kind: CertificateSigningRequest
metadata:
  annotations:
    kubectl.kubernetes.io/last - applied - configuration: |
      {"apiVersion":"certificates.k8s.io/v1","kind":"CertificateSigningRequest","metadata":
{"annotations":{},"name":"blue"},"spec":{"groups":["system:authenticated"],"request":
"LS0tLS1CRUdJTiBDRVJUSUZJQ0FURSBSRVFVRVNULS0tLS0KTUlJQ2FEQ0NBVkFDQVFBd016RU5NQXNHQTFVRUF3d
0VZbXgxWlRFU01CQUdBMVVFQ2d3S1pHVjJaV3h2Y0dWeQpNSU1CSWpBTkJna3Foa2lHOXcwQkFRRUZBQU9DQVE4QU1
JSUJDZ0tDQVFFQXUvOFlCTndFZjdaaXE3VmtmSnVkCllPVGFldlR0M0lpVFlqbmlTaWl5ek04ZTFob0paRlNwaU1DM
XFlcUY2NnZGcmVZOGNsNU9TWkx0MDBvclBMTHgKSjVXWXpEcDRxdzBtcmtPS1pzdldDQlJzY3VkZ2VvVFdPZkVGcHp
HU1NKWWxwaXFuMEtiRzdZZ2lZdk44UzRhSQp1dG9IUkVraTZqU3VkTEQ4RnFQOXB1elRjWlR3dEF3VVZaUzFJZnV1N
U1GN0diZ2plMnlHRTFYRXkzc2pLYlVyCkpMaU1WRHcrbWQyb0NrdUtuTXh4SEdodGRxNTJhKy9ja2ZqcHROeE1hVll
4OHRUUm9ZT2xlSjZqSUxqUU9Na0sKQzZwc1BiWHgrbGNvVWw2MnpmVStic0JaY2dCRGh4eTZ5dk1tOE9TVjZGZVBmZ
WZiQVZPN013bjZYWW51dWk0UQpNd01EQVFBQm9BQXdEUVlKS29aSWh2Y05BUUVMQlFBRGdnRUJBQ0hTVUkvS00zV2x
UTnRkcDNIQjlTOUVhQzE3CnJIeS9Md0N2cE9HMVNOd0NndjE5QUhvQThtbmtCc0U1SE13Y0dzUWJUSUltU0ZTbTErT
W1sTXZpdjBlR3h3K1AKNXpyRTJYaEtUNUh0SE8zdlpMVFM4VE9zczdyRllZeVlNUllMY29oWWpSUWhVcE9zSTRaUFE
3YmkyNUlGTEduUAoycUttWndUTWZxWWo1UHpWVXFkdVlOOHdKaXVlU0J1SEpJUENOdS9XSDRGRUxGdG9ZUHZ5L2lFS
GFEUDZkVkZDCjlrZFVETGNabjVFK09VN0RkQlUvWk8yQ240cXFIWEo4U0tTSHJKSk9IQldFOUhkTzI2a2ptRXNobHM
2c3Z6U3oKNEdJeHVnVFIzVVQrZlZRQUtQenRLM3lNNW5UTHlya1FHUzRWQ0VnZUVJVjBEVG04Lzl5aUViQjNvb009C
i0tLS0tRU5EIENFUlRJRklDQVRFIFJFUVVFU1QtLS0tLQo=","signerName":"kubernetes.io/kube - apiserver -
client","usages":["client auth"]}}
  creationTimestamp: "2021 - 07 - 03T03:27:15Z"
  name: blue
  resourceVersion: "5093"
  uid: 98b498a9 - dbe6 - 4dda - 97d7 - 6c56bf3cd488
spec:
  groups:
  - system:masters
  - system:authenticated
  request: LS0tLS1CRUdJTiBDRVJUSUZJQ0FURSBSRVFVRVNULS0tLS0KTUlJQ2FEQ0NBVkFDQVFBd016RU5NQXN
HQTFVRUF3d0VZbXgxWlRFU01CQUdBMVVFQ2d3S1pHVjJaV3h2Y0dWeQpNSU1CSWpBTkJna3Foa2lHOXcwQkFRRUZBQ
U9DQVE4QU1JSUJDZ0tDQVFFQXUvOFlCTndFZjdaaXE3VmtmSnVkCllPVGFldlR0M0lpVFlqbmlTaWl5ek04ZTFob0p
aRlNwaU1DMXFlcUY2NnZGcmVZOGNsNU9TWkx0MDBvclBMTHgKSjVXWXpEcDRxdzBtcmtPS1pzdldDQlJzY3VkZ2VvV
FdPZkVGcHpHU1NKWWxwaXFuMEtiRzdZZ2lZdk44UzRhSQp1dG9IUkVraTZqU3VkTEQ4RnFQOXB1elRjWlR3dEF3VVZ
aUzFJZnV1NU1GN0diZ2plMnlHRTFYRXkzc2pLYlVyCkpMaU1WRHcrbWQyb0NrdUtuTXh4SEdodGRxNTJhKy9ja2Zqc
HROeE1hVll4OHRUUm9ZT2xlSjZqSUxqUU9Na0sKQzZwc1BiWHgrbGNvVWw2MnpmVStic0JaY2dCRGh4eTZ5dk1tOE9
TVjZGZVBmZWZiQVZPN013bjZYWW51dWk0UQpNd01EQVFBQm9BQXdEUVlKS29aSWh2Y05BUUVMQlFBRGdnRUJBQ0hTV
UkvS00zV2xUTnRkcDNIQjlTOUVhQzE3CnJIeS9Md0N2cE9HMVNOd0NndjE5QUhvQThtbmtCc0U1SE13Y0dzUWJUSUl
tU0ZTbTErTW1sTXZpdjBlR3h3K1AKNXpyRTJYaEtUNUh0SE8zdlpMVFM4VE9zczdyRllZeVlNUllMY29oWWpSUWhVc
E9zSTRaUFE3YmkyNUlGTEduUAoycUttWndUTWZxWWo1UHpWVXFkdVlOOHdKaXVlU0J1SEpJUENOdS9XSDRGRUxGdG9
ZUHZ5L2lFSGFEUDZkVkZDCjlrZFVETGNabjVFK09VN0RkQlUvWk8yQ240cXFIWEo4U0tTSHJKSk9IQldFOUhkTzI2a
2ptRXNobHM2c3Z6U3oKNEdJeHVnVFIzVVQrZlZRQUtQenRLM3lNNW5UTHlya1FHUzRWQ0VnZUVJVjBEVG04Lzl5aUV
iQjNvb009Ci0tLS0tRU5EIENFUlRJRklDQVRFIFJFUVVFU1QtLS0tLQo=
```

```
  signerName: kubernetes.io/kube-apiserver-client
  usages:
  - client auth
  username: kubernetes-admin
status:
  certificate: LS0tLS1CRUdJTiBDRVJUSUZJQ0FURS0tLS0tCk1JSURDRENDQWZDZ0F3SUJBZ0lRSnRQbmRTVXZ
NczRhc093M0FKR00zREFOQmdrcWhraUc5dzBCQVFzRkFEQVYKTVJNd0VRWURWUVFERXdwcmRXSmxjbTVsZEdWek1CN
FhEVEl4TURjd016QXpNall6TlZvWERUSXlNRGN3TXpBegpNall6TlZvd016RVNNQkFHQTFVRUNoTUpaR1YyWld4dmN
HVnlNUTB3Q3dZRFZRUURFd1JpYkhWbE1JSUJJakFOCkJna3Foa2lHOXcwQkFRRUZBQU9DQVE4QU1JSUJDZ0tDQVFFQ
XUvOFlCTndFZjdaaXE3VmtmSnVkWU9UYWV2VHQKM0lpVFlqbmlTaWl5ek04ZTFob0paRlNwaU1DMXFlcUY2NnZGcmV
ZOGNsNU9TWkx0MDBvclBMTHhKNVdZekRwNApxdzBtcmtPS1pzdldDQlJzY3VkZ2VvVFdPZkVGcHpHU1NKWWxwaXFuM
EtiRzdZZ2lZdk44UzRhSXV0b0hSRWtpCjZqU3VkTEQ4RnFQOXB1elRjWlR3dEF3VVZaUzFJZnV1NU1GN0diZ2plMnl
HRTFYRXkzc2pLYlVySkxpTVZEdysKbWQyb0NrdUtuTXh4SEdodGRxNTJhKy9ja2ZqcHROeE1hVll4OHRUUm9ZT2xlS
jZqSUxqUU9Na0tDNnBzUGJYeAorbGNvWUw2MnpmVStic0JaY2dCRGh4eTZ5dkltOE9TVjZGZVBmZWZiQVZPN0l3bjZ
YWW51dWk0UU13SURBUUFCCm8wWXdSREFUQmdOVkhTVUVEREFLQmdnckJnRUZCUWNEQWpBTUJnTlZIUk1CQWY4RUFqQ
UFNQjhHQTFVZEl3UVkKTUJhQUZEU1hxYVA4Y2JmOTVyVlBXZlMxSllRMWdsM2lNQTBHQ1NxR1NJYjNEUUVCQ3dVQUE
0SUJBUUJVSFQ1RQpGYml6WTMxTUZpVU1xWmxJeUpCdlNUN0QzazduaHpvbDdpeVNVUEVPZXd5L2p6eGNXR3NibStMY
3p6UnVrMjlBCnZKcVh4amJycUVjTExqTGZCLytZSjEwSDlSd3RycnRlanJ0N2sxeUFSZE1UN3BQMnNuMldiQjZNekx
CelhxS2sKakwzOC92WlpseDhpR3Z0WHJrYzY4V3pFOU1QdVd0aDUxR2U3MmdVajBQNGNPSjlab01RZlMwMWV3ak5yL
2hHYgo0SGhUMVVxMFFwZVZLQlhMSGdQQkNoaWloWnluNXNLb3lhQ0VhOEd1cnptcDVTc3lrWlRKcUQ1WnFyWFdKcis
1ClpIbDlkM2lFbWpWc0dsa05pYUMyZ1dtUGdYaThVVVN4TDA5MUtKN0lhMy8wQkxjSWdmTm16STVIN0lkbTBwL3MKR
TRBSU1ZZVpEZGdSY09PRgotLS0tLUVORCBDRVJUSUZJQ0FURS0tLS0tCg==
  conditions:
  - lastTransitionTime: "2021-07-03T03:31:35Z"
    lastUpdateTime: "2021-07-03T03:31:35Z"
    message: This CSR was approved by kubectl certificate approve.
    reason: KubectlApprove
    status: "True"
    type: Approved
```

输出信息的.status.certificate字段的值就是签发的证书。

把证书保存到blue.crt文件，命令如下：

```
kubectl get csr blue -o jsonpath='{.status.certificate}'| base64 -d > blue.crt
```

为了授予blue用户权限，需要为用户创建Role和RoleBinding。

创建Role，命令如下：

```
kubectl create role developer --verb=create --verb=get --verb=list --verb=update
--verb=delete --resource=pods
```

运行命令后输出的信息如下：

```
role.rbac.authorization.k8s.io/developer created
```

查看 developer 角色的信息，命令如下：

```
kubectl get role developer -o yaml
```

运行命令后输出的信息如下：

```
apiVersion: rbac.authorization.k8s.io/v1
kind: Role
metadata:
  creationTimestamp: "2021-07-03T07:32:13Z"
  name: developer
  namespace: default
  resourceVersion: "23744"
  uid: 0d48d158-9bc2-490a-9a3d-e38befc2abc5
rules:
- apiGroups:
  - ""
  resources:
  - pods
  verbs:
  - create
  - get
  - list
  - update
  - delete
```

创建 RoleBinding，将 developer 角色绑定到 blue 用户，命令如下：

```
kubectl create rolebinding developer-binding-blue \
--role=developer \
--user=blue
```

运行命令后输出的信息如下：

```
rolebinding.rbac.authorization.k8s.io/developer-binding-blue created
```

接下来把用户 blue 添加到 kubeconfig，这样 blue 用户就可以访问集群了。

添加用户 blue 的凭证，命令如下：

```
kubectl config set-credentialsblue \
--client-key=blue.key \
--client-certificate=blue.crt \
--embed-certs=true
```

运行命令后输出的信息如下：

```
User "blue" set.
```

为用户设置 context，命令如下：

```
kubectl config set-context blue --cluster=kubernetes --user=blue
```

运行命令后输出的信息如下：

```
Context "blue" created.
```

测试 blue 用户是否可以访问集群，切换 context，命令如下：

```
kubectl config use-context blue
```

以 blue 用户身份创建 Pod，命令如下：

```
kubectl run test --image nginx
```

运行命令后输出的信息如下：

```
pod/test created
```

以 blue 用户身份查询 Pod 信息，命令如下：

```
kubectl get pod
```

运行命令后输出的信息如下：

```
NAME   READY   STATUS     RESTARTS   AGE
test   1/1     Running    0          6s
```

证明 blue 用户可以正常创建和查询 Pod。

由于 blue 用户绑定的角色只可以访问 default 命名空间下的 Pod 资源，所以当尝试访问其他资源时会失败，例如访问 Node 信息时会因为权限不足而失败，命令如下：

```
kubectl get node
```

运行命令后输出的信息如下：

```
Error from server (Forbidden): nodes is forbidden: User "blue" cannot list\
 resource "nodes" in API group "" at the cluster scope
```

输出提示访问被拒。

7.21 Kubernetes 授权

通过认证的请求会进入授权流程。授权即授予权限。对于权限不足的请求，Kubernetes 会拒绝请求，返回 403 HTTP 状态码。

通过 kubectl auth can-i 命令可以快速查询 API 授权结果，这个命令使用 SelfSubjectAccessReview API 来判断当前用户是否可以执行当前操作。这个命令在任何授权模式下都可以工作，示例命令如下：

```
kubectl auth can-i create deployments --namespace dev
```

如果授权通过，则运行命令后输出的信息如下：

```
yes
```

如果授权未通过，则运行命令后输出的信息如下：

```
no
```

API 请求分为资源请求和非资源请求。

对于请求路径不是/api/v1/...或者/apis/< group >/< version >类型的请求称为非资源请求。非资源请求使用小写的 HTTP 方法名作为请求动词。例如请求路径是/healthz 的请求的动词是 get。

Kubernetes 在授权流程中会检测 API 请求中的如下属性。

(1) user：认证后得到用户名称字符串。

(2) group：用户所属的用户组列表。

(3) extra：认证层提供的任意键值对。

(4) API：标志这个请求是否针对 API 资源。

(5) Request Path：非资源请求的路径，例如/api 或者/healthz。

(6) APIrequest verb：资源请求的请求动词，例如 get、list、create、update、patch、watch、delete 及 deletecollection。

(7) HTTP request verb：资源请求的小写的 HTTP 请求动词，例如 get、post、put 和 delete。

(8) Resource：被访问资源的 ID 或者名称。对于使用了 get、update、patch 和 delete 动词的资源请求，必须提供资源名称。

(9) Subresource：资源请求中被访问资源的子资源。

(10) Namespace：资源请求中被访问对象所在的命名空间。

(11) APIgroup：资源请求中被访问资源所在的 API 组。

7.21.1 授权模式

Kubernetes 支持多种授权模式，包括 Node、ABAC、RBAC 和 Webhook。

Node 模式是一种特殊用途的授权模式，根据调度到 kubelet 所在 Node 上运行的 Pod 为 kubelet 授予权限。

ABAC，即 Attribute-based access control 模式，是基于属性的访问控制。

RBAC，即 Role-based access control 模式，是基于角色的访问控制。

Webhook 模式是基于 HTTP 回调实现的事件通知模式。

7.21.2 RBAC

Role-Based Access Control(RBAC)，即基于用户的角色对访问进行控制。启用 RBAC 需要在启动 Kubernetes API 服务器的--authorization-mode 参数中加入 RBAC，例如--authorization-mode=Node,RBAC。从 Kubernetes 1.6 版本开始，默认开启 RBAC。

RBAC API 引入了 4 种 Kubernetes 对象，分别是 Role、ClusterRole、RoleBinding、ClusterRoleBinding。

Role 和 ClusterRole 代表一组权限，Role 是命名空间限定的角色，ClusterRole 是全局的角色，没有命名空间限定。

下面是示例 Role 的配置文件：

```
apiVersion: rbac.authorization.k8s.io/v1
kind: Role
metadata:
  namespace: default
  name: pod-reader
rules:
- apiGroups: [""] #""
  resources: ["pods"]
  verbs: ["get", "watch", "list"]
```

文件中定义的 Role 被命名为 pod-reader，可以用来授予读取 Pod 的权限。

ClusterRole 除了可以完成与 Role 相同的授权，还可以为全局的资源（如 Node）、为非资源端点（如/healthz）及命名空间限定的资源（如 Pod）授权。

下面是示例 ClusterRole 的配置文件：

```
apiVersion: rbac.authorization.k8s.io/v1
kind: ClusterRole
metadata:
  name: secret-reader
rules:
- apiGroups: [""]
```

```
  resources: ["secrets"]
  verbs: ["get", "watch", "list"]
```

授予权限需要把定义好的 Role 或 ClusterRole 绑定到用户上，这需要通过定义 RoleBinding 或 ClusterRoleBinding 实现。

下面是示例 RoleBinding 的配置文件的内容：

```
apiVersion: rbac.authorization.k8s.io/v1
kind: RoleBinding
metadata:
  name: read-pods
  namespace: default
subjects:
- kind: User
  name: blue
  apiGroup: rbac.authorization.k8s.io
roleRef:
  kind: Role
  name: pod-reader
  apiGroup: rbac.authorization.k8s.io
```

文件中的.subjects 指定了要进行角色绑定的对象列表，包含一个 blue 用户。.roleRef 字段定义了要引用哪个角色进行绑定。这里引用的角色是 pod-reader。.roleRef.kind 字段的值必须是 Role 或者 ClusterRole。.roleRef.name 是要绑定的角色的名称。

RoleBinding 除了可以引用 Role 还可以引用 ClusterRole 来为 RoleBinding 所在的命名空间中的资源授权。这样的机制允许定义一些集群级的通用角色，然后在不同的命名空间中复用这些角色。

下面引用了 ClusterRole 的 RoleBinding 示例配置文件，这里假设存在一个名为 secret-reader 的 ClusterRole，文件内容如下：

```
apiVersion: rbac.authorization.k8s.io/v1
kind: RoleBinding
metadata:
  name: read-secrets
  namespace: development
subjects:
- kind: User
  name: dave
  apiGroup: rbac.authorization.k8s.io
roleRef:
  kind: ClusterRole
  name: secret-reader
  apiGroup: rbac.authorization.k8s.io
```

文件中定义的 RoleBinding 引用了名为 secret-reader 的 ClusterRole，但是 dave 用户只能访问 development 命名空间中的 Secret，因为 RoleBinding 命名空间是 development。

在需要访问所有命名空间下的某类型资源的场景下，需要使用集群级的角色绑定，即 ClusterRoleBinding。下面是一个示例 ClusterRoleBinding 的配置文件：

```
apiVersion: rbac.authorization.k8s.io/v1
kind: ClusterRoleBinding
metadata:
  name: read-secrets-global
subjects:
- kind: Group
  name: manager
  apiGroup: rbac.authorization.k8s.io
roleRef:
  kind: ClusterRole
  name: secret-reader
  apiGroup: rbac.authorization.k8s.io
```

与 RoleBinding 不同，ClusterRoleBinding 没有.metadata.namespace 字段。这个文件中定义的 ClusterRoleBinding 将允许所有 manager 组的用户读取全部命名空间下的 Secret 对象。

7.21.3　常用命令

下面是一些 RBAC 授权模式中常用的命令。

创建允许对 Pod 执行 get、watch 和 list 的 Role，命名为 pod-reader，命令如下：

```
kubectl create role pod-reader \
--verb=get \
--verb=list \
--verb=watch \
--resource=pods
```

创建名为 pod-reader 的 Role，并指定 resourceName，命令如下：

```
kubectl create role pod-reader \
--verb=get \
--resource=pods \
--resource-name=readablepod \
--resource-name=anotherpod
```

创建名为 foo 的 Role，并指定 apiGroups，命令如下：

```
kubectl create role foo \
--verb=get,list,watch \
--resource=replicasets.apps
```

创建名为 foo 的 Role,并指定子资源,命令如下:

```
kubectl create role foo \
--verb=get,list,watch \
--resource=pods,pods/status
```

创建名为 bar 的 Role,并指定资源名称,命令如下:

```
kubectl create role my-component-lease-holder \
--verb=get,list,watch,update \
--resource=lease \
--resource-name=my-component
```

创建名为 monitoring 的 ClusterRole,并指定聚合规则,命令如下:

```
kubectl create clusterrole monitoring \
--aggregation-rule=rbac.example.com/aggregate-to-monitoring=true
```

创建 RoleBinding,把名为 admin 的 ClusterRole 中的权限授予名为 bob 的用户,命令如下:

```
kubectl create rolebinding bob-admin-binding --clusterrole=admin --user=bob --namespace=
acme
```

创建 RoleBinding,把名为 view 的 ClusterRole 中的权限授予 acme 命名空间下的名为 acme: myapp 的 SeviceAccount,命令如下:

```
kubectl create rolebinding myapp-view-binding \
--clusterrole=view \
--serviceaccount=acme:myapp \
--namespace=acme
```

把名为 cluster-admin 的 ClusterRole 中的权限授予名为 root 的用户,命令如下:

```
kubectl create clusterrolebinding root-cluster-admin-binding \
--clusterrole=cluster-admin --user=root
```

7.21.4 Service Account

Kubernetes 把用户分为普通用户和进程,普通用户是人类的身份标志,而 Service

Account 是 Pod 中进程的唯一身份标识，这样就可以让 APIserver 对进程进行认证，通过认证后才可以为进程授权，以满足 Pod 中的进程对 API 对象进行读写的需求。

新创建的 Pod，如果没有将.spec.serviceAccountName 字段指定为 Pod，并以此设置 Service Account，则会被指定一个名为 default 的 Service Account，并且与 Pod 处于同一个命名空间。

可以为 Service Account 设置权限。具体过程依赖于使用的授权组件。

下面演示如何为 Service Account 授权，下面的 Kubernetes 使用 RBAC 授权组件。

假设在 Pod 中访问 APIServer，获取所有的 Pod。创建这个 Pod，进入 Pod 后执行 sh，命令如下：

```
kubectl run go - it  -- image golang:1.16 - alpine  -- sh
```

运行命令后输出的信息如下：

```
If you don't see a command prompt, try pressing enter.
/go #
```

这个 Pod 由于没有指定 Service Account，所以会使用默认的 Service Account，其名为 default，可以通过命令验证，命令如下：

```
kubectl get pod  go - o jsonpath = '{.spec.serviceAccountName}'
```

运行命令后会输出 default。

在新创建的 Pod 中继续操作，进入/home 目录，创建文件 main.go，编辑内容如下：

```
package main

import (
"context"
"fmt"
"time"
    metav1 "k8s.io/apimachinery/pkg/apis/meta/v1"
"k8s.io/client - go/kubernetes"
"k8s.io/client - go/rest"
)

func main() {
    //creates the in - cluster config
    config, err : = rest.InClusterConfig()
    if err != nil {
        panic(err.Error())
    }
    //creates the clientset
```

```
    clientset, err := kubernetes.NewForConfig(config)
    if err != nil {
        panic(err.Error())
    }
    for {
        //get pods in all the namespaces by omitting namespace
        //Or specify namespace to get pods in particular namespace
        pods, err := clientset.CoreV1().Pods("").List(context.TODO(), metav1.ListOptions{})
        if err != nil {
            panic(err.Error())
        }
        fmt.Printf("There are %d pods in the cluster\n", len(pods.Items))

        time.Sleep(10 * time.Second)
    }
}
```

接着使用 go mod 命令创建 go.mod 文件，命令如下：

```
go mod init client
```

运行命令后输出的信息如下：

```
go: creating new go.mod: module client
go: to add module requirements and sums:
go mod tidy
```

安装依赖，命令如下：

```
go mod tidy
```

现在运行 main.go，命令如下：

```
go run main.go
```

运行命令后输出的错误提示信息如下：

```
panic: pods is forbidden: User "system:serviceaccount:default:default" cannot list resource
"pods" in API group "" at the cluster scope

goroutine 1 [running]:
main.main()
    /home/main.go:30 + 0x486
exit status 2
```

提示 Pod 中的进程身份，即 default：default 这个 Service Account 没有权限从 API 读取 Pod 信息。

创建一个拥有读取 Pod 信息权限的 ClusterRole，命令如下：

```
kubectl apply -f - << EOF
kind: ClusterRole
apiVersion: rbac.authorization.k8s.io/v1
metadata:
  namespace: default
  name: pod-reader
rules:
- apiGroups: [""] #"" indicates the core API group
  resources: ["pods"]
  verbs: ["get", "watch", "list"]
EOF
```

运行命令后输出的信息如下：

```
clusterrole.rbac.authorization.k8s.io/pod-reader created
```

创建 ClusterRoleBinding，为 Service Account(default)绑定 pod-reader 角色，使 default service account 获得读取 Pod 信息的权限。命令如下：

```
kubectl create clusterrolebinding pod-reader \
  --clusterrole=pod-reader  \
  --serviceaccount=default:default
```

再次运行 main.go 文件，命令如下：

```
go run main.go
```

会看到输出的类似信息如下：

```
There are 17 pods in the cluster
```

表示 Pod 已经通过 Service Account 拥有了读取 Pod 信息的权限。

7.22　Kubernetes 调度

7.22.1　调度简介

在 Kubernetes 中，调度是指把 Pod 放到合适的 Node。调度器会监视新创建且没有调度到 Node 上的 Pod，发现 Pod 之后，调度器会为 Pod 寻找最合适的 Node，然后把 Pod 调度到 Node 上。调度完成后，Node 上的 kubelet 就会把 Pod 运行起来。

Kubernetes 默认的调度器是 kube-scheduler，kube-scheduler 是 Kubernetes 控制面的重要组成部分。

Pod 和 Pod 中的容器对资源有不同的需求，所以调度器需要根据需求对集群中的 Node 进行过滤。

如果没有一个 Node 可以满足要求，则 Pod 会一直处于待调度状态，直到找到合适的 Node。

如果找到合适的 Node，则调度器会对候选 Node 打分，最后选出得分最高的 Node 并把这个调度决定通知给 API 服务器。如果出现得分相同的多个 Node，则调度器会随机从中选择一个 Node。

调度器需要在调度过程中考虑单独和整体的资源需求、硬件或软件或策略的限制、亲和性和反亲和性、数据局部性等诸多因素。

通过定义调度 Policy 和调度 Profile 可以配置调度器的过滤和打分行为。

7.22.2 约束 Node 选取

虽然调度器会把 Pod 放到合适的 Node，但是有些情况下需要对调度施加额外控制，例如确保把 Pod 调度到某个挂载了 SSD 硬盘的 Node，或者把 2 个需要频繁通信的 Pod 放到同一个可用区。

有多种方式可以限制 Pod 调度到哪些 Node，推荐的方式是使用标签选择器。

约束 Node 的最简单的方式是使用 nodeSelector。nodeSelector 是 Pod 中 spec 字段下的一个字段，它的值为键值对。

假设集群中某个 Node 名为 node1，挂载了 SSD 硬盘，为了支持 Pod 通过标签选择 node1，为 node1 添加 disktype=ssd 标签，命令如下：

```
kubectl label nodes node1 disktype = ssd
```

这样需要调度到 node1 的 Pod 就可以使用 nodeSelector 约束了。

验证标签是否添加成功，命令如下：

```
kubectl get node node1  -- show - labels
```

输出信息中的 LABELS 列会展示 node1 的所有标签。除了自定义标签，Kubernetes 中的 Node 会有一些预置的标准标签，这些标准标签也可以被 Pod 用来约束 Node 选取。

下面是使用了 nodeSelector 的 Pod 配置文件，文件内容如下：

```
apiVersion: v1
kind: Pod
metadata:
```

```
    name: nginx
  spec:
    containers:
    - name: nginx
      image: nginx
    nodeSelector:
      disktype: ssd
```

使用此文件创建的 Pod 会被调度到拥有 disktype=ssd 标签的 Node。

7.22.3　亲和性和反亲和性

与 nodeSelector 相比，亲和性和反亲和性提供了更丰富、更灵活的 Node 选取约束功能。亲和性有两种类型，一种是节点亲和性，另一种是 Pod 间的亲和性。节点亲和性与 nodeSelector 类似，利用 Node 的标签进行选择约束，Pod 间的亲和性则利用 Pod 标签进行选择约束。

下面介绍节点亲和性。

节点亲和性包括 requiredDuringSchedulingIgnoredDuringExecution 和 preferredDuringSchedulingIgnoredDuringExecution 两类。前者是硬性约束，后者是柔性约束。两者名字也体现出这一点，requiredDuringSchedulingIgnoredDuringExecution 中的前缀 required 的意思是必须，preferredDuringSchedulingIgnoredDuringExecution 的前缀 preferred 的意思是首选。两者名字的后缀都是 IgnoredDuringExecution，意思是 Node 上的标签如果在 Pod 运行过程中发生变化，致使 Node 不再满足亲和性规则，Pod 则会忽视这种状况并继续在 Node 上运行。

requiredDuringSchedulingIgnoredDuringExecution 指定了使 Pod 成功调度的 Node 必须满足的规则，不满足规则的 Pod 将不会被调度。preferredDuringSchedulingIgnoredDuringExecution 指定了使 Pod 成功调度的 Node 尽可能满足的规则。不满足规则的 Pod 仍会被调度到某个 Node。

下面是设置了 nodeAffinify 的 Pod 的示例配置文件：

```
apiVersion: v1
kind: Pod
metadata:
  name: with-node-affinity
spec:
  affinity:
    nodeAffinity:
      requiredDuringSchedulingIgnoredDuringExecution:
        nodeSelectorTerms:
        - matchExpressions:
          - key: kubernetes.io/e2e-az-name
```

```
              operator: In
              values:
               - e2e - az1
               - e2e - az2
        preferredDuringSchedulingIgnoredDuringExecution:
        - weight: 1
          preference:
            matchExpressions:
            - key: another - node - label - key
              operator: In
              values:
               - another - node - label - value
    containers:
    - name: with - node - affinity
      image: k8s.gcr.io/pause:2.0
```

文件中定义的亲和性规则表示这个 Pod 只能调度到拥有标签 kubernetes.io/e2e-az-name 并且标签值为 e2e-az1 或者 e2e-az2 的 Node。另外在满足规则的 Node 中，优先选择拥有标签 anather-node-label-key 的标签并且标签值为 another-node-label-value 的 Node。

除了此文件中使用的操作符 In，亲和性规则语法还支持 NotIn、Exists、DoesNotExist、Gt 和 Lt。其中 NotIn 和 DoesNotExist 用于实现反亲和性规则。

preferredDuringSchedulingIgnoredDuringExecution 中的 weight 字段的取值范围为 1～100，对于每个候选的 Node，调度器会遍历这个字段，把 weight 求和，这个分数会与 Node 的其他优先级函数的得分合并。最后得分最高的 Node 就是最合适的调度目标。

亲和性规则只作用于调度期间，当 Pod 完成调度后在 Node 上运行的过程中如果改变了 Node 标签或者删除了 Node 标签都不会使 Pod 被删除。

下面介绍 Pod 间的亲和性和反亲和性：

Pod 间的亲和性和反亲和性可以满足类似将两个频繁通信的 Service 中的 Pod 放置到同一个区域的调度需求。与 Node 亲和性不同，Pod 间的亲和性和反亲和性是根据 Node 上运行中的 Pod 所拥有的标签约束 Pod 调度，而不是根据 Node 拥有的标签。因为 Pod 受命名空间限定，所以 Pod 标签也隐式被命名空间限定，Pod 标签的选择器必须指定选择器生效的命名空间。

与 Node 亲和性类似，Pod 间的亲和性和反亲和性同样有两种类型，包括 requiredDuringSchedulingIgnoredDuringExecution 和 preferredDuringSchedulingIgnoredDuringExecution。前者表示硬性要求，后者表示柔性要求。Pod 间的亲和性规则通过 Pod 的.spec.affinity.podAffinity 字段指定，Pod 间的反亲和性规则通过 Pod 的.spec.affinity.podAntiAffinity 字段指定。

Pod 间亲和性和反亲和性需要消耗大量的计算资源，在大规模的集群中可能会严重拖慢调度速度，所以不建议在几百个节点以上规模的集群中使用。

下面是指定了 Pod 间亲和性的示例 Pod 的配置文件内容：

```yaml
apiVersion: v1
kind: Pod
metadata:
  name: with-pod-affinity
spec:
  affinity:
    podAffinity:
      requiredDuringSchedulingIgnoredDuringExecution:
      - labelSelector:
          matchExpressions:
          - key: security
            operator: In
            values:
            - S1
        topologyKey: topology.kubernetes.io/zone
    podAntiAffinity:
      preferredDuringSchedulingIgnoredDuringExecution:
      - weight: 100
        podAffinityTerm:
          labelSelector:
            matchExpressions:
            - key: security
              operator: In
              values:
              - S2
          topologyKey: topology.kubernetes.io/zone
  containers:
  - name: with-pod-affinity
    image: k8s.gcr.io/pause:2.0
```

此文件中定义了一个 Pod 间的亲和性规则和一个 Pod 间的反亲和性规则，Pod 间的亲和性规则 requiredDuringSchedulingIgnoredDuringExecution 要求 Pod 被调度到拥有标签 security 并且标签值为 S1 的 Node。反亲和性规则 preferredDuringScheduling-IgnoredDuringExecution 要求尽可能不把 Pod 调度到拥有标签 security 并且标签值为 S2 且与这个 Pod 处于同一区的 Node。

7.22.4　nodeName

约束 Node 选取的最简单的方式是 nodeName，通过 nodeName 字段直接为 Pod 指定被调度到哪个 Node，但是 nodeName 有许多局限性，所以实际很少使用。如果 Pod 指定了 nodeName，调度器不会对该 Pod 进行调度，名字为 nodeName 的 Node 上的 kubelet 会直接尝试运行该 Pod。与其他调度约束相比，nodeName 具有最高优先级。

下面是指定了 nodeName 的示例 Pod 的配置文件内容：

```
apiVersion: v1
kind: Pod
metadata:
  name: nginx
spec:
  containers:
  - name: nginx
    image: nginx
  nodeName: kube-01
```

文件中定义的 Pod 会在 kube-01 这个 Node 上运行。

7.22.5 污点和容忍

污点和容忍机制可以保证 Pod 不会被调度到不合适的节点上。

节点上的污点是 Node 的一个可选属性，可以为节点添加多个污点。污点用来排斥 Pod，使 Pod 不被调度到有污点的 Node 上。这与节点亲和性的作用恰恰相反。

容忍是 Pod 的一个可选属性，Pod 可以容忍将 Pod 调度到有污点的 Node。前提是容忍与污点相匹配。可以为一个 Pod 添加多个容忍。

污点的格式与标签类似，是键值对的形式。假设集群中有个节点 node1，给节点 node1 添加污点 key1=value1：NoSchedule，命令如下：

```
kubectl taint nodes node1 key1=value1:NoSchedule
```

value 中的 NoSchedule 表示这个污点的作用是不调度，即不把 Pod 调度到有这个污点的节点，除非 Pod 有与污点匹配的容忍。

运行命令后输出的信息如下：

```
node/node1 tainted
```

污点的作用除了 NoSchedule 还可以是 PreferNoSchedule 和 NoExecute。

PreferNoSchedule 会尽可能不把未容忍此污点的 Pod 调度到此节点。

NoExecute 除了会不把未容忍此污点的 Pod 调度到此节点，还会驱逐已经在节点上运行的 Pod。

删除通过上述命令添加到 node1 的污点，命令如下：

```
kubectl taint nodes node1 key1=value1:NoSchedule-
```

运行命令后输出的信息如下：

```
node/node1 untainted
```

如果希望将 Pod 调度到 node1 上，则可以为 Pod 添加的容忍如下：

```
tolerations:
- key: "key1"
  operator: "Equal"
  value: "value1"
  effect: "NoSchedule"
```

或者添加的容忍如下：

```
tolerations:
- key: "key1"
  operator: "Exists"
  effect: "NoSchedule"
```

这样 Pod 就可以被调度到 node1 节点上了。

下面是使用了容忍的完整示例 Pod 配置的文件内容：

```
apiVersion: v1
kind: Pod
metadata:
  name: nginx
spec:
  containers:
  - name: nginx
    image: nginx
  tolerations:
  - key: "example-key"
    operator: "Exists"
    effect: "NoSchedule"
```

容忍中的 operator 字段的默认值为 Equal。容忍与污点匹配的条件是 key 和 effect 相同，另外 operator 为 Exists 且没有指定 value，或者 operator 为 Equal 且 value 相等。

如果 key 为空且 operator 是 Exists，则会匹配所有的 key、value 和 effect，也就意味着容忍所有污点。

如果 effect 为空，则会匹配与 key 关联的所有 effect。

当一个节点存在多个污点并且一个 Pod 上存在多个容忍时，Kubernetes 会从所有污点中找出有匹配的容忍的污点，并忽略这部分污点。在剩下无法忽略的污点中，如果存在作用为 NoSchedule 的污点，则这个 Pod 将不会被调度到此节点。如果没有作用为 NoSchedule 的污点，但是有 PreferNoSchedule 的污点，Kubernetes 则会尽可能不把此 Pod 调度到此节点。如果有作用为 NoExecute 的污点，Kubernetes 则不会把此 Pod 调度到此节点，若此 Pod 已经在此节点上运行，则此 Pod 会被驱逐出此节点。

举个例子，假设有节点 node1，为节点 node1 添加多个污点，命令如下：

```
kubectl taint nodes node1 key1 = value1: NoSchedule
kubectl taint nodes node1 key1 = value1: NoExecute
kubectl taint nodes node1 key2 = value2: NoSchedule
```

假设一个 Pod 有以下容忍：

```
tolerations:
- key: "key1"
  operator: "Equal"
  value: "value1"
  effect: "NoSchedule"
- key: "key1"
  operator: "Equal"
  value: "value1"
  effect: "NoExecute"
```

因为此 Pod 没有容忍 node1 中的第 3 个污点，所以此 Pod 不会被调度到 node1 上。如果在添加污点前，这个 Pod 已经在 node1 上运行，它不会被驱逐出去，因为 Pod 中有匹配污点 key1＝value1：NoExecute 的容忍。

通常情况下，一个作用是 NoExecute 的污点被添加到节点上以后，节点上所有不容忍此污点的 Pod 会被立刻驱逐，与作用是 NoExecute 的污点匹配的容忍支持设置 tolerationSeconds 字段，表示在添加作用为 NoExecute 的污点后，Pod 继续保持在节点上运行的时间，单位是秒。

tolerationSeconds 的示例代码如下：

```
tolerations:
- key: "key1"
  operator: "Equal"
  value: "value1"
  effect: "NoExecute"
  tolerationSeconds: 3600
```

这个例子中的 Pod 首先会在节点中添加与容忍匹配的污点，然后继续在节点上运行 3600s，最后被驱逐出节点。如果这个污点在 tolerationSeconds 时间到达之前被删除，这个 Pod 就不会被驱逐。

7.22.6 Pod 优先级

可以为 Pod 设置优先级。当高优先级的 Pod 无法被调度时，调度器会把低优先级的

Pod 驱逐出节点，也就是抢占低优先级 Pod 以保障高优先级的 Pod 及时被调度。

使用 Pod 优先级前需要创建 PriorityClass，然后在 Pod 中将 priorityClassName 字段值指定为某个 PriorityClass 的名字。

PriorityClass 是一个非命名空间限定的对象，它定义了一个由名字到优先级数值的映射。PriorityClasss 的 name 字段用于指定名字，value 字段用于指定优先级数值。value 数值越大表示优先级越高。PriorityClass 对象的名字不可以包含 system-前缀，这个前缀是 Kubernetes 系统保留的。

下面是 PriorityClass 示例配置的文件内容：

```
apiVersion: scheduling.k8s.io/v1
kind: PriorityClass
metadata:
  name: high-priority
value: 1000000
globalDefault: false
description: "This priority class should be used for XYZ service pods only."
```

value 字段表示优先级数值，可以是一个任意的不大于 10 亿的 32 位整数。更大的数值被 Kubernetes 中重要的系统 Pod 所保留，这些 Pod 通常不应该被抢占或驱逐。

globalDefault 字段是可选的，表示是否把这个 PriorityClass 的优先级数值作为没有设置 priorityClassName 的 Pod 的默认优先级。系统中只允许存在一个 globalDefault 字段，并且可将 PriorityClass 设置为 true。

description 也是可选字段，其值是一个任意字符串，用来描述这个 PriorityClass 的用途。

在 Kubernetes 中添加一个 globalDefault 字段为 true 的 PriorityClass 并不会改变已经存在的 Pod 的优先级，只会影响这个 PriorityClass 之后创建的 Pod 的优先级。

同样地，删除一个 PriorityClass 后，已经使用了此 PriorityClass 的 Pod 优先级不会改变，但是不能新创建并使用此 PriorityClass 的 Pod。

如果不希望 Pod 抢占低优先级的 Pod，则可以将 Pod 使用的 PriorityClass 的 preemptionPolicy 值设置为 Never。

使用了 preemptionPolicy 为 Never 的 PriorityClass 的 Pod 在等待调度时，并不会抢占已经在运行的比它优先级低的 Pod，但是它会被优先级更高的 Pod 抢占。

非抢占的 PriorityClass 示例配置的文件内容如下：

```
apiVersion: scheduling.k8s.io/v1
kind: PriorityClass
metadata:
  name: high-priority-nonpreempting
value: 1000000
preemptionPolicy: Never
```

```
globalDefault: false
description: "This priority class will not cause other pods to be preempted."
```

创建好 PriorityClass 之后，在 Pod 配置文件中的 priorityClassName 字段就可以引用已经创建的 PriorityClass 的名字。

启用 Pod 优先级之后，调度器会按优先级把调度队列中的 Pod 排序，高优先级的 Pod 在低优先级的 Pod 之前。这样高优先级的 Pod 就可以比低优先级的 Pod 更快被调度。如果高优先级 Pod 无法被调度，调度器则会尝试调度其他低优先级的 Pod。

7.22.7 Pod 抢占

Pod 创建之后会进入调度队列中等待，调度器会在队列中取出 Pod 调度并尝试调度到节点上。如果没有一个节点可以满足 Pod 的需求，则会触发抢占机制来调度待调度的 Pod。

抢占机制的逻辑是寻找一个合适的节点（称为 N），当把此节点上比待调度 Pod（称为 P）优先级低的一个或多个 Pod 驱逐出节点后，节点 N 可以满足待调度 Pod P 的需求，使待调度节点可在此节点上运行。当找到这样的节点后，就会把目标节点上优先级较低的 Pod 驱逐出节点，之后把待调度节点放到节点运行。

节点能成为被抢占节点的筛选条件是当此节点上比待调度 Pod N 优先级低的所有 Pod 被驱逐出此节点后，可以把待调度 Pod N 调度到此节点上。如果满足了这个条件，这个节点就可以作为被抢占的节点。

但是抢占并不意味着驱逐所有优先级比待调度 Pod N 低的 Pod，如果只驱逐一部分优先级较低的 Pod 也可以满足 Pod N 调度到节点上的要求，则只会驱逐一部分优先级低的 Pod。

当 Pod P 抢占了节点 N 上的一个或多个 Pod 时，这个 Pod 的 status 中的 nominatedNodeName 字段会被设置为节点 N 的名字。nominatedNodeName 即被提名的节点的名称。

需要注意的是 Pod P 并不一定会被调度到被提名的节点，这是因为在被驱逐的 Pod 处于优雅终止阶段时，可能会有其他合适的节点出现，此时待调度的节点 P 就有可能被调度到其他节点上运行。这样 Pod P 的 nominatedNodeName 和 nodeName 字段值就会不一致。

在节点 N 上的 Pod 被调度器驱逐之后，如果出现了优先级比 Pod P 更高的 Pod，调度器则会优先把高优先级的 Pod 放到节点 N。这种情况下，调度器会把 Pod P 的 nominatedNodeName 字段清除，使 Pod P 可以重新抢占其他节点的 Pod。

被驱逐的 Pod 会进入默认时长为 30s 的优雅终止阶段，如果 Pod 在优雅终止阶段内没有成功终止，则将会被强制终止。

7.22.8　Pod 拓扑分布

均匀地把 Pod 分布到不同的拓扑域可以提高系统中所有 Pod 整体的运行稳定性，如果有的 Node 上运行了太多的 Pod，Pod 则可能会因为资源短缺而终止。如果有的 Node 上运行的 Pod 非常少，则又会造成资源浪费，所以调度器在调度过程中会尽量使新来的 Pod 不会造成集群中的 Pod 分布不均匀。

拓扑域的划分粒度可以是云服务商定义的地区、可用区域、节点或者用户自定义的划分标准。

使用 Pod 拓扑分布限制可以限制 Pod 在集群中如何分布，从而提高资源利用率和实现高可用。

调度器会利用 Pod 拓扑分布限制指定的 Pod 之间的关系决定如何部署待调度的 Pod。用于指定 Pod 拓扑分布限制的字段是 Pod 中的 spec.topologySpreadConstraints。

使用了 Pod 拓扑分布限制的 Pod 示例配置的文件如下：

```
apiVersion: v1
kind: Pod
metadata:
  name: mypod
spec:
  topologySpreadConstraints:
    - maxSkew: <integer>
      topologyKey: <string>
      whenUnsatisfiable: <string>
      labelSelector: <object>
```

topologySpreadConstraints 字段包括 maxSkew、topologyKey、whenUnsatisfiable、labelSelector 字段。可以为 Pod 指定一个或多个 topologySpreadConstraints。

maxSkew 用于描述 Pod 分布不均的程度。Skew 是偏差的意思，maxSkew 即指定拓扑类型中的任意 2 个拓扑域中匹配的 Pod 的最大数量差。maxSkew 的值必须大于 0。当 whenUnsatisfiable 的值为 ScheduleAnyway 时，maxSkew 无法保证任意 2 个拓扑域中匹配的 Pod 的最大数量差与之相等，调度器会尽可能使调度结果的拓扑偏差最小。

topologyKey 是 Node 的标签键，用来划分逻辑上的拓扑域。如果 2 个 Node 有相同的键为 topologyKey 的标签，并且值相同，则调度器会认为这 2 个 Node 处于同一个拓扑域。调度器会尽力把 Pod 均匀调度到每个拓扑域。

whenUnsatisfiable 用于指定调度器如何处理无法满足 Pod 分布限制的情况。whenUnsatisfiable 的默认值为 DoNotSchedule，表示在无法满足 Pod 分布限制时不进行调度。whenUnsatisfiable 还可以取值 ScheduleAnyway，表示在无法满足限制的情况下仍然进行调度，尽力使调度结果的各个拓扑域的 Pod 数量偏差最小。

labelSelector 用于匹配 Pod 的标签。与 labelSelector 匹配的 Pod 才会在计算拓扑域中的 Pod 数量时纳入考虑范围。

下面是另一个使用了 Pod 拓扑分布限制的示例 Pod 配置的文件内容：

```
kind: Pod
apiVersion: v1
metadata:
  name: mypod
  labels:
    foo: bar
spec:
  topologySpreadConstraints:
  - maxSkew: 1
    topologyKey: zone
    whenUnsatisfiable: DoNotSchedule
    labelSelector:
      matchLabels:
        foo: bar
  containers:
  - name: pause
    image: k8s.gcr.io/pause:3.1
```

下面解释文件中的指令。

定义拓扑域按照 Node 是否具有标签 zone：< value >来划分。调度器会把 Pod 均匀分布到这些拓扑域中，指令如下：

```
topologyKey: zone
```

当无法满足分布限制时，不进行调度，指令如下：

```
whenUnsatisfiable: DoNotSchedule
```

假设集群中有 4 个节点，分别是 Node1、Node2、Node3、Node4，Node1 和 Node2 有标签 zone=zoneA，Node3 和 Node4 有标签 zone=zoneB。逻辑上 Node1 与 Node2 处于同一个拓扑域 zoneA，Node3 和 Node4 处在同一个拓扑域 zoneB。集群中分布了 3 个具有标签 foo=bar 的 Pod，其中 Node1、Node2、Node3 分别运行一个 Pod，Node4 没有 Pod 在运行。当创建上述 Pod 以后，调度器会把它调度到 Node4 上运行。因为这样可以满足 Pod 拓扑分布限制，如果把它调度到 zoneA 中的 Node1 或者 Node2，则会导致 zoneA 与 zoneB 中 Pod 数量偏差变成 3－1=2，大于 maxSkew 规定的 1，从而违反分布约束，所以调度器只能把它调度到 zoneB 中的 Node3 或者 Node4。

如果想让 Pod 不仅均匀分布到不同区域，而且均匀分布到节点，则可以把上述 Pod 的分布限制中的 topologyKey 设置为 node，并将 maxSkew 保持为 1，这样上述的 Pod 就只会被调度到 Node4 上了。

下面是指定了多个分布限制的示例 Pod 配置的文件内容：

```
kind: Pod
apiVersion: v1
metadata:
  name: mypod
  labels:
    foo: bar
spec:
  topologySpreadConstraints:
  - maxSkew: 1
    topologyKey: zone
    whenUnsatisfiable: DoNotSchedule
    labelSelector:
      matchLabels:
        foo: bar
  - maxSkew: 1
    topologyKey: node
    whenUnsatisfiable: DoNotSchedule
    labelSelector:
      matchLabels:
        foo: bar
  containers:
  - name: pause
    image: k8s.gcr.io/pause:3.1
```

Pod 中指定了 2 个分布限制条件，为了满足第 1 个 topologyKey 为 zone 的限制，Pod 只能调度到 zoneB，为了同时满足第 2 个 topologyKey 为 node 的限制，Pod 只能调度到 Node 4。

7.23 Kubernetes 日志

日志对于应用的监控和排错有重要意义，容器在设计之初就对日志进行了很好的支持。容器记录日志的最佳方式是将日志输出到标准输出（stdout）和标准错误（stderr）流。

但是容器提供的日志功能还无法满足常规的日志需求。例如需要在容器崩溃、Pod 重新调度或者节点宕机等情况下查询日志。及在多节点集群环境中，需要把分布在多个节点的 Pod 产生的日志进行收集合并。

一个可靠的日志解决方案需保证日志独立于容器、Pod 和节点的生命周期。这就需要把日志存储到独立的系统。Kubernetes 并没有提供内置的日志存储解决方案，需要用户自行将其他日志处理系统集成到集群中。

7.23.1 Kubernetes 基础日志功能

kubectl 提供了基础的查看日志的命令 kubectl logs。

下面通过运行 alpine 镜像的 Pod，演示基础的日志功能。

新建 Pod 配置文件 log-pod.yaml，文件内容如下：

```
apiVersion: v1
kind: Pod
metadata:
  name: ping
spec:
  containers:
  - name: ping
    image: alpine
    args: [/bin/sh, -c,'ping github.com']
```

这个 Pod 中的容器运行 ping github.com 命令，ping 命令会持续将 sh 进程的标准输出流和标准错误流输出到容器。

使用此文件创建 Pod，命令如下：

```
kubectl apply -f log-pod.yaml
```

运行命令后输出的信息如下：

```
pod/ping created
```

查看 Pod 日志，命令如下：

```
kubectl logs ping
```

由于这个 Pod 中只有一个容器，所以使用 kubectl 查看日志不需要指定容器名，如果 Pod 中有多个容器，则需要在命令中加上-c 参数指定容器名，以查看指定容器的日志。

运行命令后输出的信息如下：

```
PING github.com (13.229.188.59): 56 data Bytes
64 Bytes from 13.229.188.59: seq=0 ttl=38 time=41.898 ms
64 Bytes from 13.229.188.59: seq=1 ttl=38 time=41.777 ms
```

如果希望持续观察日志，则可以加上-f 参数，命令如下：

```
kubectl logs -f ping
```

7.23.2　节点级日志

容器运行时 Docker 会把容器中的程序的所有输出转发到容器中应用程序的 stdout 和 stderr 流。

当使用的容器运行时为 CRI 时，每个节点上运行的 kubelet 会负责对日志进行轮转并管理存储日志的目录结构。默认情况下，kubelet 会保留终止容器的日志，而如果 Pod 被驱逐出节点，则 Pod 中容器的日志不会被保留。kubelet 会告诉容器运行时的日志目录信息，然后容器运行时把日志写入指定的日志目录。

日志文件的大小可以通过 kubelet 参数 container-log-max-size 指定，日志文件数量可以通过 kubelet 参数 container-log-max-files 指定。

当运行 kubectl logs 时，节点上的 kubelet 会从日志文件读取日志内容并返回。

节点上除了运行应用容器还运行系统组件，有的系统组件以容器形式运行，如 Kubernetes 调度器和 kube-proxy，有的不以容器形式运行，如 kubelet 和容器运行时。

当主机使用 systemd 时，不在容器中运行的系统组件日志会写入 journald，如果没有使用 systemd，则不在容器中运行的系统组件会把日志写入主机的/var/log 文件夹下。

在容器中运行的系统组件会把日志写入/var/log 文件夹下。

7.23.3　集群级日志

Kubernetes 对集群级的日志没有内置的支持，常见的集群日志处理方案包括：通过在每个节点安装日志代理、添加日志处理边车（sidecar）容器及在应用中直接把日志推送到日志处理后端。

日志代理是一个专门转发日志的工具，用于把日志持续推送到日志处理后端。一般情况下，Kubernetes 节点中日志代理以容器方式运行，可以获取节点上所有运行的容器所产生的日志文件。

因为日志代理需要在集群中的每个节点都运行一个实例，推荐以 DaemonSet 的方式部署。日志代理的部署不需要对现有的应用程序做任何改动，侵入性极低。

日志边车容器，即在应用程序容器所处的 Pod 中部署的专门转发日志的容器。一种方案是边车容器把应用容器日志发送到自己的 stdout 流，这种情况下边车容器和应用容器往往通过共享应用容器日志文件夹来使边车容器可以实时获取日志。另一种方案是在边车容器中运行专门的日志代理工具（例如 Fluentd）把日志直接发送到一个日志处理后端（例如 Elasticsearch）。

下面演示如何使用边车容器流式读取日志并发送到 stdout，创建配置文件 two-files-counter-pod-streaming-sidecar. yaml，编辑内容如下：

```
apiVersion: v1
kind: Pod
```

```
metadata:
  name: counter
spec:
  containers:
  - name: count
    image: busybox
    args:
    - /bin/sh
    - -c
    - >
      i=0;
      while true;
      do
        echo "$i: $(date)">> /var/log/1.log;
        echo "$(date) INFO $i">> /var/log/2.log;
        i=$((i+1));
        sleep 1;
      done
    volumeMounts:
    - name: varlog
      mountPath: /var/log
  - name: count-log-1
    image: busybox
    args: [/bin/sh, -c, 'tail -n+1 -f /var/log/1.log']
    volumeMounts:
    - name: varlog
      mountPath: /var/log
  - name: count-log-2
    image: busybox
    args: [/bin/sh, -c, 'tail -n+1 -f /var/log/2.log']
    volumeMounts:
    - name: varlog
      mountPath: /var/log
  volumes:
  - name: varlog
    emptyDir: {}
```

此文件中定义了一个 Pod，Pod 中有 3 个容器，1 个容器名为 count，它会将日志输出到 2 个不同的日志文件中，并且日志内容具有不同的格式，因为日志直接写入了文件，所以使用 kubectl logs 命令无法获取 count 容器的日志。另外 2 个容器是边车容器，用来收集日志。3 个容器共享了同一个卷 varlog，这样 count 容器写入 varlog 的日志就可以被另外 2 个边车容器读取。边车容器把读取的日志直接输出为标准输出流，这样就可以用 kubectl logs 命令获取日志了。

使用此文件创建 Pod，命令如下：

```
kubectl apply -f two-files-counter-pod-streaming-sidecar.yaml
```

运行命令后输出的信息如下：

```
pod/counter created
```

查看 count-log-1 容器日志，命令如下：

```
kubectl logs counter count-log-1
```

运行命令后输出的信息如下：

```
0: Fri Jun  4 07:01:01 UTC 2021
1: Fri Jun  4 07:01:02 UTC 2021
2: Fri Jun  4 07:01:03 UTC 2021
...
```

查看 count-log-1 容器日志，命令如下：

```
kubectl logs counter count-log-2
```

运行命令后输出的信息如下：

```
Fri Jun  4 07:01:01 UTC 2021 INFO 0
Fri Jun  4 07:01:02 UTC 2021 INFO 1
Fri Jun  4 07:01:03 UTC 2021 INFO 2
...
```

下面演示如何使用边车容器运行日志代理工具 Fluentd，首先创建 ConfigMap，用来保存 Fluentd 配置。创建 ConfigMap 的定义文件 fluentd-sidecar-config. yaml，编辑内容如下：

```
apiVersion: v1
kind: ConfigMap
metadata:
  name: fluentd-config
data:
  fluentd.conf: |
<source>
      type tail
      format none
      path /var/log/1.log
      pos_file /var/log/1.log.pos
      tag count.format1
</source>
```

```
<source>
       type tail
       format none
       path /var/log/2.log
       pos_file /var/log/2.log.pos
       tag count.format2
</source>

<match **>
@type file
path /var/log/fluent/access
</match>
```

为了简单起见，文件中将 Fluentd 配置为使用文件类型的输出插件，收集的日志会被写入 Fluentd 容器中的/var/log/fluent 文件夹下。

使用此文件创建 ConfigMap，命令如下：

```
kubectl apply -f fluentd-sidecar-config.yaml
```

运行命令后输出的信息如下：

```
configmap/fluentd-config created
```

创建配置文件 two-files-counter-pod-agent-sidecar.yaml，编辑内容如下：

```
apiVersion: v1
kind: Pod
metadata:
  name: counter
spec:
  containers:
   - name: count
    image: busybox
    args:
    - /bin/sh
    - -c
    - >
     i=0;
     while true;
     do
       echo "$i: $(date)">> /var/log/1.log;
       echo "$(date) INFO $i">> /var/log/2.log;
       i=$((i+1));
       sleep 1;
     done
```

```
    volumeMounts:
    - name: varlog
      mountPath: /var/log
  - name: count-agent
    image: k8s.gcr.io/fluentd-gcp:1.30
    env:
    - name: FLUENTD_ARGS
      value: -c /etc/fluentd-config/fluentd.conf
    volumeMounts:
    - name: varlog
      mountPath: /var/log
    - name: config-volume
      mountPath: /etc/fluentd-config
  volumes:
  - name: varlog
    emptyDir: {}
  - name: config-volume
    configMap:
      name: fluentd-config
```

使用此文件创建 Pod，命令如下：

```
kubectl apply -f two-files-counter-pod-agent-sidecar.yaml
```

运行命令后输出的信息如下：

```
pod/counter created
```

查看 counter Pod 中的 count-agent 容器内的日志文件，命令如下：

```
kubectl exec counter -c  count-agent -- ls /var/log/fluent
```

运行命令后输出的信息如下：

```
access.20210604.b5c3edb9250d62dd3
```

查看这个文件的内容，命令如下：

```
kubectl exec counter -c  count-agent \
-- head -n4 /var/log/fluent/access.20210604.b5c3edb9250d62dd3
```

运行命令后输出的类似信息如下：

```
2021-06-04T09:57:44+00:00  count.format1  {"message":"1: Fri Jun  4 09:57:44 UTC 2021"}
2021-06-04T09:57:44+00:00  count.format2  {"message":"Fri Jun  4 09:57:44 UTC 2021 INFO 1"}
```

```
2021-06-04T09:57:45+00:00  count.format1  {"message":"2: Fri Jun  4 09:57:45 UTC 2021"}
2021-06-04T09:57:45+00:00  count.format2  {"message":"Fri Jun  4 09:57:45 UTC 2021 INFO 2"}
```

输出信息显示 count-agent 容器中运行的 Fluentd 收集了 counter 容器中两个日志文件夹下的日志内容，并输出到了/var/log/fluent 文件夹下的文件中。与 Fluentd 的配置一致。

7.24 Kustomize

7.24.1 Kustomize 简介

Kustomize 是一个用于声明式管理 Kubernetes 对象的 CLI 工具。Kustomize 使用的配置文件称为 kustomization 文件。

kubectl 支持查看某个文件夹下所有 kustomization 文件中包含的资源，命令如下：

```
kubectl kustomize <directory>
```

例如查看当前文件夹下所有 kustomization 文件包含的资源，命令如下：

```
kubectl kustomize
```

Kustomize 可以利用其他来源创建资源，以横切的方式设置资源的字段、组合一组资源并进行定制。

7.24.2 生成 ConfigMap

ConfigMap 一般用于保存配置信息，并被其他 Kubernetes 对象(如 Pod)引用。ConfigMap 中数据的来源一般是 Kubernetes 集群外部的文件，如.properties 文件。Kustomize 中的 configMapGenerator 可以利用文件或者字面值生成 ConfigMap。

假设有一个 app.properties 文件，文件内容如下：

```
ENV=test
```

新建一个 kustomization 文件，编辑内容如下：

```
configMapGenerator:
- name: example-configmap-1
  files:
  - app.properties
```

预览生成的 ConfigMap，命令如下：

```
kubectl kustomize ./
```

运行命令后输出的信息如下：

```
apiVersion: v1
data:
  app.properties: |
    ENV = test
kind: ConfigMap
metadata:
  name: example-configmap-1-bkg9g7m579
```

输出显示 app.properties 文件作为一个整体变成了 ConfigMap 中的一个键。

利用这个文件生成 ConfigMap，命令如下：

```
kubectl apply -k ./
```

运行命令后输出的类似信息如下：

```
configmap/example-configmap-1-bkg9g7m579 created
```

如果通过.env 文件生成 ConfigMap，则需要设置 configGenerator 中的 envs 字段。假设有一个.env 文件，文件内容如下：

```
ENV = test
```

新建一个 kustomization.yaml 文件，编辑内容如下：

```
configMapGenerator:
- name: example-configmap-1
envs:
  - .env
```

预览生成的 ConfigMap，命令如下：

```
kubectl kustomize ./
```

运行命令后输出的信息如下：

```
apiVersion: v1
data:
  ENV: test
kind: ConfigMap
metadata:
  name: example-configmap-2-kg7t728m5k
```

输出显示.env文件中的每个变量都变成ConfigMap中的一个单独的键。

另外还可以通过字面值形式的键值对生成ConfigMap，需要将键值对添加到configMapGenerator的literals字段中。

假设有一个kustomization.yaml文件，内容如下：

```
configMapGenerator:
- name: example-configmap-3
  literals:
  - ENV=test
```

预览生成的ConfigMap，命令如下：

```
kubectl kustomize ./
```

运行命令后输出的信息如下：

```
apiVersion: v1
data:
  ENV: test
kind: ConfigMap
metadata:
  name: example-configmap-3-kg7t728m5k
```

从上述例子可以看出生成的ConfigMap的名字与configMapGenerator中的名字不一致，如果在其他对象(如Deployment)中引用生成的ConfigMap，只需引用configMapGenerator中的名字，Kustomize会自动使用最后生成的ConfigMap的名字进行替换。

假设有一个app.properties文件，文件内容如下：

```
ENV=test
```

假设有一个deployment.yaml文件，内容如下：

```
apiVersion: apps/v1
kind: Deployment
metadata:
  name: my-app
spec:
  selector:
    matchLabels:
      app: my-app
  template:
 metadata:
     labels:
```

```
        app: my-app
    spec:
      containers:
      - name: app
        image: my-app
        volumeMounts:
        - name: config
          mountPath: /config
      volumes:
      - name: config
        configMap:
          name: example-configmap-1
```

假设有一个 kustomization.yaml 文件，内容如下：

```
resources:
- deployment.yaml
configMapGenerator:
- name: example-configmap-1
  files:
  - app.properties
```

预览 Kustomize 生成的结果，命令如下：

```
kubectl kustomize ./
```

运行命令后输出的信息如下：

```
apiVersion: v1
data:
  app.properties: |
    ENV=test
kind: ConfigMap
metadata:
  name: example-configmap-1-bkg9g7m579
---
apiVersion: apps/v1
kind: Deployment
metadata:
  name: my-app
spec:
  selector:
    matchLabels:
      app: my-app
  template:
 metadata:
```

```
      labels:
        app: my-app
    spec:
      containers:
      - image: my-app
        name: app
        volumeMounts:
        - mountPath: /config
          name: config
      volumes:
      - configMap:
          name: example-configmap-1-bkg9g7m579
        name: config
```

可以看到 Deployment 中引用的 ConfigMap 的名字已经被自动替换为生成的 ConfigMap 的名字而不是 configMapGenerator 中的名字。

7.24.3 生成 Secret

与 ConfigMap 类似，Kustomize 可以利用文件或者字面量形式的键值对来生成 Secret。下面演示通过文件生成 Secret。

新建 password.txt 文件，编辑内容如下：

```
username=admin
password=secret
```

新建一个 kustomization.yaml 文件，编辑内容如下：

```
secretGenerator:
- name: example-secret-1
  files:
  - password.txt
```

预览 Kustomize 生成的结果，命令如下：

```
kubectl kustomize ./
```

运行命令后输出的信息如下：

```
apiVersion: v1
data:
  password.txt: dXNlcm5hbWU9YWRtaW4KcGFzc3dvcmQ9c2VjcmV0Cg==
kind: Secret
metadata:
  name: example-secret-1-2kdd8ckcc7
type: Opaque
```

下面是一个通过字面量形式的键值对生成 Secret 的例子。

新建一个 kustomization.yaml 文件，编辑内容如下：

```
secretGenerator:
- name: example-secret-2
  literals:
  - username=admin
  - password=secret
```

预览 Kustomize 生成的结果，命令如下：

```
kubectl kustomize ./
```

运行命令后输出的信息如下：

```
apiVersion: v1
data:
  password: c2VjcmV0
  username: YWRtaW4=
kind: Secret
metadata:
  name: example-secret-2-8c5228dkb9
type: Opaque
```

与 ConfigMap 类似，生成的 Secret 可以被其他资源（如 Deployment）引用。

新建 deployment.yaml 文件，编辑内容如下：

```
apiVersion: apps/v1
kind: Deployment
metadata:
  name: my-app
  labels:
    app: my-app
spec:
  selector:
    matchLabels:
      app: my-app
  template:
    metadata:
      labels:
        app: my-app
    spec:
      containers:
      - name: app
        image: my-app
```

```
        volumeMounts:
        - name: password
          mountPath: /secrets
      volumes:
      - name: password
        secret:
          secretName: example-secret-1
```

新建一个 kustomization.yaml 文件，编辑内容如下：

```
resources:
- deployment.yaml
secretGenerator:
- name: example-secret-1
  files:
  - password.txt
```

预览 Kustomize 生成的结果，命令如下：

```
kubectl kustomize ./
```

运行命令后输出的信息如下：

```
apiVersion: v1
data:
  password.txt: dXNlcm5hbWU9YWRtaW4KcGFzc3dvcmQ9c2VjcmV0Cg==
kind: Secret
metadata:
  name: example-secret-1-2kdd8ckcc7
type: Opaque
---
apiVersion: apps/v1
kind: Deployment
metadata:
  labels:
    app: my-app
  name: my-app
spec:
  selector:
    matchLabels:
      app: my-app
  template:
    metadata:
      labels:
        app: my-app
```

```
    spec:
      containers:
      - image: my-app
        name: app
        volumeMounts:
        - mountPath: /secrets
          name: password
      volumes:
      - name: password
        secret:
          secretName: example-secret-1-2kdd8ckcc7
```

输出显示 Deployment 中引用 Secret 的名字已经被替换成生成的 Secret 的名字。

7.24.4　生成器选项

Kustomize 生成的 ConfigMap 和 Secret 的名字中有一个根据内容计算的哈希值，这样可以保证当内容发生变化后会生成新的 ConfigMap 和 Secret，如果希望禁用这个功能，则可以使用 generatorOptions 实现。

新建一个 kustomization.yaml 文件，编辑内容如下：

```
configMapGenerator:
- name: example-configmap-4
  literals:
  - ENV=test
generatorOptions:
  disableNameSuffixHash: true
```

预览 Kustomize 生成的结果，命令如下：

```
kubectl kustomize ./
```

运行命令后输出的信息如下：

```
apiVersion: v1
data:
  ENV: test
kind: ConfigMap
metadata:
  name: example-configmap-4
```

输出显示 ConfigMap 的名字与 configMapGenerator 中的名字一致，没有哈希值后缀。

在 generatorOptions 中还可以设置资源的标签和注解，新建一个 kustomization.yaml 文件，编辑内容如下：

```
configMapGenerator:
- name: example-configmap-5
  literals:
  - ENV=test
generatorOptions:
  labels:
    type: generated
  annotations:
    note: generated
```

预览 Kustomize 生成的结果，命令如下：

```
kubectl kustomize ./
```

运行命令后输出的信息如下：

```
apiVersion: v1
data:
  ENV: test
kind: ConfigMap
metadata:
  annotations:
    note: generated
  labels:
    type: generated
  name: example-configmap-5-kg7t728m5k
```

7.24.5 设置横切字段

很多时候可能需要为 Kubernetes 中的资源统一设置某些通用字段，这些字段被称为横切字段，横切(cross-cutting)意味着相同，例如为所有资源设置相同的命名空间、添加相同的名字前缀或后缀、添加相同的标签及添加相同的注解等。

下面是使用 Kustomize 为一个 Deployment 设置横切字段的示例。

新建 deployment.yaml 文件，编辑内容如下：

```
apiVersion: apps/v1
kind: Deployment
metadata:
  name: nginx-deployment
  labels:
    app: nginx
spec:
  selector:
    matchLabels:
```

```
      app: nginx
  template:
    metadata:
      labels:
        app: nginx
    spec:
      containers:
      - name: nginx
        image: nginx
```

新建 kustomization. yaml 文件，编辑内容如下：

```
namespace: my-namespace
namePrefix: dev-
nameSuffix: "-001"
commonLabels:
  app: bingo
commonAnnotations:
  oncallPager: 800-555-1212
resources:
- deployment.yaml
```

预览 Kustomize 生成的结果，命令如下：

```
kubectl kustomize ./
```

运行命令后输出的信息如下：

```
apiVersion: apps/v1
kind: Deployment
metadata:
  annotations:
    oncallPager: 800-555-1212
  labels:
    app: bingo
  name: dev-nginx-deployment-001
  namespace: my-namespace
spec:
  selector:
    matchLabels:
      app: bingo
  template:
    metadata:
      annotations:
        oncallPager: 800-555-1212
```

```
      labels:
        app: bingo
    spec:
      containers:
      - image: nginx
        name: nginx
```

输出显示 Deployment 中已经添加了 kustomization.yaml 文件中指定的所有横切字段。

7.24.6 组合

Kustomize 支持把多个文件中的资源组合起来,实现对资源的批量管理。

下面是 Kustomize 组合示例。

新建文件 deployment.yaml,编辑内容如下:

```
apiVersion: apps/v1
kind: Deployment
metadata:
  name: nginx-deployment
  labels:
    app: nginx
spec:
  selector:
    matchLabels:
      app: nginx
  template:
    metadata:
      labels:
        app: nginx
    spec:
      containers:
      - name: nginx
        image: nginx
```

新建文件 service.yaml,编辑内容如下:

```
apiVersion: v1
kind: Service
metadata:
  name: my-nginx
  labels:
    run: my-nginx
spec:
  ports:
```

```
  - port: 80
    protocol: TCP
  selector:
    run: my-nginx
```

新建文件 service.yaml，编辑内容如下：

```
apiVersion: v1
kind: Service
metadata:
  name: my-nginx
  labels:
    run: my-nginx
spec:
  ports:
  - port: 80
    protocol: TCP
  selector:
    run: my-nginx
```

新建 kustomization.yaml 文件，编辑内容如下：

```
resources:
- deployment.yaml
- service.yaml
```

预览 Kustomize 生成的结果，命令如下：

```
kubectl kustomize ./
```

运行命令后输出的信息如下：

```
apiVersion: v1
kind: Service
metadata:
  labels:
    run: my-nginx
  name: my-nginx
spec:
  ports:
  - port: 80
    protocol: TCP
  selector:
    run: my-nginx
---
apiVersion: apps/v1
```

```
kind: Deployment
metadata:
  labels:
    app: nginx
  name: nginx-deployment
spec:
  selector:
    matchLabels:
      app: nginx
  template:
    metadata:
      labels:
        app: nginx
    spec:
      containers:
      - image: nginx
        name: nginx
```

输出显示 Kustomize 生成的结果中包含 Deployment 和 Service。

7.24.7 定制

Kustomize 支持对资源进行定制,也就是对一些字段进行修改或称为打补丁,在 Kustomize 生成的结果中修改的字段会覆盖资源原来的字段。修改的机制包括 patchesStrategicMerge 和 patchesJson6902。其中 patchesStrategicMerge 包含的是一个文件路径列表。每个文件对应一个补丁。建议每个补丁只做一件事。例如修改 Deployment 的 replica 字段对应一个补丁或者修改内存限制对应一个补丁。

下面是使用 Kustomize 对一个 Deployment 配置文件进行定制的示例,有两个补丁,分别用于修改 Deployment 的 replicas 字段和内存限制。

新建 deployment.yaml 文件,编辑文件内容如下:

```
apiVersion: apps/v1
kind: Deployment
metadata:
  name: my-nginx
spec:
  selector:
    matchLabels:
      run: my-nginx
  replicas: 2
  template:
    metadata:
      labels:
        run: my-nginx
```

```
    spec:
      containers:
      - name: my-nginx
        image: nginx
        ports:
        - containerPort: 80
```

新建 increase_replicas.yaml 文件，编辑内容如下：

```
apiVersion: apps/v1
kind: Deployment
metadata:
  name: my-nginx
spec:
  replicas: 3
```

新建 set_memory.yaml 文件，编辑内容如下：

```
apiVersion: apps/v1
kind: Deployment
metadata:
  name: my-nginx
spec:
  template:
    spec:
      containers:
      - name: my-nginx
        resources:
          limits:
            memory: 512Mi
```

最后新建 kustomization.yaml 文件，编辑内容如下：

```
resources:
- deployment.yaml
patchesStrategicMerge:
- increase_replicas.yaml
- set_memory.yaml
```

现在预览 Kustomize 生成的结果，命令如下：

```
kubectl kustomize ./
```

运行命令后输出的信息如下：

```
apiVersion: apps/v1
kind: Deployment
metadata:
  name: my-nginx
spec:
  replicas: 3
  selector:
    matchLabels:
      run: my-nginx
  template:
    metadata:
      labels:
        run: my-nginx
    spec:
      containers:
      - image: nginx
        name: my-nginx
        ports:
        - containerPort: 80
        resources:
          limits:
            memory: 512Mi
```

输出结果显示2个文件对应的补丁已经成功应用到了Deployment上。此Deployment中的replicas字段值已被修改为3,另外添加了内存限制512Mi。

有些字段无法使用patchesStrategicMerge进行修改,而另一种定制的机制patchesJson6902支持对资源的任意字段进行修改。下面是使用patchesJson6902修改Deployment的replicas字段的示例。

新建patch.yaml文件,编辑内容如下:

```
- op: replace
  path: /spec/replicas
  value: 3
```

新建kustomization.yaml文件,编辑内容如下:

```
resources:
- deployment.yaml

patchesJson6902:
- target:
    group: apps
    version: v1
    kind: Deployment
    name: my-nginx
  path: patch.yaml
```

现在预览 Kustomize 生成的结果，命令如下：

```
kubectl kustomize ./
```

运行命令后输出的信息如下：

```
apiVersion: apps/v1
kind: Deployment
metadata:
  name: my-nginx
spec:
  replicas: 3
  selector:
    matchLabels:
      run: my-nginx
  template:
    metadata:
      labels:
        run: my-nginx
    spec:
      containers:
      - image: nginx
        name: my-nginx
        ports:
        - containerPort: 80
```

输出显示 replicas 字段被修改为 3，说明补丁已生效。

7.24.8 变量注入

Kustomize 支持使用变量来引用其他资源，这可以解决由于资源名称发生变化时（例如资源名称被 Kustomize 加入前缀等），使用原有的名称无法正确引用资源的问题。

下面是变量注入的示例。

新建 deployment.yaml 文件，编辑内容如下：

```
apiVersion: apps/v1
kind: Deployment
metadata:
  name: my-nginx
spec:
  selector:
    matchLabels:
      run: my-nginx
  replicas: 2
  template:
```

```
    metadata:
      labels:
        run: my-nginx
    spec:
      containers:
      - name: my-nginx
        image: nginx
        command: ["start", "--host", "$(MY_SERVICE_NAME)"]
```

新建 service.yaml 文件，编辑内容如下：

```
apiVersion: v1
kind: Service
metadata:
  name: my-nginx
  labels:
    run: my-nginx
spec:
  ports:
  - port: 80
    protocol: TCP
  selector:
    run: my-nginx
```

新建 kustomization.yaml 文件，编辑内容如下：

```
namePrefix: dev-
nameSuffix: "-001"

resources:
- deployment.yaml
- service.yaml

vars:
- name: MY_SERVICE_NAME
  objref:
    kind: Service
    name: my-nginx
    apiVersion: v1
```

现在预览 Kustomize 生成的结果，命令如下：

```
kubectl kustomize ./
```

运行命令后输出的信息如下：

```
apiVersion: v1
kind: Service
metadata:
  labels:
    run: my-nginx
  name: dev-my-nginx-001
spec:
  ports:
  - port: 80
    protocol: TCP
  selector:
    run: my-nginx
---
apiVersion: apps/v1
kind: Deployment
metadata:
  name: dev-my-nginx-001
spec:
  replicas: 2
  selector:
    matchLabels:
      run: my-nginx
  template:
    metadata:
      labels:
        run: my-nginx
    spec:
      containers:
      - command:
        - start
        - --host
        - dev-my-nginx-001
        image: nginx
        name: my-nginx
```

输出显示 $(MY_SERVICE_NAME)变量被替换为 dev-my-nginx-001，说明变量注入成功。

7.24.9　基准和覆盖

不同环境下可能需要对资源进行不同的定制，例如加入不同名字的前缀。Kustomize 支持 overlay（覆盖）功能来对同一个 base（基准）进行不同定制。这样只需维护一个基准就可以产生不同的资源定义。

下面演示如果使用基准和覆盖功能。

新建文件夹 base，在 base 下新建文件 deployment.yaml，编辑内容如下：

```yaml
apiVersion: apps/v1
kind: Deployment
metadata:
  name: my-nginx
spec:
  selector:
    matchLabels:
      run: my-nginx
  replicas: 2
  template:
    metadata:
      labels:
        run: my-nginx
    spec:
      containers:
      - name: my-nginx
        image: nginx
```

在 base 下新建 service.yaml 文件，编辑内容如下：

```yaml
apiVersion: v1
kind: Service
metadata:
  name: my-nginx
  labels:
    run: my-nginx
spec:
  ports:
  - port: 80
    protocol: TCP
  selector:
    run: my-nginx
```

在 base 下新建 kustomization.yaml 文件，编辑内容如下：

```yaml
resources:
- deployment.yaml
- service.yaml
```

这样就完成了基准定义，下面定义 2 个覆盖。

新建 2 个文件夹 dev 和 prod。

在 dev 文件夹下新建文件 kustomization.yaml，编辑内容如下：

```yaml
bases:
- ../base
namePrefix: dev-
```

在 prod 文件夹下新建文件 kustomization.yaml，编辑内容如下：

```
bases:
- ../base
namePrefix: prod-
```

使用 dev/kustomization.yaml 文件预览生成的结果，命令如下：

```
kubectl kustomize dev
```

运行命令后输出的信息如下：

```
apiVersion: v1
kind: Service
metadata:
  labels:
    run: my-nginx
  name: dev-my-nginx
spec:
  ports:
  - port: 80
    protocol: TCP
  selector:
    run: my-nginx
---
apiVersion: apps/v1
kind: Deployment
metadata:
  name: dev-my-nginx
spec:
  replicas: 2
  selector:
    matchLabels:
      run: my-nginx
  template:
    metadata:
      labels:
        run: my-nginx
    spec:
      containers:
      - image: nginx
        name: my-nginx
```

输出显示生成的 Deployment 和 Service 定义的名字都加入了 dev-前缀。

使用 prod/kustomization.yaml 文件预览生成的结果，命令如下：

```
kubectl kustomize prod
```

运行命令后输出的信息如下：

```
apiVersion: v1
kind: Service
metadata:
  labels:
    run: my-nginx
  name: prod-my-nginx
spec:
  ports:
  - port: 80
    protocol: TCP
  selector:
    run: my-nginx
---
apiVersion: apps/v1
kind: Deployment
metadata:
  name: prod-my-nginx
spec:
  replicas: 2
  selector:
    matchLabels:
      run: my-nginx
  template:
    metadata:
      labels:
        run: my-nginx
    spec:
      containers:
      - image: nginx
        name: my-nginx
```

输出显示生成的 Deployment 和 Service 定义的名字都加入了 prod-前缀。

7.24.10　应用、查询和删除对象

应用 kustomization.yaml 文件中的资源，使用 kubectl apply -k 命令，-k 参数表示包含 kustomization.yaml 文件的文件夹路径。

下面假设 kustomization.yaml 在当前工作目录。

应用 kustomization.yaml 文件中的资源，命令如下：

```
kubectl apply -k ./
```

查询 kustomization.yaml 文件中的资源，命令如下：

```
kubectl get -k ./
```

查询 kustomization.yaml 文件中的资源的详细信息，命令如下：

```
kubectl describe -k ./
```

删除 kustomization.yaml 文件中的资源，命令如下：

```
kubectl delete -k ./
```

比对 Kubernetes 集群中实时的资源状态和 kustomization.yaml 文件的资源配置，命令如下：

```
kubectl diff -k ./
```

第 8 章

Kubernetes 部署应用

本章讲解如何把第 1 章介绍的用户认证项目部署到 Kubernetes 集群。

8.1 环境

查看环境,命令如下:

```
kubectl version
```

运行命令后输出的信息如下:

```
Client Version: version.Info{Major:"1", Minor:"21", GitVersion:"v1.21.0", GitCommit:
"cb303e613a121a29364f75cc67d3d580833a7479", GitTreeState:"clean", BuildDate:"2021-04-
08T16:31:21Z", \
GoVersion:"go1.16.1", Compiler:"gc", Platform:"darwin/amd64"}
Server Version: version.Info{Major:"1", Minor:"20+", \
GitVersion:"v1.20.4-eks-6b7464", GitCommit:"6b746440c04cb81db4426842b4ae65c3f7035e53",
GitTreeState:"clean", BuildDate:"2021-03-19T19:33:03Z", \
GoVersion:"go1.15.8", Compiler:"gc", Platform:"Linux/amd64"}
```

8.1.1 开发环境

使用 AWS 托管的 Kubernetes 集群作为开发环境。访问应用的域名为本地 DNS 域名 app.com。

8.1.2 生产环境

使用 AWS 托管的 Kubernetes 集群作为生产环境。访问应用的域名为真实注册的域名 unit.ink。

8.2 MySQL 服务

MySQL 数据库默认会把数据保存到/var/lib/mysql 文件夹下，需要提供一个 PV 来存储 MySQL 数据。

8.2.1 开发环境

开发环境下的 Kubernetes 集群只有一个节点，部署 MySQL 可以使用 hostPath 类型的 PV，新建 mysql-pv. yaml 文件，内容如下：

```
apiVersion: v1
kind: PersistentVolume
metadata:
  name: mysql-pv-volume
  labels:
    type: local
spec:
  storageClassName: manual
  capacity:
    storage: 20Gi
  accessModes:
     - ReadWriteOnce
  hostPath:
    path: "/mnt/data"
```

文件定义了一个大小为 20Gi 的 PV。PV 类型为 hostPath，将使用宿主机的/mnt/data 目录来存储数据。使用 mysql-pv. yaml 创建 PV，命令如下：

```
kubectl apply -f k8s/mysql-pv.yaml
```

运行命令后输出的信息如下：

```
persistentvolume/mysql-pv-volume created
```

验证 PV 创建成功，命令如下：

```
kubectl describe pv mysql-pv-volume
```

运行命令后输出的信息如下：

```
Name:            mysql-pv-volume
Labels:          type=local
Annotations:     <none>
```

```
Finalizers:        [kubernetes.io/pv-protection]
StorageClass:      manual
Status:            Available
Claim:
Reclaim Policy:    Retain
Access Modes:      RWO
VolumeMode:        Filesystem
Capacity:          20Gi
Node Affinity:     <none>
Message:
Source:
    Type:          HostPath (bare host directory volume)
    Path:          /mnt/data
    HostPathType:
Events:            <none>
```

输出信息中 Status 字段为 Available，表示 PV 创建成功，并且处于可用状态。

接下来创建将要与 mysql-pv-volume 绑定的 PVC，新建 PVC 配置文件 mysql-pvc.yaml，内容如下：

```
apiVersion: v1
kind: PersistentVolumeClaim
metadata:
  name: mysql-pv-claim
  labels:
    app: accounts
spec:
  storageClassName: manual
  accessModes:
    - ReadWriteOnce
  resources:
    requests:
      storage: 20Gi
```

使用 mysql-pvc.yaml 创建 PVC，命令如下：

```
kubectl apply -f k8s/mysql-pvc.yaml
```

运行命令后输出的信息如下：

```
persistentvolumeclaim/mysql-pv-claim created
```

查看新创建的 PVC，验证是否与 mysql-pv-volume 绑定成功，命令如下：

```
kubectl describe pvc mysql-pv-claim
```

运行命令后输出的信息如下：

```
Name:           mysql-pv-claim
Namespace:      default
StorageClass:   manual
Status:         Bound
Volume:         mysql-pv-volume
Labels:         app=accounts
Annotations:    pv.kubernetes.io/bind-completed: yes
                pv.kubernetes.io/bound-by-controller: yes
Finalizers:     [kubernetes.io/pvc-protection]
Capacity:       20Gi
Access Modes:   RWO
VolumeMode:     Filesystem
Used By:        <none>
Events:         <none>
```

输出信息中 Status 的字段值为 Bound，表示已绑定符合要求的 PV，Volume 字段显示绑定的 PV 是 mysql-pv-volume。

创建 PVC 以后，运行 MySQL 的 Pod 就可以使用此 PVC 来存储数据，这里使用 Deployment 来创建 MySQL Pod，新建 Deployment 配置文件 mysql-deployment.yaml，内容如下：

```
apiVersion: apps/v1
kind: Deployment
metadata:
  name: mysql
  labels:
    app: accounts
spec:
  selector:
    matchLabels:
      app: accounts
      tier: mysql
  strategy:
    type: Recreate
  template:
    metadata:
      labels:
        app: accounts
        tier: mysql
    spec:
      containers:
        - image: mysql:5.7
```

```
            name: mysql
            env:
              - name: MYSQL_ROOT_PASSWORD
                valueFrom:
                  secretKeyRef:
                    name: mysql - pass
                    key: password
              - name: MYSQL_DATABASE
                valueFrom:
                  configMapKeyRef:
                    name: mysql - config
                    key: db
            ports:
              - containerPort: 3306
                name: mysql
            args: [" -- character - set - server = utf8mb4"," -- default - time - zone = +08:
00"," -- ignore - db - dir = lost + found]
            volumeMounts:
              - name: mysql - persistent - storage
                mountPath: /var/lib/mysql
        volumes:
          - name: mysql - persistent - storage
            persistentVolumeClaim:
              claimName: mysql - pv - claim
```

env 字段指定了注入 MySQL 容器的环境变量，其中 MYSQL_ROOT_PASSWORD 的值会作为 MySQL root 用户的密码，MYSQL_DATABASE 的作用是 MySQL 会根据它的值自动创建数据库。环境变量 MYSQL_ROOT_PASSWORD 的值引用自名为 mysql-pass 的 Secret，环境变量 MYSQL_DATABASE 的值引用自名为 mysql-config 的 ConfigMap。

args 字段指定了 MySQL 的启动参数。其中--ignore-db-dir=lost+found 参数的作用是告诉 MySQL 忽略 lost+found 文件夹。生产环境中动态供应的卷如果使用了 ext4 文件系统，则会在根目录自动创建 lost+found 文件夹。当把卷挂载到 MySQL 存放数据的目录/var/lib/mysql 后，lost+found 文件夹会出现在 MySQL 容器中的/var/lib/mysql。如果不指定这个参数，则会导致 MySQL 启动失败。

创建名为 mysql-pass 的 Secret，用来保存 root 用户的密码，将 root 用户的密码设置为 123，命令如下：

```
kubectl create secret generic mysql - pass  -- from - literal = password = 123
```

运行命令后输出的信息如下：

```
secret/mysql - pass created
```

创建名为 mysql-config 的 Configmap，用来保存 MySQL 自动创建的数据库的名称，将数据库名称设置为 accounts，命令如下：

```
kubectl create configmap mysql-config --from-literal=db=accounts
```

运行命令后输出的信息如下：

```
configmap/mysql-config created
```

创建 mysql-pass 和 mysql-config 之后，使用 mysql-deployment.yaml 创建 Deployment，命令如下：

```
kubectl apply -f k8s/mysql-deployment.yaml
```

运行命令后输出的信息如下：

```
deployment.apps/mysql created
```

查看 Deployment 信息，验证 mysql-deployment 是否创建成功，命令如下：

```
kubectl get deploy
```

运行命令后输出的信息如下：

```
NAME    READY   UP-TO-DATE   AVAILABLE   AGE
mysql   1/1     1            1           14m
```

输出显示 MySQL 已经运行在 1 个 Pod 中，接下来把 mysql-deployment 暴露为 Service，新建 mysql-service.yaml 文件，内容如下：

```
apiVersion: v1
kind: Service
metadata:
  name: mysql
  labels:
    app: accounts
spec:
  ports:
    - port: 3306
  selector:
    app: accounts
    tier: mysql
  clusterIP: None
```

使用此文件创建 Service，命令如下：

```
kubectl apply -f k8s/mysql-service.yaml
```

运行命令后输出的信息如下：

```
service/mysql created
```

表示 Service 创建成功。

8.2.2 生产环境

生产环境可以使用自动卷供应功能提供存储卷来保存 MySQL 数据。不需要使用文件 mysql-pv.yaml 手动创建 PV，并且当 Pod 调度到其他节点之后，Pod 还可以访问之前挂载的 PV 中的数据。

自动卷供应需要用到 StorageClass，新建 StorageClass 配置文件 storageclass-ebs.yaml，文件内容如下：

```
apiVersion: storage.k8s.io/v1
kind: StorageClass
metadata:
  name: aws-ebs
  annotations:
    storageclass.kubernetes.io/is-default-class: "true"
provisioner: kubernetes.io/aws-ebs
parameters:
  type: gp2
  fsType: ext4
reclaimPolicy: Retain
allowVolumeExpansion: true
mountOptions:
  - debug
volumeBindingMode: WaitForFirstConsumer
```

文件定义了名为 aws-ebs 的 StorageClass，provisioner 字段指定了存储插件是 kubernetes.io/aws-ebs。fsType 将文件系统的类型指定为 ext4。

使用这个文件创建 StorageClass，命令如下：

```
kubectl apply -f storageclass-ebs.yaml
```

运行命令后输出的信息如下：

```
storageclass.storage.k8s.io/aws-ebs created
```

对应地，修改 mysql-pvc.yaml 中的 storageClassName 字段如下：

```
apiVersion: v1
kind: PersistentVolumeClaim
metadata:
  name: mysql - pv - claim
  labels:
    app: accounts
spec:
  storageClassName: aws - ebs
  accessModes:
    - ReadWriteOnce
  resources:
    requests:
      storage: 10Gi
```

.spec.storageClassName 字段的值是上面新创建的 StorageClass 的名字 aws-ebs。

使用 mysql-pvc.yaml 创建 PVC，命令如下：

```
kubectl apply - f k8s/mysql - pvc.yaml
```

运行命令后输出的信息如下：

```
persistentvolumeclaim/mysql - pv - claim created
```

查看 PVC 的详细信息，命令如下：

```
kubectl describe pvc mysql - pv - claim
```

运行命令后输出的信息如下：

```
Name:          mysql - pv - claim
Namespace:     default
StorageClass:  aws - ebs
Status:        Pending
Volume:
Labels:        app = accounts
Annotations:   < none >
Finalizers:    [kubernetes.io/pvc - protection]
Capacity:
Access Modes:
VolumeMode:    Filesystem
Used By:       < none >
Events:
  Type    Reason                   Age              From            Message
  ----    -------                  ----             ----            -------
  Normal  WaitForFirstConsumer1s(x18 over6s) persistentvolume - controller\
 waiting for first consumer to be created before binding
```

使用 mysql-deployment.yaml 文件创建 Deployment，命令如下：

```
kubectl apply -f k8s/mysql-deployment.yaml
```

再次查看 PVC 的详细信息，命令如下：

```
kubectl describe pvc mysql-pv-claim
```

运行命令后输出的截取信息如下：

```
Name:          mysql-pv-claim
Namespace:     default
StorageClass:  aws-ebs
Status:        Bound
Volume:        pvc-7c5fbff4-d54e-42c3-ae6a-efb42c123f63
Labels:        app=accounts
Annotations:   pv.kubernetes.io/bind-completed: yes
               pv.kubernetes.io/bound-by-controller: yes
               volume.beta.kubernetes.io/storage-provisioner: kubernetes.io/aws-ebs
               volume.kubernetes.io/selected-node: ip-192-168-81-163.ap-east-1.
               compute.internal
Finalizers:    [kubernetes.io/pvc-protection]
Capacity:      20Gi
Access Modes:  RWO
VolumeMode:    Filesystem
Used By:       mysql-5b4548bb86-z2bzs
...
```

输出信息显示 mysql-pv-claim 这个 PVC 已经绑定了 PV。

查看 PV 的信息，命令如下：

```
kubectl get pv
```

运行命令后输出的截取信息如下：

```
NAME     STATUS CLAIM
pvc-7c5fbff4-d54e-42c3-ae6a-efb42c123f63 Bound default/mysql-pv-claim
```

这个 PV 是 Kubernetes 自动创建的卷。输出信息显示这个 PV 被绑定到了 mysql-pv-claim。

3min

8.3 数据迁移

用户认证项目使用了 users 数据表，需要使用数据迁移工具生成。这里使用开源项目 goose 进行数据迁移。

在 accounts 项目根目录下的 src/database/migrations/20210304203736_create_table_user.sql 是迁移使用的文件。里面定义了对数据库执行的 SQL 语句。

为了使运行数据迁移的 Pod 可以访问数据迁移文件，把文件保存在 ConfigMap 中，命令如下：

```
kubectl create configmap migration --from-file=src/database/migrations
```

运行命令后输出的信息如下：

```
configmap/migration created
```

下面定义执行数据迁移任务的 Job。

新建 migration.yaml 文件，内容如下：

```
apiVersion: batch/v1
kind: Job
metadata:
  name: migration
spec:
  template:
    spec:
      containers:
        - name: migration
          image: bitmyth/goose:v1.0.0
          command: ["goose",  "up"]
          workingDir: /migrations
          env:
            - name: GOOSE_DRIVER
              value: mysql
            - name: GOOSE_DBSTRING
              value: root:123@tcp(mysql)/accounts
          volumeMounts:
            - name: migration
              mountPath: /migrations
      restartPolicy: Never
      initContainers:
        - name: wait-db
          image: busybox:1.28
          command: ['sh', '-c', "until telnet mysql 3306; do echo waiting for mysql; sleep 1;
done"]
      volumes:
        - name: migration
          configMap:
            name: migration
  backoffLimit: 4
```

文件中定义了名为 migration 的 Job，其中的 Pod 运行的是 bitmyth/goose：v1.0.0 镜像。容器执行的命令是 goose up。goose 使用环境变量 GOOSE_DRIVER 判断数据库驱动，使用环境变量 GOOSE_DBSTRING 的值连接数据库。

Job 中的 Pod 把 migration 的 ConfigMap 作为卷挂载到了 Pod 中/migrations 目录。

initContainer 用于等待 MySQL 服务器就绪，通过 telnet mysql 3306 判断 MySQL 的端口是否可用。如果不可用就会一直等待。在就绪之后才会运行 Job 中的容器，以便进行数据迁移。这样可以避免在 MySQL 服务还未就绪时运行迁移程序而导致失败。

使用 migration.yaml 文件创建 Job，命令如下：

```
kubectl apply -f k8s/migration.yaml
```

运行命令后输出的信息如下：

```
job.batch/migration created
```

查看 Job 信息，命令如下：

```
kubectl get job
```

运行命令后输出的信息如下：

```
NAME        COMPLETIONS   DURATION    AGE
migration   1/1           50s         14m
```

查看 Job 开始和结束时间等更多信息，命令如下：

```
kubectl describe jobs migration
```

运行命令后输出的类似信息如下：

```
Name:            migration
Namespace:       default
Selector:        controller-uid=e67f53a8-8abc-4e72-babb-406242299e3c
Labels:          controller-uid=e67f53a8-8abc-4e72-babb-406242299e3c
                  job-name=migration
Annotations:     <none>
Parallelism:     1
Completions:     1
Start Time:      Wed, 26 May 2021 17:09:46 +0800
Completed At:    Wed, 26 May 2021 17:10:36 +0800
Duration:        50s
Pods Statuses:   0 Running / 1 Succeeded / 0 Failed
```

```
Pod Template:
  Labels:  controller-uid=e67f53a8-8abc-4e72-babb-406242299e3c
           job-name=migration
  Containers:
   migration:
    Image:       bitmyth/goose:v1.0.0
    Port:        <none>
    Host Port:   <none>
    Command:
      goose
      up
    Environment:
      GOOSE_DRIVER:     mysql
      GOOSE_DBSTRING:   root:123@tcp(mysql)/accounts
    Mounts:
      /migrations from migration (rw)
  Volumes:
   migration:
    Type:       ConfigMap (a volume populated by a ConfigMap)
    Name:       migration
    Optional:   false
Events:
  Type    Reason            Age   From            Message
  ----    -------           ----  ----            --------
  Normal  SuccessfulCreate  13m   job-controller  Created pod: migration-4cxk2
  Normal  Completed         12m   job-controller  Job completed
```

为了验证 MySQL 数据库中是否已生成 users 表，下面在运行 MySQL 的 Pod 中执行 MySQL 命令查询。

首先查看运行 MySQL 的 Pod 信息，命令如下：

```
kubectl get pod -l tier=mysql
```

运行命令后输出的信息如下：

```
NAME                    READY   STATUS    RESTARTS   AGE
mysql-5b4548bb86-z2bzs  1/1     Running   0          102m
```

在 Podmysql-5b4548bb86-z2bzs 中执行命令，以便查看数据库 accounts 中有哪些数据表，命令如下：

```
kubectl exec mysql-5b4548bb86-z2bzs -- mysql -uroot -p123  -e 'show tables' accounts
```

运行命令后输出的信息如下：

```
Tables_in_accounts
goose_db_version
users
mysql: [Warning] Using a password on the command line interface can be insecure.
```

输出信息说明数据库 accounts 中已经创建了 users 数据表。goose_db_version 数据表是 goose 用来记录数据迁移的历史版本。

8.4 后端服务

3min

在第 1 章介绍的后端项目的根目录下的 k8s 文件夹下新建文件 api-deployment. yaml，内容如下：

```
apiVersion: apps/v1
kind: Deployment
metadata:
  name: api
  labels:
    app: api
spec:
  selector:
    matchLabels:
      app: api
      tier: backend
  strategy:
    type: Recreate
  template:
    metadata:
      labels:
        app: api
        tier: backend
    spec:
      containers:
        - image: bitmyth/accounts:v1.0.12
          name: api
          ports:
            - containerPort: 80
              name: api
          volumeMounts:
            - name: plain-config
              mountPath: "/config"
              readOnly: true
```

```
      volumes:
        - name: secret-config
          secret:
            secretName: secret-config
        - name: plain-config
          configMap:
            name: plain-config
```

此文件定义了一个名为 api 的 Deployment。Deployment 中运行的 Pod 使用镜像 bitmyth/accounts:v1.0.12。Pod 暴露的端口是 80。挂载了一个名为 plain-config 的卷,挂载目标是容器中的/config 目录。

创建名为 plain-config 的 ConfigMap,命令如下:

```
kubectl create configmap plain-config \
--from-file=config/plain.yaml \
--from-file=config/secret.yaml
```

运行命令后输出的信息如下:

```
configmap/plain-config created
```

创建名为 secret-config 的 Secret,命令如下:

```
kubectl create secret generic secret-config --from-file=config/secret.yaml
```

运行命令后输出的信息如下:

```
secret/secret-config created
```

使用 api-deployment 文件创建 Deployment,命令如下:

```
kubectl apply -f k8s/api-deployment.yaml
```

运行命令后输出的信息如下:

```
deployment.apps/api created
```

查看 Deployment 信息,命令如下:

```
kubectl get deploy api
```

运行命令后输出的信息如下:

```
NAME    READY    UP-TO-DATE    AVAILABLE    AGE
api     1/1      1             1            8m26s
```

输出信息显示 Deployment 中 1 个 Pod 已经启动。

新建 api-service.yaml 文件，内容如下：

```
apiVersion: v1
kind: Service
metadata:
  name: api
  labels:
    app: api
spec:
  ports:
    - port: 80
  selector:
    app: api
    tier: backend
```

使用此文件创建 Service，命令如下：

```
kubectl apply -f k8s/api-service.yaml
```

运行命令后输出的信息如下：

```
service/api created
```

查看 Web Service 信息，命令如下：

```
kubectl get svc web
```

运行命令后输出的信息如下：

```
NAME   TYPE       CLUSTER-IP      EXTERNAL-IP   PORT(S)        AGE
web    NodePort   10.100.31.208   <none>        80:32065/TCP   12h
```

3min

8.5 前端服务

新建 web-deployment.yaml 文件，内容如下：

```
apiVersion: apps/v1
kind: Deployment
```

```
metadata:
  name: web
  labels:
    app: web
spec:
  selector:
    matchLabels:
      app: web
      tier: frontend
  strategy:
    type: Recreate
  template:
    metadata:
      labels:
        app: web
        tier: frontend
    spec:
      containers:
        - image: bitmyth/accounts-frontend:v1.0.54
          name: web
          ports:
            - containerPort: 80
              name: api
```

使用此文件创建 Deployment，命令如下：

```
kubectl apply -f k8s/web-deployment.yaml
```

运行命令后输出的信息如下：

```
deployment.apps/web created
```

新建 web-service.yaml 文件，内容如下：

```
apiVersion: v1
kind: Service
metadata:
  name: web
  labels:
    app: web
spec:
  ports:
    - port: 80
  selector:
    app: web
    tier: frontend
  type: NodePort
```

使用此文件创建 Service，命令如下：

```
kubectl apply -f k8s/web-service.yaml
```

运行命令后输出的信息如下：

```
service/web created
```

8.6 Ingress

在托管的 Kubernetes 集群环境中，安装 ingress-nginx，命令如下：

```
kubectl apply -f ingress-nginx-controller-aws.yaml
```

ingress-nginx-controller-aws.yaml 文件中的内容是从网页中复制的，网址为 https://raw.bithubusercontent.com/kubernetes/ingress-nginx/controller-v0.45.0/deploy/static/provider/aws/deploy.yaml。

运行命令后输出的信息如下：

```
namespace/ingress-nginx created
serviceaccount/ingress-nginx created
configmap/ingress-nginx-controller created
clusterrole.rbac.authorization.k8s.io/ingress-nginx created
clusterrolebinding.rbac.authorization.k8s.io/ingress-nginx created
role.rbac.authorization.k8s.io/ingress-nginx created
rolebinding.rbac.authorization.k8s.io/ingress-nginx created
service/ingress-nginx-controller-admission created
service/ingress-nginx-controller created
deployment.apps/ingress-nginx-controller created
validatingwebhookconfiguration.admissionregistration.k8s.io/ingress-nginx-admission created
serviceaccount/ingress-nginx-admission created
clusterrole.rbac.authorization.k8s.io/ingress-nginx-admission created
clusterrolebinding.rbac.authorization.k8s.io/ingress-nginx-admission created
role.rbac.authorization.k8s.io/ingress-nginx-admission created
rolebinding.rbac.authorization.k8s.io/ingress-nginx-admission created
job.batch/ingress-nginx-admission-create created
job.batch/ingress-nginx-admission-patch created
```

新建 ingress.yaml 文件，内容如下：

```
apiVersion: networking.k8s.io/v1
kind: Ingress
```

```
metadata:
  name: app-ingress
  annotations:
    nginx.ingress.kubernetes.io/rewrite-target: /
spec:
  defaultBackend:
    service:
      name: web
      port:
        number: 80
  rules:
    - host: app.com
      http:
        paths:
          - path: /
            pathType: Prefix
            backend:
              service:
                name: web
                port:
                  number: 80
    - host: api.app.com
      http:
        paths:
          - path: /
            pathType: Prefix
            backend:
              service:
                name: api
                port:
                  number: 80
```

文件中定义了名为 app-ingress 的 Ingress，默认后端服务是 web。定义了 2 个规则，当请求访问的主机名是 app.com 并且请求路径的前缀匹配到路径/时会转发到后端服务 web 的 80 端口。当请求访问的主机名是 api.app.com 并且请求路径的前缀匹配到路径/时会转发到后端服务 api 的 80 端口。

使用此文件创建 Ingress，命令如下：

```
kubectl apply -f k8s/ingress-nginx.yaml
```

运行命令后输出的信息如下：

```
ingress.networking.k8s.io/app-ingress created
```

查看 app-ingress 的详细信息，命令如下：

```
kubectl describe ing app-ingress
```

运行命令后输出的信息如下：

```
Name:             app-ingress
Namespace:        default
Address:          a64fbdc3e7b0648fb99cdd870e99ac3b-8fec328a6834e3ab.elb.ap-east-1.amazonaws.com
Default backend:   web:80 (192.168.65.140:80)
Rules:
  Host          Path    Backends
  ----         ----   ---------
  app.com
                /   web:80 (192.168.65.140:80)
  api.app.com
                /   api:80 (192.168.18.243:80)
Annotations:   nginx.ingress.kubernetes.io/rewrite-target: /
Events:
  Type     Reason     Age                 From                       Message
  ----     ------     ----                ----                       -------
  Normal   Sync       48s (x2 over 57s)   nginx-ingress-controller   Scheduled for sync
```

Address 字段显示的是 Kubernetes 集群为 Ingress 分配的负载均衡器的访问地址。在浏览器网址栏可以访问这个域名。

8.7 DNS

虽然使用 Kubernetes 集群提供的负载均衡器地址可以访问项目，但是使用固定的域名访问项目会更加方便，因为域名更加简洁且容易记忆。

8.7.1 开发环境

假设项目使用的域名是 app.com。前端项目代码中会写入后端项目访问的地址。假设后端项目的域名为 api.app.com。

在域名注册商设置一条 CNAME 解析记录，把 app.com 指向 Ingress 关联的负载均衡地址。这样就可以通过 app.com 访问 Ingress 了。

在开发环境可以使用 dnsmasq 作为本地 DNS 服务器，并设置 CNAME 解析记录。

新建 resolv.conf 文件，用于指定上游 DNS 服务器，文件内容如下：

```
nameserver 8.8.8.8
```

新建 dnsmasq 配置文件 dnsmasq.conf，文件内容如下：

```
cname = app.com, a64fbdc3e7b0648fb99cdd870e99ac3b-8fec328a6834e3ab.elb.ap-east-1.amazonaws.com
cname = api.app.com, a64fbdc3e7b0648fb99cdd870e99ac3b-8fec328a6834e3ab.elb.ap-east-1.amazonaws.com
```

这里设置了 2 条 CNAME 记录。解析 app.com 和 api.app.com 返回相同的结果，即负载均衡器的地址。

运行 dnsmasq 容器，命令如下：

```
docker run --rm -d\
 -v $(pwd)/resolv.conf:/etc/resolv.conf \
 -v $(pwd)/dnsmasq.conf:/etc/dnsmasq.conf \
 -p 53:53/tcp -p 53:53/udp \
 --cap-add=NET_ADMIN \
 --name dnsmasq bitmyth/dnsmasq
```

命令中-v 参数把本地的 resolv.conf 和 dnsmasq 文件分别挂载到容器中，覆盖容器中/etc/resolv.conf 和/etc/dnsmasq.conf 文件。

dnsmasq 容器启动后，可修改本地 DNS 配置，把 DNS 服务器地址设置为本机的 IP 地址。这样本地 DNS 服务器就设置完成了。

测试 app.com 解析结果是否符合预期，命令如下：

```
nslookup app.com
```

运行命令后输出的信息如下：

```
Server:   192.168.0.168
Address:   192.168.0.168#53

app.com   canonical name = \
a64fbdc3e7b0648fb99cdd870e99ac3b-8fec328a6834e3ab.elb.ap-east-1.amazonaws.com.
```

输出信息证明解析设置已经生效，app.com 会解析到负载均衡地址。

这时打开浏览器访问 app.com 就可以访问前端项目了。

8.7.2 生产环境

生产环境设置 DNS 需要进入域名注册商后台对域名进行设置，假设在生产环境中使用域名 unit.ink 访问前端，使用 api.unit.ink 域名访问后端，并且通过 Ingress 作为 Kubernetes 集群流量入口。可以设置 2 条 DNS 解析记录，并且都解析到 Ingress 关联的负载均衡器的地址，如图 8-1 所示。

记录类型:

CNAME- 将域名指向另外一个域名

主机记录:

@ .unit.ink

解析线路:

默认 - 必填! 未匹配到智能解析线路时,返回【默认】线路设置结果

* 记录值:

a64fbdc3e7b0648fb99cdd870e99ac3b-8fec328a6834e3ab.elb.ap-east-1.amazonaws.com

图 8-1 域名 unit.ink 的 CNAME 解析记录

这样设置生效以后域名 unit.ink 会解析到 a64fbdc3e7b0648fb99cdd870e99ac3b-8fec328a6834e3ab.elb.ap-east-1.amazonaws.com。

使用通配符 * 可以匹配任意与 api.unit.ink 具有相同层级的域名,如图 8-2 所示。

记录类型:

CNAME- 将域名指向另外一个域名

主机记录:

* .unit.ink

解析线路:

默认 - 必填! 未匹配到智能解析线路时,返回【默认】线路设置结果

* 记录值:

a64fbdc3e7b0648fb99cdd870e99ac3b-8fec328a6834e3ab.elb.ap-east-1.amazonaws.com

图 8-2 域名 *.unit.ink 的 CNAME 解析记录

这样设置生效以后域名 *.unit.ink 会被解析到 a64fbdc3e7b0648fb99cdd870e99ac3b-8fec328a6834e3ab.elb.ap-east-1.amazonaws.com。

8.8　TLS

使用 HTTP 协议存在流量劫持、流量嗅探、中间人攻击等诸多安全风险。为了使应用更加安全,防范基本的安全风险,需使用 HTTPS 协议访问。尤其对于暴露在公网中的生产环境来讲,HTTPS 已经成为一项必不可少的配置。

用户认证项目使用 Ingress 接入外部流量,只需对 Ingress 启用 TLS,这样就可以支持 HTTPS 访问。利用 Ingress 的 TLS 终止功能,转发到后端服务的数据是解密后的明文数据。

8.8.1　证书管理软件

启用 HTTPS 需要配置证书,在证书过期后需要替换新证书。手动维护证书比较烦琐。可以利用一些工具提高证书管理的自动化水平。

开源项目 cert-manager 可以实现自动化证书管理。下面介绍在 Kubernetes 集群中使用 cert-manager 的步骤。

在 Kubernetes 集群中安装 cert-manager,命令如下:

```
kubectl apply -f \
https://github.com/jetstack/cert-manager/releases/download/v1.3.1/cert-manager.yaml
```

运行命令后输出的信息如下:

```
customresourcedefinition.apiextensions.k8s.io/certificaterequests.cert-manager.io created
customresourcedefinition.apiextensions.k8s.io/certificates.cert-manager.io created
customresourcedefinition.apiextensions.k8s.io/challenges.acme.cert-manager.io created
customresourcedefinition.apiextensions.k8s.io/clusterissuers.cert-manager.io created
customresourcedefinition.apiextensions.k8s.io/issuers.cert-manager.io created
customresourcedefinition.apiextensions.k8s.io/orders.acme.cert-manager.io created
namespace/cert-manager created
serviceaccount/cert-manager-cainjector created
serviceaccount/cert-manager created
serviceaccount/cert-manager-webhook created
clusterrole.rbac.authorization.k8s.io/cert-manager-cainjector created
clusterrole.rbac.authorization.k8s.io/cert-manager-controller-issuers created
clusterrole.rbac.authorization.k8s.io/cert-manager-controller-clusterissuers created
clusterrole.rbac.authorization.k8s.io/cert-manager-controller-certificates created
clusterrole.rbac.authorization.k8s.io/cert-manager-controller-orders created
clusterrole.rbac.authorization.k8s.io/cert-manager-controller-challenges created
clusterrole.rbac.authorization.k8s.io/cert-manager-controller-ingress-shim created
clusterrole.rbac.authorization.k8s.io/cert-manager-view created
clusterrole.rbac.authorization.k8s.io/cert-manager-edit created
```

```
clusterrole.rbac.authorization.k8s.io/cert-manager-controller-approve:cert-manager-io created
clusterrole.rbac.authorization.k8s.io/cert-manager-webhook:subjectaccessreviews created
clusterrolebinding.rbac.authorization.k8s.io/cert-manager-cainjector created
clusterrolebinding.rbac.authorization.k8s.io/cert-manager-controller-issuers created
clusterrolebinding.rbac.authorization.k8s.io/cert-manager-controller-clusterissuers created
clusterrolebinding.rbac.authorization.k8s.io/cert-manager-controller-certificates created
clusterrolebinding.rbac.authorization.k8s.io/cert-manager-controller-orders created
clusterrolebinding.rbac.authorization.k8s.io/cert-manager-controller-challenges created
clusterrolebinding.rbac.authorization.k8s.io/cert-manager-controller-ingress-shim created
clusterrolebinding.rbac.authorization.k8s.io/cert-manager-controller-approve:cert-manager-io created
clusterrolebinding.rbac.authorization.k8s.io/cert-manager-webhook:subjectaccessreviews created
role.rbac.authorization.k8s.io/cert-manager-cainjector:leaderelection created
role.rbac.authorization.k8s.io/cert-manager:leaderelection created
role.rbac.authorization.k8s.io/cert-manager-webhook:dynamic-serving created
rolebinding.rbac.authorization.k8s.io/cert-manager-cainjector:leaderelection created
```

查看 cert-manager 启动的相关 Pod 信息，以此来验证 cert-manager 是否已经安装成功，命令如下：

```
kubectl get pods --namespace cert-manager
```

运行命令后输出的信息如下：

```
NAME                                       READY   STATUS    RESTARTS   AGE
cert-manager-7dd5854bb4-cpd28              1/1     Running   0          5m
cert-manager-cainjector-64c949654c-fsxtd   1/1     Running   0          5m
cert-manager-webhook-6bdffc7c9d-wr6qc      1/1     Running   0          5m
```

在 cert-manager 中负责签发证书的是 Issuer。cert-manager 支持多种类型的 Issuer，包括 SelfSigned、CA、Vault、Venafi、External 和 ACME。

其中 ACME(Automated Certificate Management Environment)类型的证书一般是免费的。

8.8.2 ACME

Let's Encrypt 是一个 ACME 服务，下面使用 Let's Encrypt 提供证书，生产环境的 Let's Encrypt 有严格的请求速率限制，请求过于频繁会导致后续请求失败，为了避免超过请求速率，可以在测试环境或者开发环境中使用测试环境的 Let's Encrypt，在测试通过后切换到生产环境的 Let's Encrypt。

新建文件 issuer-letsencrypt-staging.yaml，用于创建使用测试环境 Let's Encrypt 的

Issuer。文件内容如下：

```
apiVersion: cert-manager.io/v1
kind: Issuer
metadata:
 name: letsencrypt-staging
spec:
 acme:
   # The ACME server URL
   server: https://acme-staging-v02.api.letsencrypt.org/directory
   # Email address used for ACME registration
   email: fishis@163.com
   # Name of a secret used to store the ACME account private key
   privateKeySecretRef:
     name: letsencrypt-staging
   # Enable the HTTP-01 challenge provider
   solvers:
   - http01:
       ingress:
         class:  nginx
```

使用此文件创建 Issuer，命令如下：

```
kubectl apply -f issuer-letsencrypt-staging.yaml
```

运行命令后输出的信息如下：

```
issuer.cert-manager.io/letsencrypt-staging created
```

新建文件 issuer-letsencrypt-prod.yaml，用于创建并使用生产环境 Let's Encrypt 的 Issuer。文件内容如下：

```
apiVersion: cert-manager.io/v1
kind: Issuer
metadata:
 name: letsencrypt-prod
spec:
 acme:
   # The ACME server URL
   server: https://acme-v02.api.letsencrypt.org/directory
   # Email address used for ACME registration
   email: fishis@163.com
   # Name of a secret used to store the ACME account private key
   privateKeySecretRef:
     name: letsencrypt-prod
```

```
    # Enable the HTTP-01 challenge provider
    solvers:
    - http01:
        ingress:
          class: nginx
```

使用此文件创建 Issuer，命令如下：

```
kubectl apply -f  issuer-letsencrypt-prod.yaml
```

运行命令后输出的信息如下：

```
issuer.cert-manager.io/letsencrypt-prod created
```

查看 Issuer 信息，命令如下：

```
kubectl get issuer
```

运行命令后输出的信息如下：

```
NAME                   READY   AGE
letsencrypt-prod        True    6m34s
letsencrypt-staging    True    6m34s
```

8.8.3 Ingress TLS

修改 ingress.yaml 文件，开启 TLS，修改后的 ingress.yaml 文件的内容如下：

```
apiVersion: networking.k8s.io/v1
kind: Ingress
metadata:
  name: app-ingress
  annotations:
    nginx.ingress.kubernetes.io/rewrite-target: /
    kubernetes.io/ingress.class: "nginx"
    cert-manager.io/issuer: "letsencrypt-prod"
spec:
  tls:
    - hosts:
        - unit.ink
        - api.unit.ink
      secretName: unit-tls
  defaultBackend:
    service:
```

```
          name: web
          port:
            number: 80
  rules:
    - host: api.unit.ink
      http:
        paths:
          - path: /
            pathType: Prefix
            backend:
              service:
                name: web
                port:
                  number: 80
    - host: unit.ink
      http:
        paths:
          - path: /
            pathType: Prefix
            backend:
              service:
                name: api
                port:
                  number: 80
```

文件使用的是生产环境 Let's Encrypt 的服务器地址，如果在测试阶段使用测试环境的 Let's Encrypt，则需要修改注解，需要修改的注释如下：

```
cert-manager.io/issuer: "letsencrypt-prod"
```

修改后的注释如下：

```
cert-manager.io/issuer: "letsencrypt-staging"
```

重新应用文件使修改生效，命令如下：

```
kubectl apply -f k8s/ingress.yaml
```

查看 certificate 信息，命令如下：

```
kubectl get certificate
```

运行命令后输出的信息如下：

```
NAME       READY   SECRET     AGE
unit-tls   True    unit-tls   48m
```

输出信息显示生成了 unit-tls 证书，并且保存在 unit-tls 的 Secret 对象中。

在生成证书过程中，cert-manager 还会生成其他对象。查看 cert-manager 生成的 Order 信息，命令如下：

```
kubectl get order
```

运行命令后输出的信息如下：

```
NAME                              STATE   AGE
unit-tls-fgnjm-3117089320   valid   17m
```

查看 cert-manager 生成的 CertificateRequest 信息，命令如下：

```
kubectl get certificaterequest
```

运行命令后输出的信息如下(前 5 列)：

```
NAME              APPROVED   DENIED   READY   ISSUER
unit-tls-fgnjm    True                True
```

后两列信息如下：

```
REQUESTOR                                                          AGE
letsencrypt-prod   system:serviceaccount:cert-manager:cert-manager   4m
```

查看名为 unit-tls 的 Secret 信息，命令如下：

```
kubectl get secret unit-tls
```

运行命令后输出的信息如下：

```
NAME       TYPE                DATA   AGE
unit-tls   kubernetes.io/tls   2      49m
```

查看名为 unit-tls 的 Secret 详细信息，命令如下：

```
kubectl describe secret unit-tls
```

运行命令后输出的信息如下：

```
Name:          unit-tls
Namespace:     default
Labels:        <none>
Annotations:   cert-manager.io/alt-names: unit.ink
                cert-manager.io/certificate-name: unit-tls
                cert-manager.io/common-name: unit.ink
                cert-manager.io/ip-sans:
                cert-manager.io/issuer-group: cert-manager.io
                cert-manager.io/issuer-kind: Issuer
                cert-manager.io/issuer-name: letsencrypt-prod
                cert-manager.io/uri-sans:

Type:  kubernetes.io/tls

Data
====
tls.key:  1675 Bytes
tls.crt:  5575 Bytes
```

如果需要查看 tls.key 和 tls.crt 的具体值，则可以使用的命令如下：

```
kubectl get secret unit-tls -o yaml
```

打开浏览器，访问 https://unit.ink。在网址栏左边会看到小锁图标，表示当前使用 HTTPS 协议访问，如图 8-3 所示。

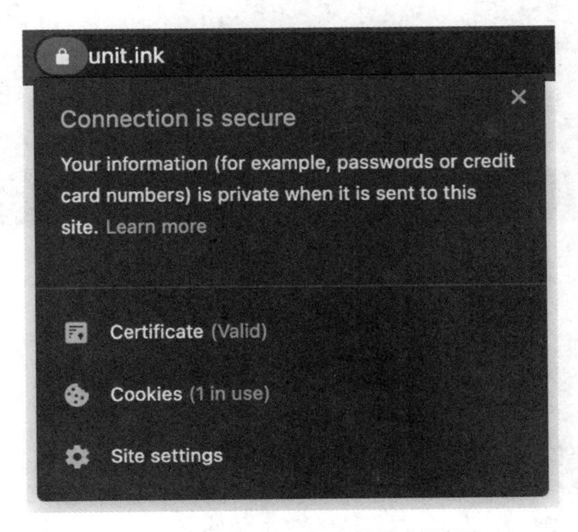

图 8-3　域名 unit.ink 的访问页面

单击小锁图标，在弹出的窗口中单击 Certificate。可以查看证书信息，如图 8-4 所示。

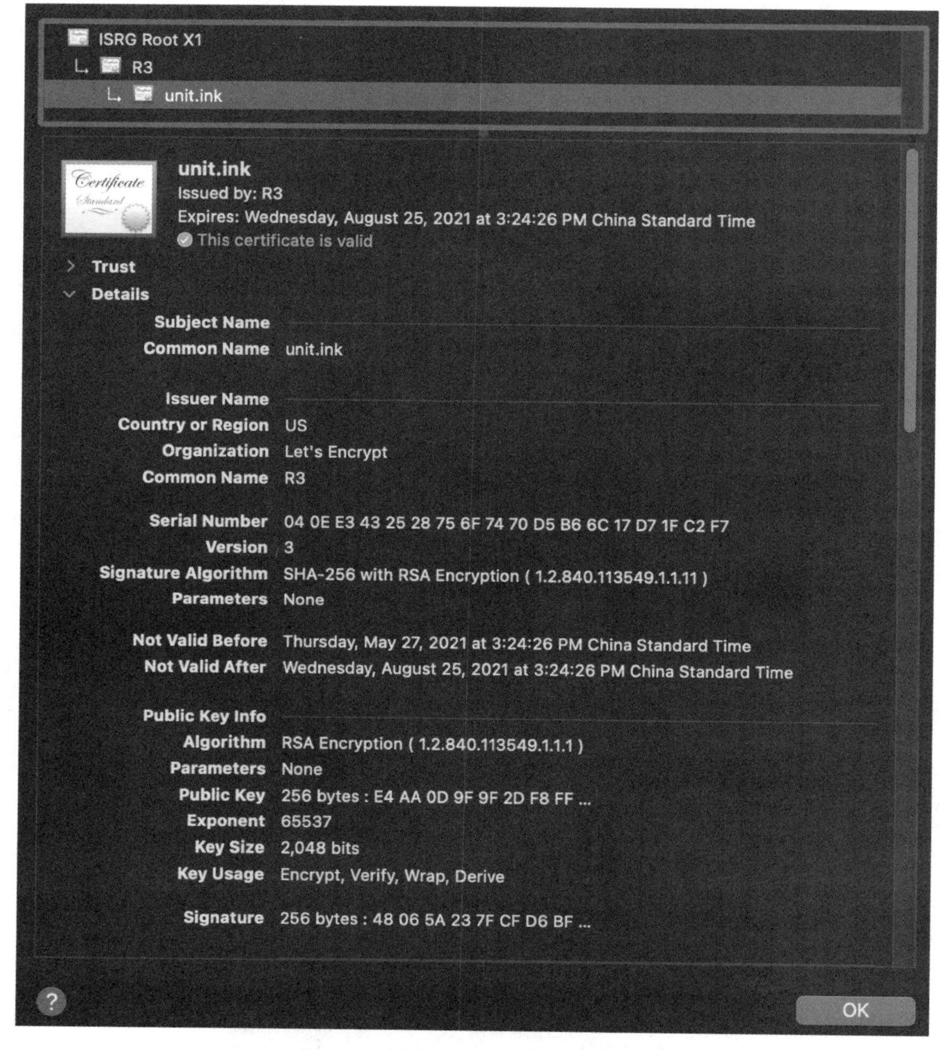

图 8-4　域名 unit. ink 的证书信息

8.9　日志

8.9.1　方案简介

项目使用 EFK 技术栈作为日志解决方案。使用开源项目 Fluentd 收集日志，使用 Elasticsearch 存储和索引日志。使用 Kibana 提供日志查询的 Web 界面。

Fluentd 在每个节点上都运行一个实例，用来收集日志，使用 DaemonSet 来部署。

Fluentd 需要把日志数据转发到 Elasticsearch,所以需要先启动 Elasticsearch。如果先启动 Fluentd 后启动 Elasticsearch,则 Fluentd 会在启动后的一段时间无法连接到 Elasticsearch。在 Fluentd Pod 日志中将会看到类似下面的警告信息:

```
2021-06-02 10:00:17 +0000 [warn]: #0 [out_es] Could not communicate to Elasticsearch, resetting
connection and trying again. no address for elasticsearch.default (Resolv::ResolvError)
```

8.9.2 ElasticSearch

在项目根目录下的 k8s 文件夹中新建文件 elasticsearch.yaml,编辑内容如下:

```
apiVersion: v1
kind: Service
metadata:
  name: elasticsearch
  labels:
    app: elasticsearch
spec:
  ports:
    - port: 9200
  selector:
    app: elasticsearch
    tier: logging
  type: ClusterIP
---
apiVersion: apps/v1 #for versions before 1.9.0 use apps/v1beta2
kind: Deployment
metadata:
  name: elasticsearch
  labels:
    app: elasticsearch
spec:
  selector:
    matchLabels:
      app: elasticsearch
      tier: logging
  strategy:
    type: Recreate
  template:
    metadata:
      labels:
        app: elasticsearch
        tier: logging
    spec:
      containers:
```

```
          - image: docker.elastic.co/elasticsearch/elasticsearch:7.10.2
            name: elasticsearch
            ports:
              - containerPort: 9200
            env:
              - name: discovery.type
                value: single-node
```

文件包含两个 YAML,分别定义了一个 Service 和一个 Deployment。

第一部分定义了 Service,命名为 elasticsearch,用于暴露 TCP 9200 端口。Service 类型为 ClusterIP。

第二部分定义了 Deployment。命名为 elasticsearch。由于没有指定 replicas 字段,所以默认只运行 1 个 Pod。使用的容器镜像为 docker.elastic.co/elasticsearch/elasticsearch:7.10.2。容器用于暴露 TCP 9200 端口。

使用此文件创建 Service,命令如下:

```
kubectl apply -f elasticsearch.yaml
```

运行命令后输出的信息如下:

```
service/elasticsearch created
deployment.apps/elasticsearch created
```

创建好 ElasticSearch 后,接着创建 Fluentd。

8.9.3 Fluentd

Fluentd 是一个开源的日志收集、处理和转发工具。在 Kubernetes 中使用 Fluentd 之前需要为 Fluentd 创建一个 Service Account,以便获取访问某些资源的权限。

创建文件 fluentd-rbac.yaml,以便定义 Service Account 和需要与 Service Account 绑定的权限,编辑内容如下:

```
apiVersion: v1
kind: ServiceAccount
metadata:
  labels:
    k8s-app: fluentd
  name: fluentd
  namespace: kube-system

---
kind: ClusterRole
apiVersion: rbac.authorization.k8s.io/v1
```

```yaml
metadata:
  name: fluentd-clusterrole
rules:
  - apiGroups:
      - ""
    resources:
      - "namespaces"
      - "pods"
    verbs:
      - "list"
      - "get"
      - "watch"

---
kind: ClusterRoleBinding
apiVersion: rbac.authorization.k8s.io/v1
metadata:
  name: fluentd-clusterrole
roleRef:
  apiGroup: rbac.authorization.k8s.io
  kind: ClusterRole
  name: fluentd-clusterrole
subjects:
  - kind: ServiceAccount
    name: fluentd
    namespace: kube-system
```

使用此文件创建与权限相关的对象，命令如下：

```
kubectl apply -f fluentd-rbac.yaml
```

运行命令后输出的信息如下：

```
serviceaccount/fluentd created
clusterrole.rbac.authorization.k8s.io/fluentd-clusterrole created
clusterrolebinding.rbac.authorization.k8s.io/fluentd-clusterrole created
```

接着在项目根目录下的 k8s 文件夹中新建文件 fluentd-daemonset-elasticsearch.yaml，编辑内容如下：

```yaml
apiVersion: apps/v1
kind: DaemonSet
metadata:
  name: fluentd
  namespace: kube-system
```

```yaml
  labels:
    k8s-app: fluentd-logging
    version: v1
spec:
  selector:
    matchLabels:
      k8s-app: fluentd-logging
      version: v1
  template:
    metadata:
      labels:
        k8s-app: fluentd-logging
        version: v1
    spec:
      serviceAccountName: fluentd
      tolerations:
      - key: node-role.kubernetes.io/master
        effect: NoSchedule
      containers:
      - name: fluentd
        image: fluent/fluentd-kubernetes-daemonset:v1-debian-elasticsearch
        env:
          - name:  FLUENT_ELASTICSEARCH_HOST
            value: "elasticsearch.default"
          - name:  FLUENT_ELASTICSEARCH_PORT
            value: "9200"
          - name: FLUENT_ELASTICSEARCH_SCHEME
            value: "http"
          # Option to configure elasticsearch plugin with self signed certs
          # ================================================================
          - name: FLUENT_ELASTICSEARCH_SSL_VERIFY
            value: "true"
          # Option to configure elasticsearch plugin with tls
          # ================================================================
          - name: FLUENT_ELASTICSEARCH_SSL_VERSION
            value: "TLSv1_2"
          # X-Pack Authentication
          # =====================
          - name: FLUENT_ELASTICSEARCH_USER
            value: "elastic"
          - name: FLUENT_ELASTICSEARCH_PASSWORD
            value: "changeme"
        resources:
          limits:
            memory: 200Mi
          requests:
```

```
          cpu: 100m
          memory: 200Mi
        volumeMounts:
        - name: varlog
          mountPath: /var/log
        - name: varlibdockercontainers
          mountPath: /var/lib/docker/containers
          readOnly: true
      terminationGracePeriodSeconds: 30
      volumes:
      - name: varlog
        hostPath:
          path: /var/log
      - name: varlibdockercontainers
        hostPath:
          path: /var/lib/docker/containers
```

文件定义了一个 DaemonSet，命名为 fluentd，DaemonSet 中的 Pod 使用的镜像是 fluent/fluentd-kubernetes-daemonset：v1-debian-elasticsearch，这个镜像对将日志输出到 elasticsearch 进行了适配。

```
serviceAccountName: fluentd
```

这里指定 Pod 使用 fluentd 这个 Service Account。这样 Pod 就有权限从 Kubernetes API 读取需要的资源信息了。

文件中 env 字段设置了几个环境变量，用于配置 Fluentd，内容如下：

```
env:
  - name:  FLUENT_ELASTICSEARCH_HOST
    value: "elasticsearch.default"
```

环境变量 FLUENT_ELASTICSEARCH_HOST 用于设置 ElasticSearch 的主机名。主机名为 ElasticSearch 服务在 Kubernetes 集群中的 DNS 域名，即 elasticsearch.default，由于上面创建的 ElasticSearch 服务在 default 命名空间，而这个文件中定义的 DaemonSet 命名空间是 kube-system，与 default 在不同的命名空间，kube-system 命名空间中的 Pod 访问 default 命名空间的 Service 需要指定命名空间，不加命名空间无法访问 ElasticSearch 服务，所以使用 elasticsearch.default 而不是 elasticsearch。

```
  - name:  FLUENT_ELASTICSEARCH_PORT
    value: "9200"
```

环境变量 FLUENT_ELASTICSEARCH_PORT 将 ElasticSearch 服务的端口号设置

为 9200。

```
- name: FLUENT_ELASTICSEARCH_SCHEME
  value: "http"
```

环境变量 FLUENT_ELASTICSEARCH_SCHEME 将连接 ElasticSearch 服务使用的协议设置为 HTTP。

使用此文件创建 DaemonSet,命令如下:

```
kubectl apply -f fluentd-daemonset-elasticsearch.yaml
```

运行命令后输出的信息如下:

```
daemonset.apps/fluentd created
```

查看 DaemonSet 创建的 Pod 信息,验证 Fluentd 是否启动正常。命令如下:

```
kubectl get pod -n kube-system -l k8s-app=fluentd-logging
```

运行命令后输出的信息如下:

```
NAME             READY   STATUS     RESTARTS   AGE
fluentd-9blcj    1/1     Running    0          31s
fluentd-fg2bx    1/1     Running    0          31s
```

因为集群中有 2 个工作节点,所以这个 DaemonSet 创建了 2 个 Pod。分别运行在不同的节点上。

8.9.4 Kibana

创建文件 kibana.yaml,内容如下:

```
apiVersion: v1
kind: Service
metadata:
  name: kibana
  labels:
    app: kibana
spec:
  ports:
    - port: 5601
  selector:
    app: kibana
    tier: logging
  type: NodePort
```

```
---
apiVersion: apps/v1 # for versions before 1.9.0 use apps/v1beta2
kind: Deployment
metadata:
  name: kibana
  labels:
    app: kibana
spec:
  selector:
    matchLabels:
      app: kibana
      tier: logging
  strategy:
    type: Recreate
  template:
    metadata:
      labels:
        app: kibana
        tier: logging
    spec:
      containers:
        - image: kibana:7.10.1
          name: kibana
          ports:
            - containerPort: 5601
              name: api
```

使用此文件创建 Service,命令如下:

```
kubectl apply -f kibana.yaml
```

运行命令后输出的信息如下:

```
service/kibana created
deployment.apps/kibana created
```

配置 Ingress 规则,通过 Ingress 访问 Kibana。

测试环境下使用 kibana.app.com 域名访问 Kibana,编辑 ingress.yaml 文件,以便添加一条规则,代码如下:

```
- host: kibana.app.com
  http:
    paths:
      - path: /
        pathType: Prefix
```

```
          backend:
            service:
              name: kibana
              port:
                number: 5601
```

应用修改后的 ingress. yaml 文件，使其生效，命令如下：

```
kubectl apply -f ingress.yaml
```

接下来设置 dnsmasq 解析，把 kibana. com 域名解析到 Ingress 控制器关联的负载均衡器的地址。在配置文件 dnsmasq. conf 中添加一条记录，记录如下：

```
cname=kibana.app.com,a7d86ad7638c849e8ae8e5d9ab81d054-4c4c3011147415e7.elb.ap-east-1.amazonaws.com
```

重启 dnsmasq 容器，命令如下：

```
docker restart dnsmasq
```

打开浏览器访问 kibana. app. com，会看到 Kibana 的默认首页。

单击左边栏中的菜单 Stack Management，接着单击 Index patterns 菜单，会看到如图 8-5 所示的页面。

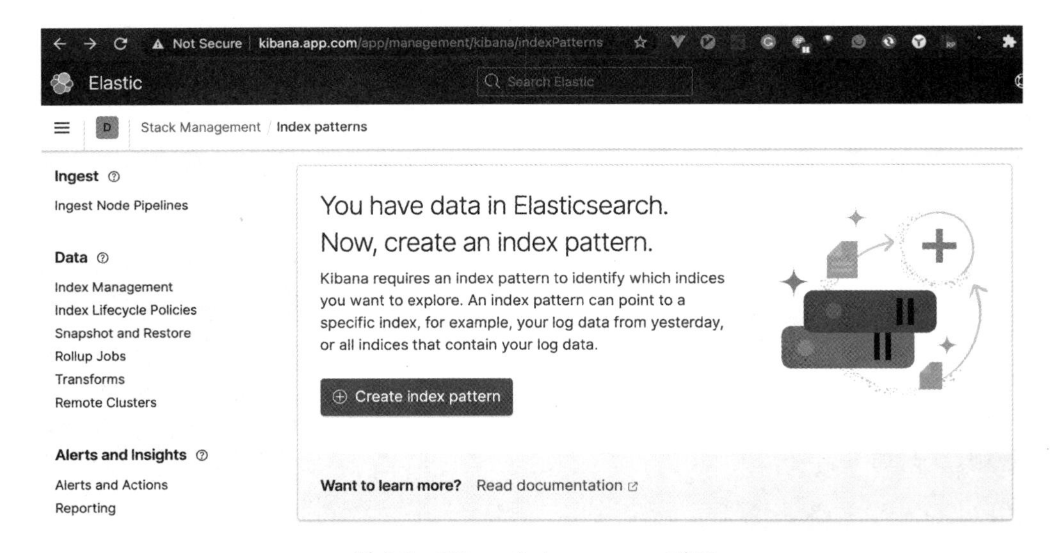

图 8-5　Kibana Index patterns 页面

单击 Create index pattern 按钮，会跳转到如图 8-6 所示的页面。

在输入框中输入 logstash*，单击右侧按钮 Next step，会跳转到如图 8-7 所示的页面。

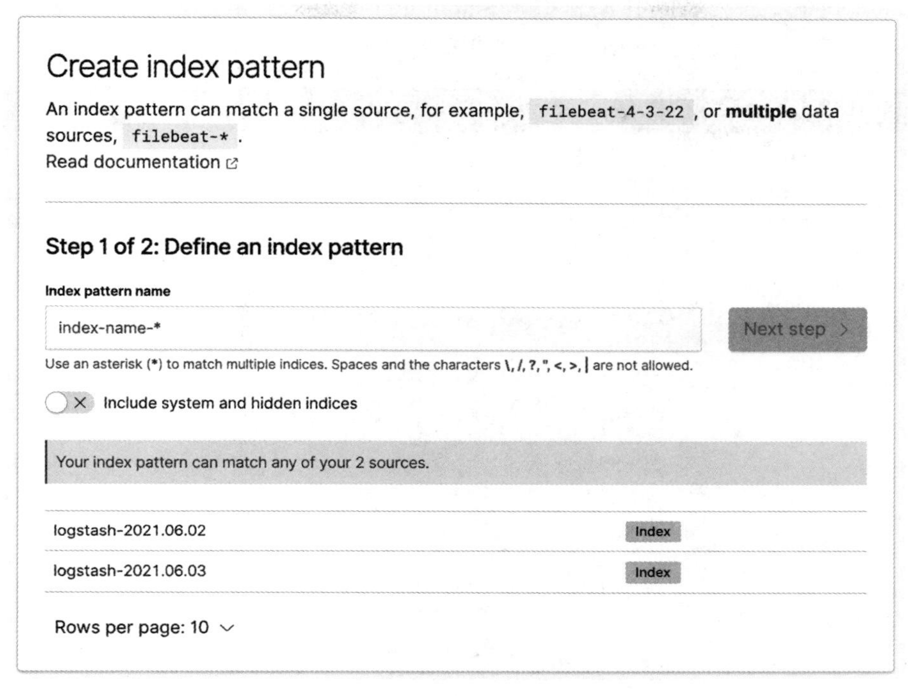

图 8-6　Kibana Create index pattern 页面(1)

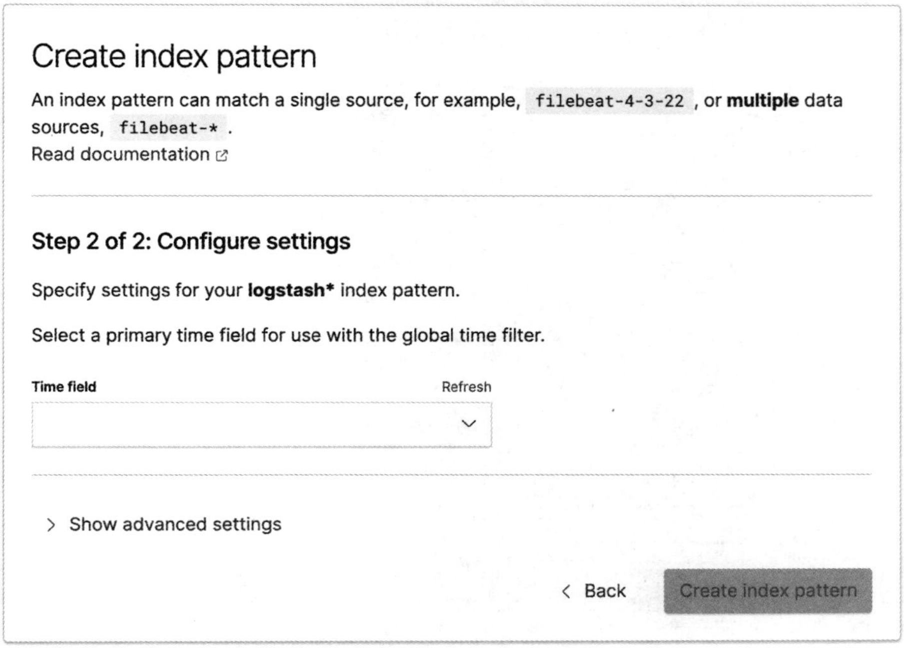

Create index pattern

An index pattern can match a single source, for example, `filebeat-4-3-22` , or **multiple** data sources, `filebeat-*` .

Read documentation

Step 2 of 2: Configure settings

Specify settings for your **logstash*** index pattern.

Select a primary time field for use with the global time filter.

Time field　Refresh

Show advanced settings

Back　Create index pattern

图 8-7　Kibana Create index pattern 页面(2)

在 Time field 下拉列表框中选择@timestamp，接着单击右下角 Create index pattern 按钮。在左边栏菜单中单击 Discover，进入如图 8-8 所示的页面。

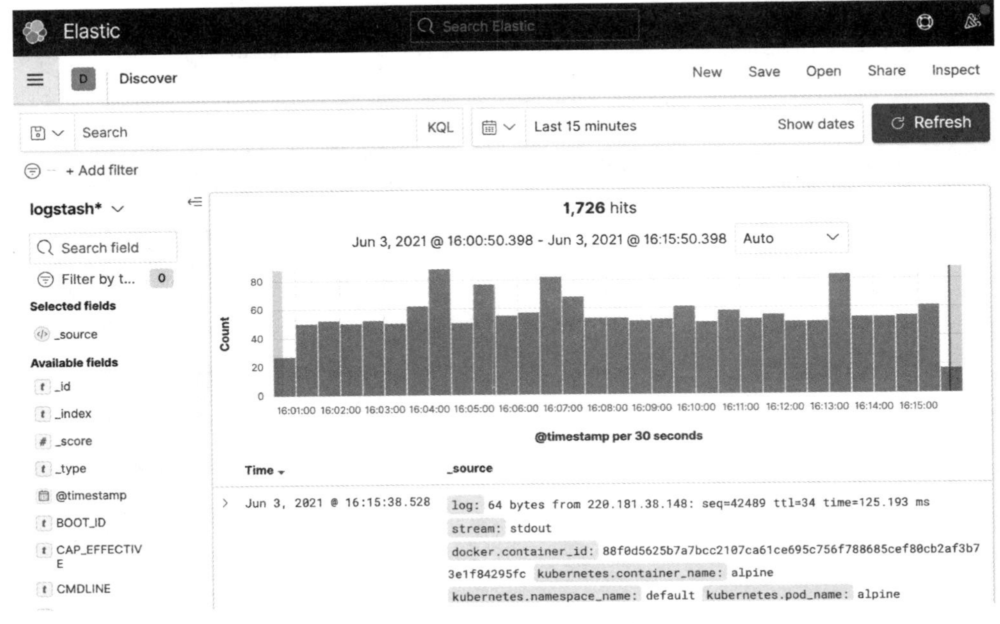

图 8-8　Kibana Discover 页面

为了测试应用日志是否可以被正常收集并保存到 ElasticSearch，可以发送一个请求到后端服务，命令如下：

```
curl api.app.com/api/version
```

现在回到 Kibana 页面查询日志，在搜索框输入"/api/version"，按回车键，会看到日志记录如图 8-9 所示。

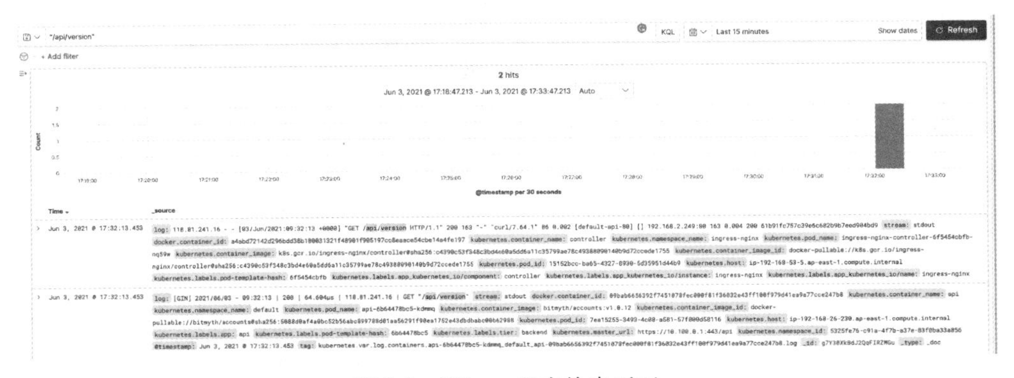

图 8-9　Kibana 日志检索页面

第 1 条日志的 kubernetes.pod_name 的字段值为 ingress-nginx-controller-6f5454cbfb-nq59w，字段 kubernetes.container_image 的值为 k8s.gcr.io/ingress-nginx/controller@sha256：c4390c53f348c3bd4e60a5dd6a11c35799ae78c49388090140b9d72ccede1755。说明这条日志是从 Ingress 控制器所在的容器中产生的。

第 2 条日志的 kubernetes.pod_name 的字段值为 api-6b64478bc5-kdmmq。字段 kubernetes.container_image 的值为 bitmyth/accounts：v1.0.12。说明这条日志是从 api 服务所在的容器中产生的。

说明日志的收集和存储都运行正常。

8.10 Kustomize

新建 kustomization.yaml 文件，内容如下：

```
configMapGenerator:
  - name: plain-config
    files:
      - config/plain.yaml
      - config/secret.yaml
  - name: mysql-config
    literals:
      - db=accounts
secretGenerator:
  - name: mysql-pass
    literals:
      - password=123
  - name: secret-config
    files:
      - config/secret.yaml
resources:
  - k8s/mysql-deployment.yaml
  - k8s/mysql-pvc.yaml
  - k8s/mysql-service.yaml
  - k8s/migration.yaml
  - k8s/api-deployment.yaml
  - k8s/api-service.yaml
  - k8s/web-deployment.yaml
  - k8s/web-service.yaml
  - k8s/ingress.yaml
  - k8s/ingress.yaml
  - k8s/ingress-nginx-controller-aws.yaml
  - k8s/elasticsearch.yaml
  - k8s/fluentd-daemonset-elasticsearch.yaml
  - k8s/fluentd-rbac.yaml
  - k8s/kibana.yaml
```

使用 kustomization. yaml 文件创建所有资源，命令如下：

```
kubectl apply -k .
```

使用 kustomization. yaml 文件删除所有资源，命令如下：

```
kubectl delete -k .
```

第9章 Helm

Helm 是一个开源的 Kubernetes 包管理器，使用 Helm 可以便捷地分享和使用运行在 Kubernetes 中的软件。

Helm 在 2015 年面世，在 2018 年成为 CNCF 项目。Helm 客户端和库使用 Go 语言编写，使用 Kubernetes 客户端库与 Kubernetes 通信。Helm 使用 Kubernetes 中的 Secret 存储信息，并没有使用独立的数据库。

2020 年 CNCF 在中国进行的第 4 次云原生调研结论表明 Helm 是最流行的 Kubernetes 应用打包工具，约 64%的 Kubernetes 用户使用 Helm。

9.1 安装 Helm

使用 Helm 前需要准备一个可用的 Kubernetes 集群环境，然后安装和配置 Helm。

Heml 项目的 GitHub release 页面提供了常见操作系统下安装 Helm 的二进制文件。根据需要下载相应的版本后，解压文件(例如 tar -zxvf helm-v3.0.0-Linux-amd64.tar.gz)，然后把解压后得到的文件夹中的 helm 可执行文件移动到系统的 PATH 目录下(例如 mv linux-amd64/helm /usr/local/bin/helm)。

除了使用二进制文件安装 Helm，Helm 还提供了脚本安装方式。

下面以 Ubuntu 操作系统为例，使用脚本安装 Helm。

下载并安装脚本，命令如下：

```
curl - fsSL - o get_helm.sh \
https://raw.bithubusercontent.com/helm/helm/master/scripts/get - helm - 3
```

这样会将脚本下载到本地的 get_helm.sh 文件。

改变脚本文件的权限，命令如下：

```
chmod 700 get_helm.sh
```

运行脚本，命令如下：

```
./get_helm.sh
```

运行命令后输出的信息如下：

```
Downloading https://get.helm.sh/helm-v3.6.1-Linux-amd64.tar.gz
Verifying checksum... Done.
Preparing to install helm into /usr/local/bin
helm installed into /usr/local/bin/helm
```

输出信息提示安装完成。

除此之外，还可以使用包管理器安装 Helm。在 macOS 上安装 Helm，命令如下：

```
brew install helm
```

在 Windows 上安装 Helm，命令如下：

```
choco install kubernetes-helm
```

在 Ubuntu/Debian 上安装 Helm，命令如下：

```
curl https://baltocdn.com/helm/signing.asc | sudo apt-key add -
sudo apt-get install apt-transport-https --yes
echo "deb https://baltocdn.com/helm/stable/debian/ all main" | sudo tee /etc/apt/sources.
list.d/helm-stable-debian.list
sudo apt-get update
sudo apt-get install helm
```

在 FreeBSD 上安装 Helm，命令如下：

```
pkg install helm
```

使用 snap 安装 Helm，命令如下：

```
sudo snap install helm --classic
```

还可以下载 Helm 项目源码自行编译安装，命令如下：

```
git clone https://github.com/helm/helm.git
cd helm
make
```

安装后，可以使用 helm help 命令验证安装是否成功。

运行 helm help 命令后输出的类似信息如下（只截取部分输出）：

```
The Kubernetes package manager

Common actions for Helm:

 - helm search:     search for charts
 - helm pull:       download a chart to your local directory to view
 - helm install:   upload the chart to Kubernetes
 - helm list:       list releases of charts
Environment variables:
...
Usage:
  helm [command]

Available Commands:
  completion   generate auto
...
```

输出表示 Helm 安装成功了。通常情况下，使用 Helm 项目预编译的二进制文件安装 Helm 是最简单的方式，可以满足绝大部分 Helm 的使用需求。

查看 Helm 版本，命令如下：

```
helm version
```

运行命令后输出的信息如下：

```
version.BuildInfo{Version: "v3.5.3", GitCommit: "041ce5a2c17a58be0fcd5f5e16fb3e7e95fea622",
GitTreeState: "dirty", GoVersion: "go1.16"}
```

输出显示 Helm 版本号是 v3.5.3。后续的演示过程都使用这个 Helm 版本。

9.2 Helm Chart

9.2.1 Chart 简介

Chart 是 Helm 打包资源的格式，Chart 中包含了描述一组相关的 Kubernetes 资源的文件。Chart 可以部署一个简单的 Pod，也可以部署一个完整的应用，例如一个包含了 HTTP 服务器和数据库及缓存的 Web 应用。

Chart 是一个包含了一组文件的文件夹，文件夹的名称就是 Chart 的名称。例如 WordPress 的 Chart 就是一个名为 wordpress 的文件夹。文件夹结构如下：

```
wordpress/
  Chart.yaml                          # 关于这个 YAML 的信息
```

```
LICENSE                        # 可选,包含许可信息
README.md                      # 可选
values.yaml                    # Chart 的默认配置文件
values.schema.json             # 可选,包含配置的结构信息
charts/                        # 包含 Chart 依赖
crds/                          # Custom Resource Definitions
templates/                     # 用于生成 Kubernetes 配置文件的模板
templates/NOTES.txt
```

其中 charts、crds 和 templates 目录是 Helm 保留的目录。

9.2.2 安装 Chart

安装 Chart 前需要添加 Chart 仓库,可以在 Artifact Hub 中查找有效的 Chart 仓库。下面以仓库 https://charts.bitnami.com/bitnami 为例。

安装 Chart 仓库 https://charts.bitnami.com/bitnami,命令如下:

```
helm repo add bitnami https://charts.bitnami.com/bitnami
```

运行命令后输出的信息如下:

```
"bitnami" has been added to your repositories
```

查看 repo 中包含的 Chart,命令如下:

```
helm search repo bitnami
```

运行命令后输出的信息如下:

```
NAME                                        CHART VERSION   APP VERSION
    DESCRIPTION
bitnami/bitnami-common                      0.0.9           0.0.9
    DEPRECATED Chart with custom templates used in ...
bitnami/airflow                             10.2.0          2.1.0
    Apache Airflow is a platform to programmaticall...
bitnami/apache                              8.5.5           2.4.48
...
```

获取最新的 Helm Chart 库,命令如下:

```
helm repo update
```

运行命令后输出的信息如下:

```
Hang tight while we grab the latest from your chart repositories...
...Successfully got an update from the "bitnami" chart repository
Update Complete.  Happy Helming!
```

输出信息显示仓库更新成功。

假设需要安装的程序是 MySQL，首先在仓库中搜索相关 Chart，命令如下：

```
helm search repo mysql
```

运行命令后输出的信息如下：

```
NAME            CHART VERSIONAPP VERSIONDESCRIPTION
bitnami/mysql 8.7.0 8.0.25 Chart to create a Highly available MySQL cluster
...
```

接下来通过 Chart(bitnami/mysql)安装 MySQL，命令如下：

```
helm install db bitnami/mysql
```

运行命令后输出的信息如下：

```
NAME: db
LAST DEPLOYED: Wed Jul  7 16:05:50 2021
NAMESPACE: default
STATUS: deployed
REVISION: 1
TEST SUITE: None
NOTES:
** Please be patient while the chart is being deployed **

Tip:

  Watch the deployment status using the command: kubectl get pods -w --namespace default

Services:

  echo Primary: db-mysql.default.svc.cluster.local:3306

Administrator credentials:

  echo Username: root
  echo Password : $(kubectl get secret --namespace default db-mysql -o jsonpath="{.data.
mysql-root-password}" | base64 --decode)

To connect to your database:
```

```
  1. Run a pod that you can use as a client:

      kubectl run db-mysql-client --rm --tty -i --restart='Never' --image  docker.
io/bitnami/mysql:8.0.25-debian-10-r37 --namespace default --command -- bash

  2. To connect to primary service (read/write):

      mysql -h db-mysql.default.svc.cluster.local -uroot -p my_database

To upgrade this helm chart:

  1. Obtain the password as described on the 'Administrator credentials' section and set the
'root.password' parameter as shown below:

      ROOT_PASSWORD=$(kubectl get secret --namespace default db-mysql -o jsonpath=
"{.data.mysql-root-password}" | base64 --decode)
      helm upgrade --namespace default db bitnami/mysql --set auth.rootPassword=$ROOT
_PASSWORD
```

输出信息提示 Chart 状态为 deployed(已部署),此时相关的 Kubernetes 资源会开始创建。输出信息中还提供了有用的信息,包括后续的配置步骤等。

查看 Chart 生成的 Pod 信息,命令如下:

```
kubectl get pod
```

运行命令后输出的信息如下:

```
NAME          READY   STATUS    RESTARTS   AGE
db-mysql-0    1/1     Running   0          1m35s
```

输出信息显示部署的 Pod 已经处于运行状态。

使用 Helm 安装 Kubernetes 资源的顺序如下:

```
NameSpace
NetworkPolicy
ResourceQuota
LimitRange
PodSecurityPolicy
PodDisruptionBudget
ServiceAccount
Secret
SecretList
ConfigMap
StorageClass
```

```
PersistentVolume
PersistentVolumeClaim
CustomResourceDefinition
ClusterRole
ClusterRoleList
ClusterRoleBinding
ClusterRoleBindingList
Role
RoleList
RoleBinding
RoleBindingList
Service
DaemonSet
Pod
ReplicationController
ReplicaSet
Deployment
HorizontalPodAutoscaler
StatefulSet
Job
CronJob
Ingress
APIService
```

Helm 不会等到所有资源全部创建完成才退出。因为安装过程包含一些耗时操作(例如下载镜像),所以有可能持续较长时间。

追踪发布状态,可以使用 helm status 命令查询。

除了从仓库安装 Chart,Helm 还支持从其他来源安装 Chart。

从 Chart archive 安装 Chart,命令如下:

```
helm install foo foo-0.1.1.tgz
```

从 URL 安装 Chart,命令如下:

```
helm install foo https://example.com/charts/foo-1.2.3.tgz
```

从 archive 解压缩后的 Chart 文件夹安装 Chart,命令如下:

```
helm install foo path/to/foo
```

9.2.3 定制 Chart

当使用上述命令安装 Chart 时会使用该 Chart 的默认配置,Helm 支持按需定制某些配置项。

以 MySQL 为例，查看 Chart 可配置的选项，命令如下：

```
helm show values bitnami/mysql
```

运行命令后输出的信息如下：

```
## Global Docker image parameters
## Please, note that this will override the image parameters, including dependencies, configured to
use the global value
## Current available global Docker image parameters: imageRegistry and imagePullSecrets
##
# global:
# imageRegistry: myRegistryName
# imagePullSecrets:
#    - myRegistryKeySecretName
# storageClass: myStorageClass

## Bitnami mysql image
## ref: https://hub.docker.com/r/bitnami/mysql/tags/
##
image:
  registry: docker.io
  repository: bitnami/mysql
  tag: 8.0.25-debian-10-r37
   ## Specify a imagePullPolicy
   ## Defaults to 'Always' if image tag is 'latest', else set to 'IfNotPresent'
   ## ref: http://kubernetes.io/docs/user-guide/images/#pre-pulling-images
...
```

输出信息中包含了此 Chart 支持的配置项，Helm 支持两种方式覆盖默认配置项。

一种是使用 YAML 文件来保存要定制的配置，然后在安装 Chart 时通过--values(或者-f)选项传入，以覆盖默认配置。如果指定多个--values 选项，最后的参数具有最高优先级；另一种是通过--set 选项直接设置覆盖配置项。

如果同时指定了--values 和--set，则--set 的值会与--values 的值合并，并且--set 的值具有更高的优先级。

--set 选项的值是 0 到多个键值对，简单的示例如下：

```
--set name=value
```

与之等价的 YAML 如下：

```
name: value
```

如果值是多个键值对，则它们之间用逗号分隔，示例代码如下：

```
--set a=b,c=d
```

与之等价的 YAML 如下：

```
a: b
c: d
```

此外--set 的值还支持更复杂的表达式，示例代码如下：

```
--set outer.inner=value
```

与之等价的 YAML 如下：

```
outer:
  inner: value
```

可以看到嵌套的数据结构使用--set 表达比较困难。

--set 的值还可以是列表，示例代码如下：

```
--set name={a,b,c}
```

与之等价的 YAML 如下：

```
name:
  - a
  - b
  - c
```

从 Helm 2.5.0 版本开始，支持使用数组索引的语法访问列表元素，示例代码如下：

```
--set servers[0].port=80
```

与之等价的 YAML 如下：

```
servers:
  - port: 80
```

如果--set 中的值包含特殊字符，则可以使用反斜杠\来转义，示例代码如下：

```
--set name=value1\,value2
```

与之等价的 YAML 如下：

```
name: "value1,value2"
```

9.2.4 Release

每次执行 helm install 的时候，Helm 都会创建一个新的 Release，即发布版本，同一个 Chart 可以被多次安装，每个发布版本都可以独立进行管理和升级。

Helm 提供了命令，用于查看已经安装的 Chart，命令如下：

```
helm ls
```

运行命令后输出的类似信息如下(前 4 列)：

```
NAME  NAMESPACE  REVISION  UPDATED
db    default    1          2021-07-07 16:05:50.95343 +0800 CST
```

后 3 列信息如下：

```
STATUS   CHART         APP VERSION
deployed  mysql-8.7.0   8.0.25
```

在上述例子中发布了 bitnami/mysql，并手动将发布名字指定为 db。如果希望 Helm 自动生成一个名字，则可以使用--generate-name 参数，命令如下：

```
helm install bitnami/mysql  --generate-name
```

运行命令后输出的类似信息如下：

```
NAME: mysql-1625648726
...
```

再次执行命令后输出的类似信息如下：

```
NAME: mysql-1625651443
...
```

可以看到每次发布相同的 Chart 都生成了不同的名字。

9.2.5 升级和回滚

发布新版本的 Chart 或者修改已发布版本的配置，可以使用 helm upgrade 命令，假设有一个名为 happy-panda 的发布，发布的 Chart 是 wordpress，现在需要对它进行升级，使用的值文件为 panda.yaml，升级命令如下：

```
helm upgrade -f panda.yaml happy-panda bitnami/wordpress
```

panda.yaml 文件的内容如下：

```
mariadb.auth.username: user1
```

升级后可以查看新的配置是否生效，命令如下：

```
helm get values happy-panda
```

如果已经生效，则可以看到的输出如下：

```
mariadb:
  auth:
    username: user1
```

如果发布过程出现了问题，则可以回滚这次发布，命令如下：

```
helm rollback happy-panda 1
```

其中 happy-panda 是发布的名称，1 是修订版本号。修订版本号是自增的，初始值是 1。每次运行 install、upgrade、rollback 都会使修订版本号加 1。使用 helm history 命令可以查看某个发布的修订历史。

Helm 支持用一行命令实现安装或者更新 Chart，也就是在 Chart 没有安装的情况下进行安装操作，在 Chart 已经安装的情况下进行升级操作，命令如下：

```
Helm upgrade --install <release name> --values <values file><chart directory>
```

9.2.6 卸载 Release

以上述发布的版本 db 为例，卸载 db 这个 Release，命令如下：

```
helm uninstall db
```

运行命令后输出的信息如下：

```
release "db" uninstalled
```

该命令会删除和 Chart 发布版本 db 相关的所有 Kubernete 对象及版本历史。如果在 helm uninstall 命令中指定--keep-history 选项，则 Helm 将保留版本历史。

查看此发布版本的信息，命令如下：

```
helm helm status db
```

运行命令后输出的信息如下：

```
Error: release: not found
```

假设删除 db 发布版本时指定了--keep-history 选项，那么查看版本信息会看到发布版本状态为 uninstalled。信息如下：

```
NAME: db
LAST DEPLOYED: Wed Jul  7 16:05:50 2021
NAMESPACE: default
STATUS: uninstalled
...
```

保留历史记录可以很方便地对发布的生命周期进行审查。

9.2.7 搜索 Chart

Helm 提供了强大的 Chart 搜索功能，根据仓库来源可以分为两类，一类是从 Artifact Hub 中包含的仓库搜索；另一类是从添加到 Helm 本地客户端的仓库搜索。

以 wordpress 应用程序为例，从 Artifact Hub 搜索 Chart，命令如下：

```
helm search hub wordpress
```

运行命令后输出的信息如下(前 4 列)：

```
URL CHART VERSION
https://artifacthub.io/packages/helm/bitnami/wo     1.1.2
```

后 2 列信息如下：

```
APPVERSION     DESCRIPTION
5.7.2          Web publishing platform for building blogs and ...
```

从本地搜索，命令如下：

```
helm search repo wordpress
```

运行命令后输出的信息如下：

```
NAME                  CHART VERSION   APP VERSION   DESCRIPTION
bitnami/wordpress     11.0.18         5.7.2         Web publishing platform for building blogs and ...
stable/wordpress      9.0.3           5.3.2         DEPRECATED Web publishing platform for building...
```

搜索功能支持模糊匹配，所以搜索 wordpress 的关键词可以是 word，命令如下：

```
helm search repo word
```

或者命令如下：

```
helm search repo press
```

9.3 Chart 模板

2min

9.3.1 模板示例

创建示例 Chart,命令如下：

```
helm create mychart
```

运行命令后输出的信息如下：

```
Creating mychart
```

命令会创建文件夹 mychart。为了简单起见,删除 mychart/templates 文件夹下的所有文件,命令如下：

```
rm -rf mychart/templates/*
```

下面开始创建第 1 个模板。

在生成的 mychart/templates 文件夹下创建文件 configmap.yaml,编辑内容如下：

```
apiVersion: v1
kind: ConfigMap
metadata:
  name: mychart-configmap
data:
  myvalue: "Hello World"
```

文件定义了一个 ConfigMap,包含了 ConfigMap 中最基础的字段。

这样就有了一个可安装的 Chart,安装这个 Chart,命令如下：

```
helm install hello-world ./mychart
```

命令中的 hello-world 参数的作用是命名此次发布,运行命令后输出的信息如下：

```
NAME: hello-world
LAST DEPLOYED: Sun Jul 11 10:30:49 2021
NAMESPACE: default
STATUS: deployed
```

```
REVISION: 1
TEST SUITE: None
```

查看生成的 ConfigMap，命令如下：

```
kubectl get configmap mychart-configmap
```

运行命令后输出的信息如下：

```
NAME                DATA   AGE
mychart-configmap   1      50s
```

安装 Chart 会生成一个发布版本，查看发布的内容，命令如下：

```
helm get manifest hello-world
```

命令中的 hello-word 参数即此次发布的名称。

运行命令后输出的信息如下：

```
---
# Source: mychart/templates/configmap.yaml
apiVersion: v1
kind: ConfigMap
metadata:
  name: mychart-configmap
data:
  myvalue: "Hello World"
```

运行命令后输出此次发布中 Helm 实际加载的模板文件，Helm 会把这些文件上传到 Kubernetes。每个文件开头都是---，用来标记 YAML 文件的开始。接着是一行生成的注释，说明此文件是基于哪个模板文件生成的。

卸载这次发布，命令如下：

```
helm uninstall hello-world
```

运行命令后输出的信息如下：

```
release "hello-world" uninstalled
```

9.3.2 模板调用

到目前为止上述 ConfigMap 模板中没有使用模板调用，下面使用模板调用功能动态生成 ConfigMap 的 name 字段的值，代码如下：

```
apiVersion: v1
kind: ConfigMap
metadata:
  name: {{ .Release.Name }}-configmap
data:
  myvalue: "Hello World"
```

双花括号中的内容是模板指令。.Release.Name 即发布的名称。Release 之前的句点表示最顶层的命名空间。.Release.Name 即在顶层命名空间中找到 Release 对象，接着在 Release 对象中找到 Name 对象。Release 对象是 Helm 的内置对象之一。

现在重新安装 Chart，观察模板指令的作用，命令如下：

```
helm install hello-world-2 ./mychart
```

安装完成后，查看创建的 ConfigMap 信息，命令如下：

```
kubectl get configmap
```

运行命令后输出的信息如下：

```
NAME                          DATA    AGE
hello-world-2-configmap         1     5s
```

输出显示 ConfigMap 的名称由 Release 的名称和后缀-configmap 构成，与模板中的定义一致。

Helm 提供了模拟安装，以此来预览模板渲染结果的功能，而不会真正安装到 Kubernetes，命令如下：

```
helm install --dry-run hello-world-2 ./mychart
```

运行命令后输出的信息如下：

```
NAME: hello-world-2
LAST DEPLOYED: Sun Jul 11 16:10:51 2021
NAMESPACE: default
STATUS: pending-install
REVISION: 1
TEST SUITE: None
HOOKS:
MANIFEST:
---
# Source: mychart/templates/configmap.yaml
apiVersion: v1
kind: ConfigMap
```

```
metadata:
  name: hello-world-2-configmap
data:
  myvalue: "Hello World"
```

如果希望显示更多细节,则可以加上--debug 参数。--dry-run 使测试模板变得简单,但是需要注意的是尽管--dry-run 可以正常运行,不代表生成的结果一定可以被 Kubernetes 接受。

9.3.3 内置对象

在渲染模板的过程中,模板引擎会把对象传入模板中,对象简单到只包含一个值,也可以包含其他对象或者函数。

在模板中可以访问的内置对象有 Release、Values、Chart、Files、Capabilities 和 Template。

Release 对象用于描述发布信息,它包括的其他对象如下。

(1) Release.Name:表示发布名称。

(2) Release.Namespace:表示发布到哪个命名空间。

(3) Release.IsUpgrade:布尔值,表示当前操作是否为升级,值是 false 表示回滚。

(4) Release.IsInstall:布尔值,true 表示当前操作是安装。

(5) Release.Revision:表示当前发布的修订版本数字,当第一次安装完成后,这个值是 1,之后每次升级或者回滚,这个值会加 1。

(6) Release.Service:表示渲染当前模板的服务,在 Helm 中,这个值是 Helm。

Values 对象用于保存 values.yaml 文件和其他用户提供的文件中的值及通过--set 选项传递的独立参数,Values 默认为空。

Chart 对象用于保存 Chart.yaml 文件中的内容,Chart.yaml 文件中的任何数据都可以从 Chart 对象中访问。

Files 对象用于访问 Chart 中所有除模板文件以外的其他普通文件。Files 对象包含多个函数。

(1) Files.Get 函数用于读取文件,例如(.Files.Get config.ini)。

(2) Files.GetBytes 用于以字节数组的形式读取文件,适用于读取图像等二进制文件。

(3) Files.Lines 用于逐行读取文件。

(4) Files.AsSecrets 用于读取文件并返回 Base64 格式编码后的字符串。

(5) Files.AsConfig 用于读取文件并以 YAML map 的形式返回。

Capabilities 对象用于描述 Kubernetes 集群支持的功能信息。Capabilities 对象包含如下对象:

(1) Capabilities.APIVersions 是一个版本集合。

(2) Capabilities.APIVersions.Has $version 用于表示 Kubernetes 中某个版本或者某个资源是否可用。

（3）Capabilities. KubeVersion 和 Capabilities. KubeVersion. Version 表示 Kubernetes 版本。

（4）Capabilities. KubeVersion. Major 是 Kubernetes 的主版本号。

（5）Capabilities. KubeVersion. Minor 是 Kubernetes 的小版本号。

Template 对象包含当前被渲染的模板的信息。

（1）Template. Name 是当前模板的路径，例如 mychart/templates/mytemplate. yaml。

（2）Template. BasePath 是当前模板所在的文件夹，例如 mychart/template。

9.3.4 值文件

值文件（Values Files）是 YAML 格式的文件。值文件的来源包括 Chart 中的 values. yaml 文件，父 Chart 中的 values. yaml，helm install 或者 helm upgrade 命令中的-f 选项指定的文件，以及--set 选项传递的参数（例如 helm install --set foo=bar . /mychart）。

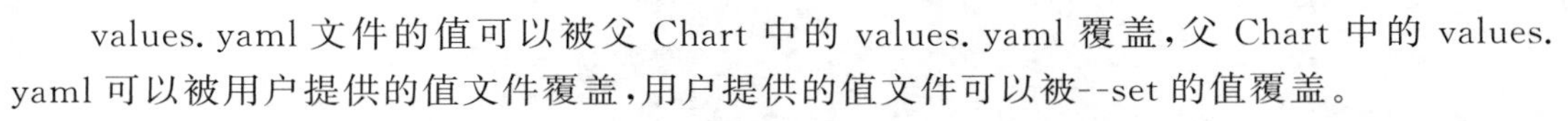

values. yaml 文件的值可以被父 Chart 中的 values. yaml 覆盖，父 Chart 中的 values. yaml 可以被用户提供的值文件覆盖，用户提供的值文件可以被--set 的值覆盖。

值文件中的变量名字应该以小写字母开头，使用小驼峰式，命名示例如下：

```
chicken: true
chickenNoodleSoup: true
```

不正确的命名示例如下：

```
Chicken: true                          #首字母不应该大写
chicken-noodle-soup: true              #名字中不应该使用连字符
```

下面演示如何使用 values. yaml 文件，修改 mychart/values. yaml 文件，编辑内容如下：

```
favoriteDrink: tea
```

然后在 mychart/templates/configmap. yaml 文件中使用这个值，编辑 configmap. yaml 文件如下：

```
apiVersion: v1
kind: ConfigMap
metadata:
  name: {{ .Release.Name }}-configmap
data:
  myvalue: "Hello World"
  drink: {{ .Values.favoriteDrink }}
```

通过 Values 对象可以访问 values. yaml 文件中的值。

下面预览模板生成的结果，命令如下：

```
helm install --dry-run --debug hello-world-3 ./mychart
```

运行命令后输出的信息如下：

```
install.go:173: [debug] Original chart version: ""
install.go:190: [debug] CHART PATH: /Users/gsh/projects/k8s/helm/mychart

NAME: hello-world-3
LAST DEPLOYED: Sun Jul 11 18:12:19 2021
NAMESPACE: default
STATUS: pending-install
REVISION: 1
TEST SUITE: None
USER-SUPPLIED VALUES:
{}

COMPUTED VALUES:
favoriteDrink: tea

HOOKS:
MANIFEST:
---
# Source: mychart/templates/configmap.yaml
apiVersion: v1
kind: ConfigMap
metadata:
  name: hello-world-3-configmap
data:
  myvalue: "Hello World"
  drink: tea
```

输出信息显示 favoriteDrink 的值是 tea，也就是 values.yaml 中指定的值。通过优先级更高的--set 选项可以覆盖这个值，命令如下：

```
helm install --dry-run --debug hello-world-3 . --set favoriteDrink=coffee
```

运行命令后输出的信息如下：

```
install.go:173: [debug] Original chart version: ""
install.go:190: [debug] CHART PATH: /Users/gsh/projects/k8s/helm/mychart

NAME: hello-world-3
LAST DEPLOYED: Sun Jul 11 18:15:42 2021
NAMESPACE: default
STATUS: pending-install
```

```
REVISION: 1
TEST SUITE: None
USER-SUPPLIED VALUES:
favoriteDrink: coffee

COMPUTED VALUES:
favoriteDrink: coffee

HOOKS:
MANIFEST:
---
#Source: mychart/templates/configmap.yaml
apiVersion: v1
kind: ConfigMap
metadata:
  name: hello-world-3-configmap
data:
  myvalue: "Hello World"
  drink: coffee
```

可以看到生成结果中的 favoriteDrink 的值是 coffee，说明覆盖成功。

值文件中除了可以包含上述键值对，还可以包含更加结构化的数据，编辑 values.yaml 文件如下：

```
favorite:
  drink: tea
  food: noodle
```

接着修改模板文件 configmap.yaml，内容如下：

```
apiVersion: v1
kind: ConfigMap
metadata:
  name: {{ .Release.Name }}-configmap
data:
  myvalue: "Hello World"
  drink: {{ .Values.favorite.drink }}
  food: {{ .Values.favorite.food }}
```

再次预览模板生成的结果，命令如下：

```
helm install --dry-run --debug hello-world-4 ./mychart
```

运行命令后输出的信息如下：

```
install.go:173: [debug] Original chart version: ""
install.go:190: [debug] CHART PATH: /Users/gsh/projects/k8s/helm/mychart

NAME: hello-world-4
LAST DEPLOYED: Sun Jul 11 18:23:00 2021
NAMESPACE: default
STATUS: pending-install
REVISION: 1
TEST SUITE: None
USER-SUPPLIED VALUES:
{}

COMPUTED VALUES:
favorite:
  drink: tea
  food: noodle

HOOKS:
MANIFEST:
---
# Source: mychart/templates/configmap.yaml
apiVersion: v1
kind: ConfigMap
metadata:
  name: hello-world-4-configmap
data:
  myvalue: "Hello World"
  drink: tea
  food: noodle
```

虽然支持结构化数据，但是建议结构树层次不要过多，因为当把值赋给子 Chart 时，值的命名会使用树结构。树过深会导致命名比较复杂。

删除值文件中的默认键需要把对应的键设置为 null。下面以 Drupal Chart 为例，演示如何删除键。下面是默认的值：

```
livenessProbe:
  httpGet:
    path: /user/login
    port: http
  initialDelaySeconds: 120
```

然后希望使用--set 把 livenessProbe 的 httpGet 覆盖为 exec。--set 设置如下：

```
--set livenessProbe.exec.command=[cat,docroot/CHANGELOG.txt]
```

而实际上 Helm 处理的结果如下：

```
livenessProbe:
  httpGet:
    path: /user/login
    port: http
  exec:
    command:
    - cat
    - docroot/CHANGELOG.txt
  initialDelaySeconds: 120
```

这个结果与期望不符，并没有覆盖 httpGet，而是合并在一起了。Kubernetes 在应用这个 YAML 时会报错，因为 Kubernetes 不支持多个 livenessProbe 处理器。

为了实现覆盖的效果，需要把 livenessProbe.httpGet 设置为 null，以此来使 Helm 删除这个键，命令如下：

```
helm install stable/drupal \
 --set image=my-registry/drupal:0.1.0 \
 --set livenessProbe.exec.command=[cat,docroot/CHANGELOG.txt] \
 --set livenessProbe.httpGet=null
```

9.3.5 模板函数和管道

前面介绍了如何在模板中注入数据，有时候需要修改注入的数据，这就需要使用 Helm 中的模板函数，Helm 提供了超过 60 种模板函数，其中一部分来自 Go 模板语言，其他大部分来自 Sprig 模板库。

1. quote

例如为字符串数据加入双引号，这可以通过模板函数中的 quote 函数实现，示例如下：

```
apiVersion: v1
kind: ConfigMap
metadata:
  name: {{ .Release.Name }}-configmap
data:
  myvalue: "Hello World"
  drink: {{ quote .Values.favorite.drink }}
  food: {{ quote .Values.favorite.food }}
```

模板函数的语法如下：

```
functionName arg1 arg2...
```

上述 YAML 中的{{ quote . Values. favorite. food }}表示使用 1 个参数调用 quote 函数。

2. 管道

模板语言中的管道是另一个强大的特性，管道的概念来自 UNIX，使用管道可以方便地链式调用模板函数，以此来处理数据格式，并且可以简洁地表达出数据的顺序处理流程。

下面使用管道实现上述加括号的示例：

```
apiVersion: v1
kind: ConfigMap
metadata:
  name: {{ .Release.Name }} - configmap
data:
  myvalue: "Hello World"
  drink: {{ .Values.favorite.drink | quote }}
  food: {{ .Values.favorite.food | quote }}
```

文件中使用管道符(|)把参数传送到了 quote 函数，而不是使用 quote ARGUMENT 这种格式。使用管道可以链式调用多个函数，示例如下：

```
apiVersion: v1
kind: ConfigMap
metadata:
  name: {{ .Release.Name }} - configmap
data:
  myvalue: "Hello World"
  drink: {{ .Values.favorite.drink | quote }}
  food: {{ .Values.favorite.food | upper | quote }}
```

这个模板生成的 YAML 如下：

```
apiVersion: v1
kind: ConfigMap
metadata:
  name: hello - world - 4 - configmap
data:
  myvalue: "Hello World"
  drink: "tea"
  food: "NOODLE"
```

可以看到 noodle 经过 upper 函数和 quote 函数按一定的顺序处理后变成了"NOODLE"。

3. repeat

如果需要重复字符串，则可以使用 repeat 函数，下面是使用 repeat 函数的示例：

```
apiVersion: v1
kind: ConfigMap
metadata:
  name: {{ .Release.Name }}-configmap
data:
  myvalue: "Hello World"
  drink: {{ .Values.favorite.drink | repeat 5 | quote }}
  food: {{ .Values.favorite.food | upper | quote }}
```

该模板生成的 YAML 如下：

```
apiVersion: v1
kind: ConfigMap
metadata:
  name: melting-porcup-configmap
data:
  myvalue: "Hello World"
  drink: "coffeecoffeecoffeecoffeecoffee"
  food: "PIZZA"
```

4. default

模板函数中比较常用的还有一个 default 函数，用于指定模板中的默认值。default 函数的语法如下：

```
default DEFAULT_VALUE GIVEN_VALUE
```

使用 default 函数的示例模板如下：

```
apiVersion: v1
kind: ConfigMap
metadata:
  name: {{ .Release.Name }}-configmap
data:
  myvalue: "Hello World"
  drink: {{ .Values.favorite.drink | default "coffee" | quote }}
  food: {{ .Values.favorite.food | upper | quote }}
```

保持 values.yaml 文件内容不变：

```
favorite:
  drink: tea
  food: noodle
```

模板生成的 YAML 如下：

```
apiVersion: v1
kind: ConfigMap
metadata:
  name: hello-world-4-configmap
data:
  myvalue: "Hello World"
  drink: "tea"
  food: "NOODLE"
```

修改 values.yaml 文件，注释掉 drink：tea，示例如下：

```
favorite:
#drink: tea
  food: noodle
```

模板生成的 YAML 如下：

```
apiVersion: v1
kind: ConfigMap
metadata:
  name: hello-world-4-configmap
data:
  myvalue: "Hello World"
  drink: "coffee"
  food: "NOODLE"
```

输出信息显示 drink 的值为 coffee，也就是 default 函数指定的默认值。

5. include

Go 语言提供了 template 函数，用于实现在模板中包含另一个模板，但是 template 函数不支持用于管道操作，这样就无法继续操作 template 函数的输出结果了。

而 include 函数支持管道，这样就可以对模板输出继续进行修改，示例如下：

```
{{ include "toYaml" $value | indent 2 }}
```

这个模板中的 include 指令包含了 toYaml 命名模板，传递了 $value 参数，接着输出结果会传入 indent 函数。

6. tpl

tpl 用于在模板中把字符串当作模板继续解析，从而实现模板的嵌套。

例如 values.yaml 文件的内容如下：

```
template: "{{ .Values.name }}"
name: "Tom"
```

模板如下：

```
data: {{ tpl .Values.template . }}
```

模板渲染结果如下：

```
data: Tom
```

考虑模板的执行过程，模板中引用的值.Values.template 为"{{ .Values.name }}"，所以模板首先应该被替换，示例如下：

```
{{ tpl "{{ .Values.name }}" . }}
```

接着 tpl 函数会把"{{ .Values.name }}"这个字符串当作模板继续解析，解析结果就是 Tom。

下面的示例演示如何利用 tpl 函数渲染模板外部的配置文件。

假设有一个配置文件 conf/app.conf，其内容如下：

```
firstName = {{ .Values.firstName }}
lastName = {{ .Values.lastName }}
```

模板文件 templates/demo.yaml 的内容如下：

```
data: |
{{- tpl (.Files.Get "conf/app.conf") . | nindent 2 }}
```

值文件 values.yaml 的内容如下：

```
firstName: Peter
lastName: Parker
```

现在渲染模板，然后查看输出结果，命令如下：

```
helm template .
```

运行命令后输出的信息如下：

```
data: |
  firstName = Peter
  lastName = Parker
```

9.3.6 流程控制

控制结构提供了控制模板生成流程的能力。Helm 模板语言提供了如下控制结构：

(1) if/else 用于创建条件结构。

(2) with 用于创建范围。

(3) range 用于创建循环结构。

条件结构示例如下:

```
{{ if PIPELINE }}
  # Do something
{{ else if OTHER PIPELINE }}
  # Do something else
{{ else }}
  # Default case
{{ end }}
```

这个条件结构会根据条件包含不同的文本块。条件可以是管道而不仅仅是一个简单的值。

在之前的 configmap.yaml 文件中加入条件结构,代码如下:

```
apiVersion: v1
kind: ConfigMap
metadata:
  name: {{ .Release.Name }}-configmap
data:
  myvalue: "Hello World"
  drink: {{ .Values.favorite.drink | default "tea" | quote }}
  food: {{ .Values.favorite.food | upper | quote }}
  {{ if eq .Values.favorite.drink "coffee" }}mug: "true"{{ end }}
```

修改 values.yaml 文件,代码如下:

```
favorite:
  drink:coffee
  food: noodle
```

模板生成的 YAML 文件的内容如下:

```
apiVersion: v1
kind: ConfigMap
metadata:
  name: eyewitness-elk-configmap
data:
  myvalue: "Hello World"
  drink: "coffee"
  food: "PIZZA"
  mug: "true"
```

为了使模板更易读，修改模板中的条件结构，代码如下：

```
apiVersion: v1
kind: ConfigMap
metadata:
  name: {{ .Release.Name }}-configmap
data:
  myvalue: "Hello World"
  drink: {{ .Values.favorite.drink | default "tea" | quote }}
  food: {{ .Values.favorite.food | upper | quote }}
  {{ if eq .Values.favorite.drink "coffee" }}
    mug: "true"
  {{ end }}
```

预览模板生成的结果，命令如下：

```
helm install --dry-run hello-world-5 ./mychart
```

运行命令后输出的信息如下：

```
Error: YAML parse error on mychart/templates/configmap.yaml: error converting YAML to JSON:
yaml: line 9: did not find expected key
```

这是由于缩进导致的 YAML 格式错误，上述模板生成的 YAML 文件的内容如下：

```
apiVersion: v1
kind: ConfigMap
metadata:
  name: eyewitness-elk-configmap
data:
  myvalue: "Hello World"
  drink: "coffee"
  food: "PIZZA"
    mug: "true"
```

mug 缩进错误，修改后的内容如下：

```
apiVersion: v1
kind: ConfigMap
metadata:
  name: {{ .Release.Name }}-configmap
data:
  myvalue: "Hello World"
  drink: {{ .Values.favorite.drink | default "tea" | quote }}
  food: {{ .Values.favorite.food | upper | quote }}
  {{ if eq .Values.favorite.drink "coffee" }}
```

```
  mug: "true"
  {{ end }}
```

再次预览生成的 YAML，会看到以下结果：

```
apiVersion: v1
kind: ConfigMap
metadata:
  name: hello-world-4-configmap
data:
  myvalue: "Hello World"
  drink: "coffee"
  food: "NOODLE"

  mug: "true"
```

food 与 mug 之间多了一个空行。这是因为模板引擎只会移除{{和}}之间的内容，并不会移除空白字符。换行符也是空白字符，空白字符在 YAML 中有重要意义，所以管理空白字符十分重要。Helm 提供了处理空白的几个工具。

模板语法中的{{-可以去除左边的空白，-}}可以去除右边的空白。需要注意的是连字符-和数据要用空格隔开，例如{{- 3}}会去掉左边的空格，输出 3，而{{-3}}会输出－3。

使用这个语法修改，修改后的模板如下：

```
apiVersion: v1
kind: ConfigMap
metadata:
  name: {{ .Release.Name }}-configmap
data:
  myvalue: "Hello World"
  drink: {{ .Values.favorite.drink | default "tea" | quote }}
  food: {{ .Values.favorite.food | upper | quote }}
  {{- if eq .Values.favorite.drink "coffee" }}
  mug: "true"
  {{- end }}
```

模板生成的 YAML 文件的内容如下：

```
apiVersion: v1
kind: ConfigMap
metadata:
  name: hello-world-4-configmap
data:
  myvalue: "Hello World"
  drink: "coffee"
```

```
  food: "NOODLE"
  mug: "true"
```

为了方便理解上述模板中被去掉的空白字符，用星号 * 代替会被去掉的空白字符，代码如下：

```
apiVersion: v1
kind: ConfigMap
metadata:
  name: {{ .Release.Name }}-configmap
data:
  myvalue: "Hello World"
  drink: {{ .Values.favorite.drink | default "tea" | quote }}
  food: {{ .Values.favorite.food | upper | quote }}*
**{{- if eq .Values.favorite.drink "coffee" }}
  mug: "true"*
**{{- end }}
```

下面开始介绍可以改变范围的 with。范围(scope)也可以理解为上下文。之前提到过的句号(.)用于引用当前的范围，所以常用的.Values 用于在当前作用范围寻找 Values 对象。

with 的语法如下：

```
{{ with PIPELINE }}
  # restricted scope
{{ end }}
```

使用 with 可以把当前范围从(.)修改为其他对象，例如把当前范围设置为.Values.favorite，代码如下：

```
apiVersion: v1
kind: ConfigMap
metadata:
  name: {{ .Release.Name }}-configmap
data:
  myvalue: "Hello World"
  {{- with .Values.favorite }}
  drink: {{ .drink | default "tea" | quote }}
  food: {{ .food | upper | quote }}
  {{- end }}
```

模板中使用.drink 和.food 来引用值而不需要使用.Values.favorite 来限定。这是由于 with 把范围从(.)修改为了(.Values.favorite)。(.)会在{{- end }}指令之后被重置到之前的范围。

需要注意的是在受限的范围内，不再允许访问父范围的其他对象，代码如下：

```
{{- with .Values.favorite }}
drink: {{ .drink | default "tea" | quote }}
food: {{ .food | upper | quote }}
release: {{ .Release.Name }}
{{- end }}
```

这样会报错，因为在受限范围.Values.favorite内并不存在Release.Name。

调整后的代码如下：

```
{{- with .Values.favorite }}
  drink: {{ .drink | default "tea" | quote }}
  food: {{ .food | upper | quote }}
  {{- end }}
  release: {{ .Release.Name }}
```

或者使用$符号访问根范围。根范围在整个模板执行过程中都不会改变。使用$修改模板后的代码如下：

```
{{- with .Values.favorite }}
  drink: {{ .drink | default "tea" | quote }}
  food: {{ .food | upper | quote }}
  release: {{ $.Release.Name }}
  {{- end }}
```

这样模板就可以正常解析了。

下面介绍循环结构。

在很多编程语言中支持循环结构，例如for循环或者foreach循环等。Helm模板语言中使用range来遍历一个集合。

下面演示如何使用range，修改后values.yaml文件的代码如下：

```
favorite:
  drink: coffee
  food: pizza
pizzaToppings:
   - mushrooms
   - cheese
   - peppers
   - onions
```

现在修改模板，打印pizzaToppings集合中的值。修改后configmap.yaml文件的代码如下：

```
apiVersion: v1
kind: ConfigMap
metadata:
  name: {{ .Release.Name }}-configmap
data:
  myvalue: "Hello World"
  {{- with .Values.favorite }}
  drink: {{ .drink | default "tea" | quote }}
  food: {{ .food | upper | quote }}
  {{- end }}
  toppings: |-
    {{- range .Values.pizzaToppings }}
    - {{ . | title | quote }}
    {{- end }}
```

或者使用 $ 从父范围中访问 Values.pizzaToppings，代码如下：

```
apiVersion: v1
kind: ConfigMap
metadata:
  name: {{ .Release.Name }}-configmap
data:
  myvalue: "Hello World"
  {{- with .Values.favorite }}
  drink: {{ .drink | default "tea" | quote }}
  food: {{ .food | upper | quote }}
  toppings: |-
    {{- range $.Values.pizzaToppings }}
    - {{ . | title | quote }}
    {{- end }}
  {{- end }}
```

模板中的 range 会遍历 pizzaToppings 列表，与 with 类似，range 也可以改变(.)指向的范围。在每轮遍历中，(.)都会指向当前正在遍历的元素，第 1 轮指向 pizzaToppings 列表中的第 1 个元素 mashrooms，第 2 轮指向第 2 个元素 cheese，以此类推。

模板生成的 YAML 文件的内容如下：

```
# Source: mychart/templates/configmap.yaml
apiVersion: v1
kind: ConfigMap
metadata:
  name: hello-world-4-configmap
data:
```

```
myvalue: "Hello World"
drink: "coffee"
food: "NOODLE"
toppings: |-
  - "Mushrooms"
  - "Cheese"
  - "Peppers"
  - "Onions"
```

如果 range 遍历的列表是固定长度的，则可以使用 tupple 在模板中快速创建一个列表，代码如下：

```
sizes: |-
 {{- range tuple "small""medium""large" }}
 - {{ . }}
 {{- end }}
```

模板生成的 YAML 文件的内容如下：

```
sizes: |-
    - small
    - medium
    - large
```

除了支持遍历列表和元组，range 还支持遍历键值对类型的数据结构，例如 map 或者 dict。

9.3.7 变量

Helm 中的变量是对其他对象的一个命名引用。变量格式为 $name，变量赋值操作符为 :=。使用变量修改下面的模板，模板内容如下：

```
apiVersion: v1
kind: ConfigMap
metadata:
  name: {{ .Release.Name }}-configmap
data:
  myvalue: "Hello World"
  {{- with .Values.favorite }}
  drink: {{ .drink | default "tea" | quote }}
  food: {{ .food | upper | quote }}
  release: {{ $.Release.Name }}
  {{- end }}
```

重写后模板的内容如下：

```
apiVersion: v1
kind: ConfigMap
metadata:
  name: {{ .Release.Name }}-configmap
data:
  myvalue: "Hello World"
  {{- $relname := .Release.Name -}}
  {{- with .Values.favorite }}
  drink: {{ .drink | default "tea" | quote }}
  food: {{ .food | upper | quote }}
  release: {{ $relname }}
  {{- end }}
```

预览生成的 YAML 文件，命令如下：

```
helm install --dry-run  hello-world-6 ./mychart
```

运行命令后输出的信息如下：

```
NAME: hello-world-6
LAST DEPLOYED: Tue Jul 13 14:11:29 2021
NAMESPACE: default
STATUS: pending-install
REVISION: 1
TEST SUITE: None
HOOKS:
MANIFEST:
---
# Source: mychart/templates/configmap.yaml
apiVersion: v1
kind: ConfigMap
metadata:
  name: hello-world-6-configmap
data:
  myvalue: "Hello World"
  drink: "coffee"
  food: "NOODLE"
  release: hello-world-6
```

在 range 中，变量可以用于获取列表或者与列表类似的对象的索引和值，代码如下：

```
toppings: |-
  {{- range $index, $topping := .Values.pizzaToppings }}
    {{ $index }}: {{ $topping }}
  {{- end }}
```

索引从 0 开始计数，生成的 YAML 文件的内容如下：

```
toppings: |-
    0: mushrooms
    1: cheese
    2: peppers
    3: onions
```

对于键值对形式的数据结构，可以在 range 中使用变量获取键值对，代码如下：

```
apiVersion: v1
kind: ConfigMap
metadata:
  name: {{ .Release.Name }}-configmap
data:
  myvalue: "Hello World"
  {{- range $key, $val := .Values.favorite }}
  {{ $key }}: {{ $val | quote }}
  {{- end }}
```

模板生成的 YAML 文件的内容如下：

```
apiVersion: v1
kind: ConfigMap
metadata:
  name: hello-world-7
data:
  myvalue: "Hello World"
  drink: "coffee"
  food: "pizza"
```

变量的作用域通常不是全局的，而是被限制在声明变量的语句块中，在这个示例中，$key 和 $val 的作用域是{{range}}和{{end}}块。

唯一例外的变量的 $，$ 总是指向根上下文。示例代码如下：

```
{{- range .Values.tlsSecrets }}
apiVersion: v1
kind: Secret
metadata:
  name: {{ .name }}
  labels:
    app.kubernetes.io/name: {{ template "fullname" $ }}
    helm.sh/chart: "{{ $.Chart.Name }}-{{ $.Chart.Version }}"
    app.kubernetes.io/instance: "{{ $.Release.Name }}"
    app.kubernetes.io/version: "{{ $.Chart.AppVersion }}"
```

```
    app.kubernetes.io/managed-by: "{{ $.Release.Service }}"
type: kubernetes.io/tls
data:
  tls.crt: {{ .certificate }}
  tls.key: {{ .key }}
---
{{- end }}
```

到目前为止所有 Helm 的示例 Chart 都只声明了一个模板文件，其实 Helm 支持声明和使用多个模板文件。下面介绍如何使用多个模板文件。

9.3.8 命名模板

命名模板是指定了名字的模板，模板名字必须全局唯一，如果声明了 2 个名字相同的模板，则后加载的模板会生效。一个受欢迎的命名惯例是使用 Chart 的名字做前缀，例如 {{ define "mychart.labels" }}，这样可以尽量避免在不同的 Chart 中使用相同的模板名字，从而导致冲突。

在 templates 文件夹下，除了 NOTES.txt 和名字以下画线开头的文件通常包含 Kubernetes 对象。以下画线开头的文件一般只用于被其他模板引用。

在模板文件中创建和命名一个模板需要使用 define 函数，语法如下：

```
{{ define "MY.NAME" }}
  # body of template here
{{ end }}
```

例如定义一个包含一组 Kubernetes 标签的模板，示例代码如下：

```
{{- define "mychart.labels" }}
  labels:
    generator: helm
    date: {{ now | htmlDate }}
{{- end }}
```

接下来就可以把它嵌入其他模板中，代码如下：

```
{{- define "mychart.labels" }}
  labels:
    generator: helm
    date: {{ now | htmlDate }}
{{- end }}
apiVersion: v1
kind: ConfigMap
metadata:
  name: {{ .Release.Name }}-configmap
```

```
  {{- template "mychart.labels" }}
data:
  myvalue: "Hello World"
  {{- range $key, $val := .Values.favorite }}
  {{ $key }}: {{ $val | quote }}
  {{- end }}
```

模板生成的 YAML 文件的内容如下：

```
apiVersion: v1
kind: ConfigMap
metadata:
  name: running-panda-configmap
  labels:
    generator: helm
    date: 2016-11-02
data:
  myvalue: "Hello World"
  drink: "coffee"
  food: "pizza"
```

按照惯例，命名模板通常放到部分文件中，如_helpers.tpl 文件中。另外建议在 define 函数中加入描述用途的文档块({{/* ... */}})。

当把命名模板 mychart.labels 移到_helpers.tpl 文件中以后，在 configmap.yaml 文件中仍然可以访问这个命名模板，修改后 configmap.yaml 文件的内容如下：

```
apiVersion: v1
kind: ConfigMap
metadata:
  name: {{ .Release.Name }}-configmap
  {{- template "mychart.labels" }}
data:
  myvalue: "Hello World"
  {{- range $key, $val := .Values.favorite }}
  {{ $key }}: {{ $val | quote }}
  {{- end }}
```

下面在命名模板 mychart.labels 中加入 Chart 名称和 Chart 版本号。

预览生成的 YAML，命令如下：

```
helm install --dry-run  hello-world-7 ./mychart
```

运行命令后输出的信息如下：

```
Error: unable to build kubernetes objects from release manifest: error validating "": error
validating data: [unknown object type "nil" in ConfigMap.metadata.labels.chart, unknown
object type "nil" in \
ConfigMap.metadata.labels.version]
```

报错了，为了查看生成的 YAML，使用--disable-openapi-validation 参数再次运行，命令如下：

```
helm install \
--dry-run \
--disable-openapi-validation  hello-world-7 ./mychart
```

运行命令后输出的信息如下：

```
NAME: hello-world-7
LAST DEPLOYED: Tue Jul 13 18:25:11 2021
NAMESPACE: default
STATUS: pending-install
REVISION: 1
TEST SUITE: None
HOOKS:
MANIFEST:
---
# Source: mychart/templates/configmap.yaml
apiVersion: v1
kind: ConfigMap
metadata:
  name: hello-world-7-configmap
  labels:
    generator: helm
    date: 2021-07-13
    chart:
    version:
data:
  myvalue: "Hello World"
  drink: "coffee"
  food: "noodle"
```

可以看到生成的结果不符合预期，chart 和 version 字段的值都是空的。这是因为在调用 template 的时候，没有传入范围，代码如下：

```
{{- template "mychart.labels" }}
```

所以在命名模板中无法访问(.)中的对象。

需要在 configmap.yaml 中调用 template 的时候加入范围参数，修改后的代码如下：

```
{{- template "mychart.labels" . }}
```

因为 Release 对象是顶级范围中的对象,所以传入了顶级范围(.)。

再次预览生成的 YAML,命令如下:

```
helm install --dry-run  hello-world-7 ./mychart
```

生成的 YAML 文件的内容如下:

```
apiVersion: v1
kind: ConfigMap
metadata:
  name: hello-world-7-configmap
  labels:
    generator: helm
    date: 2021-07-13
    chart: mychart
    version: 0.1.0
data:
  myvalue: "Hello World"
  drink: "coffee"
  food: "noodle"
```

可以看到{{ .Chart.Name }}被替换为 mychart,{{ .Chart.Version }}被替换为 0.1.0。

下面介绍另一个模板函数 include。

假设有一个命名模板 mychart.app,代码如下:

```
{{- define "mychart.app" -}}
app_name: {{ .Chart.Name }}
app_version: "{{ .Chart.Version }}"
{{- end -}}
```

假设需要在 configmap.yaml 中把 mychart.app 命名模板插入 labels 和 data 这 2 个字段下,代码如下:

```
apiVersion: v1
kind: ConfigMap
metadata:
  name: {{ .Release.Name }}-configmap
  labels:
    {{ template "mychart.app" . }}
data:
  myvalue: "Hello World"
  {{- range $key, $val := .Values.favorite }}
```

```
    {{ $key }}: {{ $val | quote }}
    {{- end }}
  {{ template "mychart.app" . }}
```

渲染这个模板会看到以下错误：

```
Error: unable to build kubernetes objects from release manifest: error validating "": error
validating data: [ValidationError(ConfigMap): unknown field "app_name" in io.k8s.api.core.
v1.ConfigMap, \
ValidationError(ConfigMap): unknown field "app_version" in \
io.k8s.api.core.v1.ConfigMap]
```

为了查看生成的YAML，使用--disable-openapi-validation参数再次运行，命令如下：

```
helm install \
--dry-run \
--disable-openapi-validation  hello-world-8 ./mychart
```

运行命令后输出的信息如下：

```
apiVersion: v1
kind: ConfigMap
metadata:
  name: hello-world-8-configmap
  labels:
    app_name: mychart
app_version: "0.1.0"
data:
  myvalue: "Hello World"
  drink: "coffee"
  food: "noodle"
app_name: mychart
app_version: "0.1.0"
```

生成的YAML与预期不符，这是因为嵌入的模板替换完成后向左对齐了，而由于template不是函数，无法把template的输出传入其他函数进行处理，使数据被简单地插入YAML中了。

为了处理这个问题，可以使用include来替换template，并且使用indent进行缩进，代码如下：

```
apiVersion: v1
kind: ConfigMap
metadata:
```

```
  name: {{ .Release.Name }}-configmap
  labels:
{{ include "mychart.app" . | indent 4 }}
data:
  myvalue: "Hello World"
  {{- range $key, $val := .Values.favorite }}
  {{ $key }}: {{ $val | quote }}
  {{- end }}
{{ include "mychart.app" . | indent 2 }}
```

重新渲染模板后得到的 YAML 的内容如下：

```
apiVersion: v1
kind: ConfigMap
metadata:
  name: hello-world-8-configmap
  labels:
    app_name: mychart
    app_version: "0.1.0"
data:
  myvalue: "Hello World"
  drink: "coffee"
  food: "noodle"
  app_name: mychart
  app_version: "0.1.0"
```

由于 include 是函数，所以可以对输出结果进行后续处理，正因如此，Helm 推荐使用 include 而不是 template。

9.3.9 访问文件

有时候需要在模板中访问非模板文件，并把文件内容直接注入渲染结果，而不需通过模板引擎处理。

Helm 通过.Files 对象提供文件访问功能。出于安全考虑，templates 文件夹下的文件和.helmignore 中列出的文件无法被访问。

尽管可以将文件添加到 Helm Chart 中，但是由于 Kubernetes 对象的存储限制，Chart 的大小不可以超过 1MB。

Charts 不会保留文件的 UNIX 模式信息，所以对于.Files 对象来讲，文件的权限不会影响文件的可用性。

1. 基本使用示例

下面使用一个读取 3 个文件的 ConfigMap 模板演示模板内如何访问文件。首先在 mychart 文件夹中创建 3 个文件。

创建 config1. toml 文件，编辑内容如下：

```
message = Hello from config 1
```

创建 config2. toml 文件，编辑内容如下：

```
message = Hello from config 2
```

创建 config3. toml 文件，编辑内容如下：

```
message = Hello from config 3
```

接下来在模板中使用 range 遍历这 3 个文件，并注入文件内容，编辑模板如下：

```
apiVersion: v1
kind: ConfigMap
metadata:
  name: {{ .Release.Name }}-configmap
data:
  {{- $files := .Files }}
  {{- range tuple "config1.toml""config2.toml""config3.toml" }}
  {{ . }}: |-
        {{ $files.Get . }}
  {{- end }}
```

模板中使用 tuple 生成了一个包含 3 个元素的列表，然后把每个元素对应的文件名打印出来，并通过{{ $files. Get . }}注入文件内容。

模板生成的 YAML 文件的内容如下：

```
# Source: mychart/templates/configmap.yaml
apiVersion: v1
kind: ConfigMap
metadata:
  name: quieting-giraf-configmap
data:
  config1.toml: |-
        message = Hello from config 1

  config2.toml: |-
        message = This is config 2

  config3.toml: |-
        message = Goodbye from config 3
```

2. 路径帮助函数

处理文件的时候往往需要对文件路径进行操作，Helm 从 Go 中的 path 包引入了许多与文件路径相关的函数，以此来满足这个需求，引入的函数包括 Base、Dir、Ext、IsAbs 和 Cleaen。

3. Glob

Files 对象提供了 Glob 方法，用于匹配多个文件。Glob 方法返回值的类型是 Files 类型，所以可以在方法返回对象上调用 Files 的所有方法。

假设有以下文件结构：

```
foo/:
  foo.txt foo.yaml

bar/:
  bar.go bar.conf baz.yaml
```

在模板中使用 Glob 的示例代码如下：

```
{{ $currentScope := .}}
{{ range $path, $_ :=  .Files.Glob  "**.yaml" }}
    {{- with $currentScope}}
        {{ .Files.Get $path }}
    {{- end }}
{{ end }}
```

或者示例代码如下：

```
{{ range $path, $_ :=  .Files.Glob  "**.yaml" }}
      {{ $.Files.Get $path }}
{{ end }}
```

模板会找到符合 **.yaml 模式的文件，并把每个文件的内容打印出来。

为了观察已找到的所有文件的文件名，可以将文件路径打印出来，代码如下：

```
apiVersion: v1
kind: ConfigMap
metadata:
  name: {{ .Release.Name }}-configmap
data:
  {{ range $path, $_ :=  .Files.Glob  "**.yaml" }}
      {{- $path }}
  {{ end }}
```

模板生成的 YAML 文件的内容如下：

```
apiVersion: v1
kind: ConfigMap
metadata:
  name: hello-world-8-configmap
data:
  bar/bar.yaml
  foo/foo.yaml
```

Glob 匹配了 bar/bar.yaml 和 foo/foo.yaml 这两个文件。

4. 编码

Helm 支持导入一个文件并对文件内容进行 base64 编码，示例模板如下：

```
apiVersion: v1
kind: Secret
metadata:
  name: {{ .Release.Name }}-secret
type: Opaque
data:
  token: |-
        {{ .Files.Get "config1.toml" | b64enc }}
```

模板生成的 YAML 文件的内容如下：

```
apiVersion: v1
kind: Secret
metadata:
  name: hello-world-8-secret
type: Opaque
data:
  token: |-
        bWVzc2FnZSA9IEhlbGxvIGZyb20gY29uZmlnIDEK
```

5. ConfigMap 和 Secret 帮助函数

有时需要把文件内容导入 ConfigMap 或者 Secret，Files 对象提供了相关的方法 AsConfig 和 AsSecrets 以实现这个功能。

使用 AsConfig 模板的示例代码如下：

```
apiVersion: v1
kind: ConfigMap
metadata:
  name: conf
data:
{{ (.Files.Glob "foo/*").AsConfig | indent 2 }}
```

模板生成的 YAML 文件的内容如下：

```yaml
apiVersion: v1
kind: ConfigMap
metadata:
  name: conf
data:
  foo.txt: ""
  foo.yaml: ""
```

使用 AsSecrets 模板的示例代码如下：

```yaml
apiVersion: v1
kind: Secret
metadata:
  name: very-secret
type: Opaque
data:
{{ (.Files.Glob "bar/*").AsSecrets | indent 2 }}
```

模板生成的 YAML 文件的内容如下：

```yaml
apiVersion: v1
kind: Secret
metadata:
  name: very-secret
type: Opaque
data:
  bar.conf: ""
  bar.go: ""
  bar.yaml: ""
```

6. 逐行读取文件

Files 对象还提供了 Lines 方法，用于实现逐行读取文件，示例代码如下：

```yaml
data:
  some-file.txt: {{ range .Files.Lines "foo/bar.txt" }}
    {{ . }}{{ end }}
```

9.3.10 NOTES.txt

为了帮助用户更好地使用 Chart，Helm 支持在 Chart 中添加一些帮助信息文本，使用户在运行 helm install 或者 helm upgrade 命令的时候可以看到帮助信息。

帮助信息中往往包含一些与该 Chart 相关的常用指令，保存在 templates/NOTES. txt 文件中。虽然 NOTES. txt 文件是普通的文本文件，但是在 NOTES. txt 中可以使用所有的模板函数和对象。

在 templates 文件夹下创建 NOTES. txt 文件，编辑内容如下：

```
Thank you for installing {{ .Chart.Name }}.

Your release is named {{ .Release.Name }}.

To learn more about the release, try:

  $ helm status {{ .Release.Name }}
  $ helm get all {{ .Release.Name }}
```

预览生成的 YAML 文件，命令如下：

```
helm install --dry-run  hello-world-10 ./mychart
```

在输出末尾可以看到以下帮助信息：

```
NOTES:
Thank you for installing mychart.

Your release is named hello-world-10.

To learn more about the release, try:

  $ helm status hello-world-10
  $ helm get all hello-world-10
```

说明 NOTES. txt 文件已经生效，虽然为 Chart 创建 NOTES. txt 不是必需的，但是推荐创建这个文件，以帮助用户更好地使用 Chart。

9.3.11 helmignore 文件

类似 Git 中的. gitignore 文件，Helm 中使用. helmignore 文件来忽略不想包含在 Chart 中的文件。

当运行 helm package 命令对 Chart 打包时，如果存在. helmignore 文件，则所有与. helmignore 文件内容匹配的文件会被忽略。

. helmignore 文件支持多种文件匹配模式，例如 glob 匹配、相对路径匹配及排除匹配（使用!前缀）。. helmignore 文件每一行只支持设置一个模式。

下面是一个. helmignore 文件示例，代码如下：

```
#注释前使用#标记

#匹配名字为.git的文件或者文件夹
.git

#匹配任意文本文件
*.txt

#匹配 mydir 文件夹
mydir/

#匹配根文件夹下的文本文件
/*.txt

#匹配根文件夹下的 foo.txt 文件
/foo.txt

#匹配 ab.txt、ac.txt 或 ad.txt 文件
a[b-d].txt

#匹配任意子文件夹下的 temp* 文件
*/temp*

#匹配任意子文件夹下的子文件夹中的 temp* 文件
*/*/temp*

#匹配名字长度为 5 且以 temp 开头的文件
temp?
```

.helmignore 中不支持 ** 语法。

9.3.12 Debug

当 Helm 模板报错时排查错误有时候比较困难，因为模板生成的 YAML 在发送到 Kubernetes 之后可能会因为格式错误以外的问题而被拒绝执行。

Helm 提供了几个对排查错误非常有用的命令。

命令 helm lint 可以验证模板是否遵循了最佳实践。

命令 helm install --dry-run --debug 或者 helm template --debug 可以在报错的情况下查看生成的 YAML。

命令 helm get manifest 可以查看最终发送到 Kubernetes 的 YAML。

当模板发生解析错误时，可以先把有问题的部分加上注释，然后重新运行 helm install

--dry-run --debug 命令,命令如下:

```
apiVersion: v2
#有问题的部分
#{{ .Values.foo | quote }}
```

生成的 YAML 中会原封不动地包含模板中的注释,输出信息如下:

```
apiVersion: v2
#有问题的部分
# "bar"
```

这样可以快速预览生成的 YAML。

9.3.13 最佳实践

1. templates 文件夹组织结构

生成的 YAML 格式的模板文件应该使用.yaml 后缀命名,而生成的非格式化输出的模板文件应该使用.tpl 后缀命名。

模板文件命名应该使用短横线和小写字母,不建议使用驼峰式。

每个资源定义都应该放到单独的模板文件中。

模板文件名最好可以反映出定义的资源类型,例如 foo-pod.yaml、bar-svc.yaml。

2. 命名模板的名称

命名模板的命名应该使用命名空间限定,下面是正确命名示例:

```
{{- define "nginx.fullname" }}
{{/* ... */}}
{{ end -}}
```

下面是错误命名示例:

```
{{- define "fullname" -}}
{{/* ... */}}
{{ end -}}
```

3. 模板格式化

模板采用 2 个空格进行缩进,模板指令应该与两边的花括号使用空格隔开。

下面是正确格式示例:

```
{{ .foo }}
{{ print "foo" }}
{{- print "bar" -}}
```

下面是错误格式示例：

```
{{.foo}}
{{print "foo"}}
{{-print "bar"-}}
```

模板中应该尽量去掉不必要的空白字符，代码如下：

```
foo:
  {{- range .Values.items }}
  {{ . }}
  {{ end -}}
```

控制结构等指令块应该适当进行缩进以提升可读性和更好地表达模板指令的执行流程。代码如下：

```
{{ if $foo -}}
  {{- with .Bar }}Hello{{ end -}}
{{- end -}}
```

模板生成的 YAML 中应避免出现连续的空白行，内容如下：

```
apiVersion: batch/v1
kind: Job

metadata:
  name: example






  labels:
    first: first

    second: second
```

上述文件出现了连续空行，不推荐以这种形式生成文件，好的 YAML 示例如下：

```
apiVersion: batch/v1
kind: Job
metadata:
  name: example
  labels:
    first: first
    second: second
```

虽然推荐在生成的 YAML 中尽量保留少量空白，但是在模板的不同逻辑块之前保留空白行可以提高可读性，示例代码如下：

```
apiVersion: batch/v1
kind: Job

metadata:
  name: example

  labels:
    first: first
    second: second
```

4. 注释

YAML 和 Helm 模板语言都支持添加注释，YAML 中的注释语法如下：

```
#这是注释
type: sprocket
```

Helm 模板注释的语法如下：

```
{{- /*
这是注释
*/}}
type: frobnitz
```

模板注释可用于描述模板的功能，示例代码如下：

```
{{- /*
mychart.shortname 提供 6 个字符长度的截断后的发布名称
*/}}
{{ define "mychart.shortname" -}}
{{ .Release.Name | trunc 6 }}
{{- end -}}
```

也可以用于模板内部，为用户 Debug 提供更多帮助信息，示例代码如下：

```
#这个值如果超过 100Gi 可能会造成严重问题
memory: {{ .Values.maxMem | quote }}
```

上面这行注释会在用户使用 helm install --debug 命令时看到。如果是{{- /* */}}格式的注释，则不会显示。

5. 使用 JSON

YAML 是 JSON 的超集，所以在 YAML 中可以使用 JSON 语法，有时候使用 JSON 可以一定程度上提升可读性。例如有以下 YAML 文件内容：

```
arguments:
  - "--dirname"
  - "/foo"
```

可以使用 JSON 改写，改写后代码如下：

```
arguments: ["--dirname", "/foo"]
```

对于复杂结构的数据，不建议使用 JSON 语法表达。除非使用 JSON 可以显著降低格式错误的风险。

9.4 Chart 依赖

9.4.1 简介

被依赖的 Chart 称为子 Chart。子 Chart 拥有独立的模板和值。下面演示如何创建子 Chart。

首先进入 mychart/charts 文件夹，然后运行 helm create，以便创建 Chart，命令如下：

```
helm create mysubchart
```

运行命令后输出的信息如下：

```
Creating mysubchart
```

这时在 mychart/charts 文件夹中会出现 mysubchart 文件夹。现在 mychart 就是 mysubchart 的父 Chart，父子关系由文件夹的层级决定。

为简单起见，删除 templates 文件夹下的所有文件，命令如下：

```
rm -rf templates/
```

在 mysubchart/templates 文件夹中新建文件 configmap.yaml，编辑内容如下：

```
apiVersion: v1
kind: ConfigMap
metadata:
  name: {{ .Release.Name }}-cfgmap2
data:
  dessert: {{ .Values.dessert }}
```

接下来编辑 values. yaml 文件,修改后的内容如下:

```
dessert: cake
```

因为子 Chart 是独立的 Chart,可以单独安装,测试 mysubchart 是否可用,命令如下:

```
helm install --generate-name --dry-run mychart/charts/mysubchart
```

运行命令后输出的信息如下:

```
NAME: chart-1626231573
LAST DEPLOYED: Wed Jul 14 10:59:36 2021
NAMESPACE: default
STATUS: pending-install
REVISION: 1
TEST SUITE: None
HOOKS:
MANIFEST:
---
# Source: mysubchart/templates/configmap.yaml
apiVersion: v1
kind: ConfigMap
metadata:
  name: chart-1626231573-cfgmap2
data:
  dessert: cake
```

9.4.2 值覆盖

父 Chart 可以覆盖子 Chart 中的值,mychat 是 mysubchart 的父 Chart,修改 mychart/values. yaml 文件,覆盖 dessert 的值,代码如下:

```
favorite:
  drink: coffee
  food: pizza
pizzaToppings:
   - mushrooms
   - cheese
   - peppers
   - onions

mysubchart:
  dessert: ice cream
```

最下面的 mysubchart 字段下的值会被发送到子 Chart(mysubchart)中,现在预览 mycharts 生成的 YAML,会看到子 Chart 对应的 YAML 文件的内容如下:

```
# Source: mychart/charts/mysubchart/templates/configmap.yaml
apiVersion: v1
kind: ConfigMap
metadata:
  name: chart - 1626233040 - cfgmap2
data:
  dessert: ice cream
```

dessert 的值由原来的 cake 变成了 ice cream,说明值被父 Chart 覆盖了。

9.4.3 全局值

全局值可以被任何 Chart 包括子 Chart 访问,全局值需要显式地声明。

修改 mychart/values.yaml 文件,添加一个全局值,代码如下:

```
favorite:
  drink: coffee
  food: pizza
pizzaToppings:
   - mushrooms
   - cheese
   - peppers
   - onions

mysubchart:
  dessert: ice cream

global:
  salad: caesar
```

然后在 mychart/templates/configmap.yaml 和 mysubchart/templates/configmap.yaml 中都可以使用{{ .Values.global.salad }}访问这个全局值。

修改 mychart/templates/configmap.yaml 文件,修改后的内容如下:

```
apiVersion: v1
kind: ConfigMap
metadata:
  name: {{ .Release.Name }} - configmap
data:
  salad: {{ .Values.global.salad }}
```

模板生成的 YAML 文件的内容如下：

```
---
# Source: mychart/templates/configmap.yaml
apiVersion: v1
kind: ConfigMap
metadata:
  name: chart-1626256431-configmap
data:
  salad: caesar
```

修改 mysubchart/templates/configmap.yaml 文件，修改后的内容如下：

```
apiVersion: v1
kind: ConfigMap
metadata:
  name: {{ .Release.Name }}-cfgmap2
data:
  dessert: {{ .Values.dessert }}
  salad: {{ .Values.global.salad }}
```

模板生成的 YAML 文件的内容如下：

```
---
# Source: mychart/charts/mysubchart/templates/configmap.yaml
apiVersion: v1
kind: ConfigMap
metadata:
  name: chart-1626256431-cfgmap2
data:
  dessert: ice cream
  salad: caesar
```

9.5 Chart Hook

9.5.1 简介

Helm 提供了钩子(Hook)机制，使用户可以介入 Chart Release 生命周期中的特定节点并执行某些操作。例如在加载 Chart 之前加载 ConfigMap、Secret 或在安装 Chart 之前执行一个备份数据库的 Job 等。

钩子的工作机制与普通的 Helm 模板类似，只不过钩子拥有特定的注解。Helm 中可用的钩子如表 9-1 所示。

表 9-1 Helm 中可用的钩子

注 解 值	描 述
pre-install	在模板渲染完成后且在 Kubernetes 创建相关资源前执行
post-install	在 Kubernetes 创建相关资源结束后执行
pre-delete	发送删除请求后在 Kubernetes 删除相关资源前执行
post-delete	发送删除请求后在 Kubernetes 删除相关资源后执行
pre-upgrade	发送升级请求后在 Kubernetes 更新资源前执行
post-upgrade	发送升级请求后在 Kubernetes 更新资源后执行
pre-rollback	发送回滚请求后在 Kubernetes 回滚资源前执行
post-rollback	发送回滚请求后在 Kubernetes 回滚资源后执行
test	在执行 helm test 命令时执行

9.5.2 Hook 示例

Hook 除了在 metadata 中包含了特殊的注解外，与普通的模板文件并无区别，在 Hook 中可以使用所有模板的功能特性，如读取.Values、.Release 和.Template 中的值。

创建 Hook 示例文件 templates/post-install-job.yaml，编辑内容如下：

```
apiVersion: batch/v1
kind: Job
metadata:
  name: "{{ .Release.Name }}"
  labels:
    app.kubernetes.io/managed-by: {{ .Release.Service | quote }}
    app.kubernetes.io/instance: {{ .Release.Name | quote }}
    app.kubernetes.io/version: {{ .Chart.AppVersion }}
    helm.sh/chart: "{{ .Chart.Name }}-{{ .Chart.Version }}"
  annotations:
    # This is what defines this resource as a hook. Without this line, the
    # job is considered part of the release.
"helm.sh/hook": post-install
"helm.sh/hook-weight": "-5"
"helm.sh/hook-delete-policy": hook-succeeded
spec:
  template:
    metadata:
      name: "{{ .Release.Name }}"
      labels:
        app.kubernetes.io/managed-by: {{ .Release.Service | quote }}
        app.kubernetes.io/instance: {{ .Release.Name | quote }}
        helm.sh/chart: "{{ .Chart.Name }}-{{ .Chart.Version }}"
    spec:
      restartPolicy: Never
```

```
containers:
- name: post-install-job
  image: "alpine:3.3"
  command: ["/bin/sleep","{{ default "10" .Values.sleepyTime }}"]
```

文件中的注解 helm.sh/hook 是使这个模板变成 Hook 的关键,代码如下:

```
annotations:
"helm.sh/hook": post-install
```

一个资源可以用于实现多个 Hook,只需要在 helm.sh/hook 注解值中定义多个 Hook 类型,用逗号分隔,代码如下:

```
annotations:
"helm.sh/hook": post-install,post-upgrade
```

在子 Chart 中也可以定义 Hook,父 Chart 中无法禁用子 Chart 中声明的 Hook。

9.5.3 Hook 权重

当 Hook 中定义了多个资源时,多个资源会按顺序执行,Hook 支持设置权重来决定不同资源的执行顺序,权重通过 helm.sh/hook-weight 注解定义,代码如下:

```
annotations:
"helm.sh/hook-weight": "5"
```

当资源定义了权重后,它们会按照权重排序并执行。定义 Hook 权重是一个好的实践,如果权重不重要,则可以把它设置为 0。权重可以是正数或负数,以字符串形式表示。

9.5.4 Hook 删除策略

Helm 支持不同的删除策略,通过 helm.sh/hook-delete-policy 注解指定,示例代码如下:

```
annotations:
"helm.sh/hook-delete-policy": before-hook-creation,hook-succeeded
```

支持的删除策略如表 9-2 所示。

表 9-2 Hook 支持的删除策略

注解值	描述
before-hook-creation	在下一个 Hook 执行前删除 Hook 资源,此为默认值
hook-succeed	在 Hook 执行结束后删除 Hook 资源
hook-failed	在 Hook 执行失败后删除 Hook 资源

在没有指定删除策略的情况下，会使用默认值 before-hook-creation。

9.6 Chart 测试

9.6.1 测试简介

Chart 测试可以验证安装 Chart 之后创建的诸多 Kubernetes 资源是否工作正常，这对于开发 Chart 非常有用，另外也可以使 Chart 用户更好地理解 Chart 的预期行为。

Chart 测试是在 templates 文件夹下定义一个任务资源文件，任务中的容器会运行特定的命令，根据命令的返回值是否为 0 来判断是否成功通过测试。如果返回值为 0，则表示测试成功。

Chart 测试对应的 Job 资源必须包含的注解如下：

```
helm.sh/hook: test
```

Chart 测试本质上是个 Chart Hook，所以 Hook 支持的注解都可以用到 Chart 测试对应的资源，例如 helm.sh/hook-weight 和 helm.sh/hook-delete-policy 注解。

9.6.2 测试示例

下面以 wordpress Chart 为例，了解 Chart 测试。

首页安装 wordpress 这个 Chart，命令如下：

```
helm repo add bitnami https://charts.bitnami.com/bitnami
helm pull bitnami/wordpress --untar
```

在 wordpress/templates/tests/test-mariadb-connection.yaml 文件中定义了一个测试，内容如下：

```
{{- if .Values.mariadb.enabled }}
apiVersion: v1
kind: Pod
metadata:
  name: "{{ .Release.Name }}-credentials-test"
  annotations:
"helm.sh/hook": test-success
spec:
  {{- if .Values.podSecurityContext.enabled }}
  securityContext: {{- omit .Values.podSecurityContext "enabled" | toYaml | nindent 4 }}
  {{- end }}
  containers:
    - name: {{ .Release.Name }}-credentials-test
```

```
        image: {{ template "wordpress.image" . }}
        imagePullPolicy: {{ .Values.image.pullPolicy | quote }}
        {{- if .Values.containerSecurityContext.enabled }}
        securityContext: {{- omit .Values.containerSecurityContext "enabled" | toYaml | nindent 8 }}
        {{- end }}
        env:
          - name: MARIADB_HOST
            value: {{ include "wordpress.databaseHost" . | quote }}
          - name: MARIADB_PORT
            value: "3306"
          - name: WORDPRESS_DATABASE_NAME
            value: {{ default "" .Values.mariadb.auth.database | quote }}
          - name: WORDPRESS_DATABASE_USER
            value: {{ default "" .Values.mariadb.auth.username | quote }}
          - name: WORDPRESS_DATABASE_PASSWORD
            valueFrom:
              secretKeyRef:
                name: {{ include "wordpress.databaseSecretName" . }}
                key: mariadb-password
        command:
          - /bin/bash
          - -ec
          - |
            mysql --host=$MARIADB_HOST --port=$MARIADB_PORT --user=$WORDPRESS_
DATABASE_USER --password=$WORDPRESS_DATABASE_PASSWORD
  restartPolicy: Never
{{- end }}
```

9.6.3 运行示例测试

首页安装 wordpress 这个 Chart,命令如下:

```
helm helm install --generate-name wordpress --namespace default
```

运行命令后输出的信息如下:

```
NAME: wordpress-1626328513
LAST DEPLOYED: Thu Jul 15 13:51:53 2021
NAMESPACE: default
STATUS: deployed
REVISION: 1
...
```

输出显示生成的 Release 名字为 wordpress-1626328513。
运行测试,命令如下:

```
helm test wordpress-1626328513
```

运行命令后输出的信息如下：

```
NAME: wordpress-1626328513
LAST DEPLOYED: Thu Jul 15 13:55:18 2021
NAMESPACE: default
STATUS: deployed
REVISION: 1
TEST SUITE:     wordpress-1626328513-credentials-test
Last Started:   Thu Jul 15 13:56:17 2021
Last Completed: Thu Jul 15 13:56:25 2021
Phase:          Succeeded
...
```

Phase 的字段值为 Succeeded，表示测试通过。

为了隔离测试文件与普通模板文件，可以把测试文件放到 templates/tests 文件夹下。

9.7 库 Chart

9.7.1 简介

Helm Chart 有两种类型，分别是 application 和 library。library 即库 Chart，它定义了可以被其他 Chart 复用的模板原语。库 Chart 并不定义任何 Kubernetes 资源。使用库 Chart 可以有效降低模板代码的重复，简化 Chart 开发和维护。

在 9.3.8 节介绍了可以复用的命名模板，命名模板只可以在一个 Chart 中使用，而库 Chart 可以被多个 Chart 使用。只需要在目标 Chart 中把库 Chart 声明为一个依赖。

9.7.2 示例

库 Chart 的创建方式与普通 Chart 类似，创建示例库 Chart，命令如下：

```
helm create mylibchart
```

运行命令后输出的信息如下：

```
Creating mylibchart
```

为简单起见，删除 templates 文件夹下的所有文件，命令如下：

```
rm -rf mylibchart/templates/*
```

删除值文件，命令如下：

```
rm -f mylibchart/values.yaml
```

接下来创建一个通用的 ConfigMap，创建文件 mylibchart/templates/_configmap.yaml，编辑内容如下：

```
{{- define "mylibchart.configmap.tpl" -}}
apiVersion: v1
kind: ConfigMap
metadata:
  name: {{ .Release.Name | printf "%s-%s" .Chart.Name }}
data: {}
{{- end -}}
{{- define "mylibchart.configmap" -}}
{{- include "mylibchart.util.merge" (append . "mylibchart.configmap.tpl") -}}
{{- end -}}
```

文件定义了 2 个命名模板，其中 mylibchart.configmap.tpl 中定义了一个空 ConfigMap，也就是 data 字段值为空。另一个是 mylibchart.configmap，这个命名模板中包含了名为 mylibchart.util.merge 的命名模板，mylibchart.util.merge 模板传入了 2 个参数，分别是 mylibchart.configmap 和 mylibchart.configmap.tpl。

现在创建命名模板 mylibchart.util.merge，新建文件 mylibchart/templates/_util.yaml，编辑内容如下：

```
{{- /*
mylibchart.util.merge will merge two YAML templates and output the result.
This takes an array of three values:
- the top context
- the template name of the overrides (destination)
- the template name of the base (source)
*/}}
{{- define "mylibchart.util.merge" -}}
{{-  $top := first . -}}
{{-  $overrides := fromYaml (include (index . 1) $top) | default (dict ) -}}
{{-  $tpl := fromYaml (include (index . 2) $top) | default (dict ) -}}
{{- toYaml (merge $overrides $tpl) -}}
{{- end -}}
```

最后修改 mylibchart/Chart.yaml，把 type 值变更为 library，代码如下：

```
apiVersion: v2
name: mylibchart
description: A Helm chart for Kubernetes
```

```
# A chart can be either an 'application' or a 'library' chart.
#
# Application charts are a collection of templates that can be packaged into versioned archives
# to be deployed.
#
# Library charts provide useful utilities or functions for the chart developer. They're
included as
# a dependency of application charts to inject those utilities and functions into the rendering
# pipeline. Library charts do not define any templates and therefore cannot be deployed.
type: library

# This is the chart version. This version number should be incremented each time you
make changes
# to the chart and its templates, including the app version.
# Versions are expected to follow Semantic Versioning (https://semver.org/)
version: 0.1.0

# This is the version number of the application being deployed. This version number should be
# incremented each time you make changes to the application. Versions are not expected to
# follow Semantic Versioning. They should reflect the version the application is using.
# It is recommended to use it with quotes.
appVersion: "1.16.0"
```

这个步骤非常重要，至此这个库 Chart 就可以被分享到其他 Chart 中了，使此库 Chart 中定义的 ConfigMap 可以被复用。

在继续之前，验证 Helm 是否可以识别出这个 Chart 是库 Chart，命令如下：

```
helm install mylibchart mylibchart/
```

运行命令后输出的信息如下：

```
Error: library charts are not installable
```

库 Chart 无法被安装。输出信息证明 Helm 正确地识别出这个库 Chart。

9.7.3 使用库 Chart

下面演示如何使用上述库 Chart。

创建 Chart，命令如下：

```
helm create mychart
```

运行命令后输出的信息如下：

```
Creating mychart
```

为了简单起见，同样删除 mychart/templates 下的文件，命令如下：

```
rm -rf mychart/templates/*
```

创建 mychart/templates/configmap.yaml 文件，编辑内容如下：

```
{{- include "mylibchart.configmap" (list . "mychart.configmap") -}}
{{- define "mychart.configmap" -}}
data:
  myvalue: "Hello World"
{{- end -}}
```

修改 mychart/Chart.yaml 文件，在文件末尾添加 dependencies 声明，代码如下：

```
dependencies:
- name: mylibchart
  version: 0.1.0
  repository: file://../mylibchart
```

更新依赖，命令如下：

```
helm dependency update mychart/
```

运行命令后输出的信息如下：

```
Hang tight while we grab the latest from your chart repositories...
...Successfully got an update from the "bitnami" chart repository
...Successfully got an update from the "stable" chart repository
Update Complete. Happy Helming!
Saving 1 charts
Deleting outdated charts
```

命令会把库 Chart 复制到 charts 文件夹。

现在预览 mychart 生成的 YAML 文件，命令如下：

```
helm install demo --dry-run --debug mychart
```

运行命令后输出的信息如下：

```
install.go:173: [debug] Original chart version: ""
install.go:190: [debug] CHART PATH: /Users/gsh/projects/k8s/helm/mychart
```

```
NAME: demo
LAST DEPLOYED: Fri Jul 16 21:30:00 2021
NAMESPACE: default
STATUS: pending-install
REVISION: 1
TEST SUITE: None
USER-SUPPLIED VALUES:
{}

COMPUTED VALUES:
affinity: {}
autoscaling:
  enabled: false
  maxReplicas: 100
  minReplicas: 1
  targetCPUUtilizationPercentage: 80
fullnameOverride: ""
image:
  pullPolicy: IfNotPresent
  repository: nginx
  tag: ""
imagePullSecrets: []
ingress:
  annotations: {}
  enabled: false
  hosts:
  - host: chart-example.local
    paths:
    - backend:
        serviceName: chart-example.local
        servicePort: 80
      path: /
  tls: []
mylibchart:
  global: {}
nameOverride: ""
nodeSelector: {}
podAnnotations: {}
podSecurityContext: {}
replicaCount: 1
resources: {}
securityContext: {}
service:
  port: 80
  type: ClusterIP
serviceAccount:
```

```
  annotations: {}
  create: true
  name: ""
tolerations: []

HOOKS:
MANIFEST:
---
# Source: mychart/templates/configmap.yaml
apiVersion: v1
data:
  myvalue: Hello World
kind: ConfigMap
metadata:
  name: mychart-demo
```

9.8 创建自己的 Chart

下面讲解如何为第 1 章介绍的用户认证项目创建一个 Chart，最后使用 Chart 部署。新建文件夹 helm，接着进入新建的文件夹，后续的操作以这个文件夹为工作目录。

创建 Chart.yaml 文件，编辑内容如下：

```
apiVersion: v2
name: accounts
description: A Helm chart for Kubernetes

# A chart can be either an 'application' or a 'library' chart.
#
# Application charts are a collection of templates that can be packaged into versioned archives
# to be deployed.
#
# Library charts provide useful utilities or functions for the chart developer. They're
included as
# a dependency of application charts to inject those utilities and functions into the rendering
# pipeline. Library charts do not define any templates and therefore cannot be deployed.
type: application

# This is the chart version. This version number should be incremented each time you
make changes
# to the chart and its templates, including the app version.
# Versions are expected to follow Semantic Versioning (https://semver.org/)
version: 0.1.0
```

```
# This is the version number of the application being deployed. This version number should be
# incremented each time you make changes to the application. Versions are not expected to
# follow Semantic Versioning. They should reflect the version the application is using.
# It is recommended to use it with quotes.
appVersion: "1.16.0"
```

9.8.1 后端服务

为用户认证项目中的后端服务 api 创建 Chart 模板，创建文件 templates/api-deployment.yaml，编辑内容如下：

```
apiVersion: apps/v1 # for versions before 1.9.0 use apps/v1beta2
kind: Deployment
metadata:
  name: api
  labels:
    {{- include "api.labels" . | nindent 4}}
spec:
  selector:
    matchLabels:
      {{- include "api.selectorLabels" . |nindent 6}}
  strategy:
    type: Recreate
  template:
    metadata:
      labels:
        {{- include "api.selectorLabels" . |nindent 8}}
    spec:
      containers:
        - image: "{{ .Values.api.image.repository }}:{{ .Values.api.image.tag | default .
Chart.AppVersion }}"
          name: api
          ports:
            - containerPort: {{ required "A valid .Values.api.server.port required" .Values.api.
server.port }}
              name: web
          volumeMounts:
            - name: secret-config
              mountPath: "/config"
              readOnly: true
      volumes:
        - name: secret-config
          secret:
            secretName: api-secret
```

此文件中使用了2个命名模板，分别是api.selectorLabels和api.labels，用于复用与标签相关的定义。按照惯例，它们需要在templates下的_helpers.tpl文件中定义。

新建templates/_helpers.tpl文件，编辑内容如下：

```
{{/*
Selector labels
*/}}
{{- define "api.selectorLabels" -}}
app: api
tier: backend
{{- end }}

{{/*
labels
*/}}
{{- define "api.labels" -}}
{{ include "api.selectorLabels" . }}
{{ include "common.labels" . }}
{{- end }}
{{/*

{{/*
Create chart name and version as used by the chart label.
*/}}
{{- define "api.chart" -}}
{{ printf "%s-%s" .Chart.Name .Chart.Version | replace "+" "_" | trunc 63 | trimSuffix "-"
}}
{{- end }}

Common labels
*/}}
{{- define "common.labels" -}}
helm.sh/chart: {{ include "api.chart" . }}
{{- if .Chart.AppVersion }}
app.kubernetes.io/version: {{ .Chart.AppVersion | quote }}
{{- end }}
app.kubernetes.io/managed-by: {{ .Release.Service }}
{{- end }}
```

api-deployment.yaml中通过值文件读取容器镜像配置，代码如下：

```
        - image: "{{ .Values.api.image.repository }}:{{ .Values.api.image.tag | default .
Chart.AppVersion }}"
```

另外通过值文件读取容器暴露的端口，代码如下：

```
  - containerPort: {{ required "A valid .Values.api.server.port required" .Values.api.server.port }}
```

新建 values. yaml 文件,设置模板中引用的值,编辑内容如下:

```
api:
  server:
    port: 80
  image:
    repository: bitmyth/accounts
    pullPolicy: IfNotPresent
    tag: "v1.1.7"
```

api-deployment. yaml 中定义了一个名为 secret-config 的卷,卷的数据来源是名为 api-secret 的 Secret 对象,代码如下:

```
volumes:
  - name: secret-config
    secret:
      secretName: api-secret
```

这个 Secret 中包含了后端服务所需要的配置文件 plain. yaml 和 secret. yaml。

新建文件 templates/api-secret. yaml,编辑内容如下:

```
apiVersion: v1
kind: Secret
metadata:
  name: api-secret
  namespace: {{ .Release.Namespace | quote }}
type: Opaque
stringData:
  plain.yaml: |
    server:
      port: {{ required "A valid .Values.api.server.port required" .Values.api.server.port }}
    locale: {{ .Values.api.locale }}
  secret.yaml: |-
    database:
      {{- .Values.mysql.auth | toYaml | nindent 6 }}
```

此文件中引用了值文件中定义的 mysql. auth 字段。toYamll 指令用于把 mysql. auth 定义的 map 转换为 YAML 格式的字符串。如果不转换为字符串,模板渲染时则会报错,报错信息如下:

```
wrong type for value; expected string; got map[string]interface {}
```

编辑 values.yaml 文件,添加的 mysql 字段如下:

```
mysql:
  auth:
    host: mysql
    port: 3306
    username: root
    password: 123
    schema: accounts
```

预览模板文件生成的 YAML,命令如下:

```
helm install --generate-name  --dry-run .
```

运行命令后输出的信息如下:

```
# Source: accounts/templates/api-secret.yaml
apiVersion: v1
kind: Secret
metadata:
  name: api-secret
  namespace: "default"
type: Opaque
stringData:
  plain.yaml: |
    server:
      port: 80
    locale: zh-CN
  secret.yaml: |-
    database:
      host: mysql
      password: 123
      port: 3306
      schema: accounts
      username: root
---
# Source: accounts/templates/api-deployment.yaml
apiVersion: apps/v1 # for versions before 1.9.0 use apps/v1beta2
kind: Deployment
metadata:
  name: api
  labels:
    app: api
    tier: backend
    helm.sh/chart: accounts-0.1.0
    app.kubernetes.io/version: "1.16.0"
```

```
      app.kubernetes.io/managed-by: Helm
  spec:
    selector:
      matchLabels:
        app: api
        tier: backend
    strategy:
      type: Recreate
    template:
      metadata:
        labels:
          app: api
          tier: backend
      spec:
        containers:
          - image: "bitmyth/accounts:v1.1.7"
            name: api
            ports:
              - containerPort: 80
                name: web
            volumeMounts:
              - name: secret-config
                mountPath: "/config"
                readOnly: true
        volumes:
          - name: secret-config
            secret:
              secretName: api-secret
```

创建文件 templates/api-service.yaml，编辑内容如下：

```
apiVersion: v1
kind: Service
metadata:
  name: api
  labels:
    {{- include "api.labels" . | nindent 4}}
spec:
  ports:
    - port: {{ required "A valid .Values.api.server.port required" .Values.api.server.port }}
  selector:
    {{- include "api.selectorLabels" . |nindent 6}}
```

其中 required 指令用于校验 port 值是否被指定。如果值为空，则会在渲染过程中报错，并提示 required 指令中指定的错误信息"A valid .Values.api.server.port required"。

api-service.yaml 生成的 YAML 文件的内容如下：

```
---
# Source: accounts/templates/api-service.yaml
apiVersion: v1
kind: Service
metadata:
  name: api
  labels:
    app: api
    tier: backend
    app.kubernetes.io/version: "1.16.0"
    app.kubernetes.io/managed-by: Helm
spec:
  ports:
    - port: 80
  selector:
      app: api
      tier: backend
```

9.8.2 MySQL 服务

后端服务依赖于 MySQL 存储数据，接下来创建 templates/mysql-deployment.yaml 文件，编辑内容如下：

```
apiVersion: apps/v1 # for versions before 1.9.0 use apps/v1beta2
kind: Deployment
metadata:
  name: mysql
  labels:
    {{- include "mysql.labels" . | nindent 4}}
spec:
  selector:
    matchLabels:
      {{- include "mysql.selectorLabels" . |nindent 6}}
  strategy:
    type: Recreate
  template:
    metadata:
      labels:
        {{- include "mysql.selectorLabels" . |nindent 8}}
    spec:
      containers:
        - image: "{{ .Values.mysql.image.repository }}:{{ .Values.mysql.image.tag | default
.Chart.AppVersion }}"
```

```
            name: mysql
            env:
              - name: MYSQL_ROOT_PASSWORD
                valueFrom:
                  secretKeyRef:
                    name: mysql-secret
                    key: password
              - name: MYSQL_DATABASE
                valueFrom:
                  secretKeyRef:
                    name: mysql-secret
                    key: database
            ports:
              - containerPort: 3306
                name: mysql
            args: ["--character-set-server=utf8mb4","--default-time-zone=+08:
00","--ignore-db-dir=lost+found"]
            volumeMounts:
              - name: mysql-persistent-storage
                mountPath: /var/lib/mysql
        volumes:
          - name: mysql-persistent-storage
            persistentVolumeClaim:
              claimName: mysql-pv-claim
```

与 api-deployment.yaml 类似，此模板中也引用了命名模板，分别是 mysql.labels 和 mysql.selectorLabels。在 templates/_helpers.tpl 中添加命名模板 mysql.selectorLabels，代码如下：

```
{{/*
Selector labels
*/}}
{{- define "mysql.selectorLabels" -}}
app: mysql
tier: mysql
{{- end }}
{{/*
Common labels
*/}}
```

在 templates/_helpers.tpl 中添加命名模板 mysql.labels，代码如下：

```
{{- define "mysql.labels" -}}
helm.sh/chart: {{ include "api.chart" . }}
{{ include "mysql.selectorLabels" . }}
```

```
{{- include "common.labels" . }}
{{- end }}
```

mysql-deployment. yaml 中使用了 mysql-secret 这个 Secret,以此获取数据库配置。

新建文件 mysql-secret. yaml,编辑内容如下:

```
apiVersion: v1
kind: Secret
metadata:
  name: mysql-secret
  namespace: {{ .Release.Namespace | quote }}
type: Opaque
stringData:
  {{- with .Values.mysql.auth }}
  host: {{ .host | quote }}
  username: {{ .username | quote }}
  password: {{ .password | quote }}
  database: {{ .schema | quote }}
  port: {{ .port | quote }}
  {{- end }}
```

mysql-secret. yaml 模板生成的 YAML 的文件内容如下:

```
# Source: accounts/templates/mysql-secret.yaml
apiVersion: v1
kind: Secret
metadata:
  name: mysql-secret
  namespace: "default"
type: Opaque
stringData:
  host: "mysql"
  username: "root"
  password: "123"
  database: "accounts"
  port: "3306"
```

mysql-deployment. yaml 中使用了名为 mysql-pv-claim 的 PVC。下面定义 mysql-pv-claim 的模板文件。

新建 mysql-pvc. yaml 文件,编辑内容如下:

```
apiVersion: v1
kind: PersistentVolumeClaim
metadata:
```

```
    name: mysql-pv-claim
    labels:
      app: accounts
spec:
  storageClassName: {{ .Values.storage.class }}
  accessModes:
    - ReadWriteOnce
  resources:
    requests:
      storage: {{ .Values.storage.size | quote }}
```

相应地，在 values. yaml 文件添加被引用的值，代码如下：

```
storage:
  class: aws-ebs
  size: 20Gi
```

新建文件 templates/ebs-storageclass. yaml，编辑内容如下：

```
apiVersion: storage.k8s.io/v1
kind: StorageClass
metadata:
  name: aws-ebs
  annotations:
    storageclass.kubernetes.io/is-default-class: "true"
provisioner: kubernetes.io/aws-ebs
parameters:
  type: gp2
  fsType: ext4
reclaimPolicy: Retain
allowVolumeExpansion: true
mountOptions:
  - debug
volumeBindingMode: WaitForFirstConsumer
```

最后创建 templates/mysql-service. yaml 文件，编辑内容如下：

```
apiVersion: v1
kind: Service
metadata:
  name: mysql
  labels:
    {{- include "mysql.labels" . | nindent 4}}
spec:
  ports:
    - port: {{ required "A valid .Values.mysql.auth.port required" .Values.mysql.auth.port }}
```

```
  selector:
    {{- include "mysql.selectorLabels" . |nindent 6}}
```

9.8.3 前端服务

新建文件 templates/web-deployment.yaml，编辑内容如下：

```
apiVersion: apps/v1 # for versions before 1.9.0 use apps/v1beta2
kind: Deployment
metadata:
  name: web
  labels:
    {{- include "web.labels" . | nindent 4}}
spec:
  selector:
    matchLabels:
      {{- include "web.selectorLabels" . |nindent 6}}
  strategy:
    type: Recreate
  template:
    metadata:
      labels:
        {{- include "web.selectorLabels" . |nindent 8}}
    spec:
      containers:
        - image: "{{ .Values.web.image.repository }}:{{ .Values.web.image.tag | default .Chart
.AppVersion }}"
          name: web
          ports:
            - containerPort: 80
              name: web
          readinessProbe:
            exec:
              command:
                - curl
                - api
            initialDelaySeconds: 5
            periodSeconds: 30
```

模板中引用了 2 个命名模板，分别是 web.labels 和 web.selectorLabels。

编辑 templates/_helpers.tpl 文件，添加这 2 个命名模板定义，代码如下：

```
{{/*
Selector labels
*/}}
```

```
{{- define "web.selectorLabels" -}}
app: web
tier: frontend
{{- end }}

{{/*
Web labels
*/}}
{{- define "web.labels" -}}
{{ include "web.selectorLabels" . }}
{{ include "common.labels" . }}
{{- end }}
```

编辑 values.yaml 文件，添加 templates/web-deployment.yaml 模板中引用的值，代码如下：

```
web:
  port: 80
  image:
    repository: bitmyth/accounts-frontend
    pullPolicy: IfNotPresent
    tag: "v1.1.2"
```

创建 templates/web-service.yaml 文件，编辑内容如下：

```
apiVersion: v1
kind: Service
metadata:
  name: web
  labels:
    {{- include "web.labels" . | nindent 4}}
spec:
  ports:
    - port: {{ required "Values.web.port required" .Values.web.port }}
  selector:
    {{- include "web.selectorLabels" . |nindent 6}}
  type: NodePort
```

9.8.4 数据迁移任务

新建文件 templates/migration-job.yaml，编辑内容如下：

```
apiVersion: batch/v1
kind: Job
metadata:
```

```
    name: migration
  spec:
    template:
      spec:
        containers:
          - name: migration
            image: bitmyth/goose:v1.0.0
            command: ["goose",  "up"]
            workingDir: /migrations
            env:
              - name: GOOSE_DRIVER
                value: mysql
              - name: GOOSE_DBSTRING
                value: {{ .Values.mysql.auth.username }}:{{ .Values.mysql.auth.password  }}
@tcp({{ .Values.mysql.auth.host }})/accounts
            volumeMounts:
              - name: migration
                mountPath: /migrations
        restartPolicy: Never
        initContainers:
          - name: wait-db
            image: busybox:1.28
            command: ['sh', '-c', "until telnet mysql 3306; do echo waiting for mysql; sleep 1;
done"]
        volumes:
          - name: migration
            configMap:
              name: migration
    backoffLimit: 4
```

此文件引用了名为 migration 的 ConfigMap 并作为卷挂载到容器中的/migrations 目录，这个 ConfigMap 中包含了迁移数据的 SQL 文件。

新建文件 templates/migration-configmap.yaml，编辑内容如下：

```
apiVersion: v1
kind: ConfigMap
metadata:
  name: migration
data:
{{ (.Files.Glob "migrations/*.sql").AsConfig | indent 2 }}
```

这里使用了 AsConfig 帮助函数，把.Files.Glob 函数返回的文件转换为 ConfigMap 中的数据项。

接下来把后端项目中 src/database/migrations 目录复制到当前 Chart 的根目录下，这样上述模板中.Files.Glob "migrations/*.sql"就可以返回包含了迁移数据的 SQL 文件。

9.8.5 Ingress

新建文件 templates/ingress.yaml，编辑内容如下：

```
{{- if .Values.ingress.enabled -}}
apiVersion: networking.k8s.io/v1
kind: Ingress
metadata:
  name: app-ingress
  {{- with .Values.ingress.annotations }}
  annotations:
    {{- toYaml . | nindent 4 }}
  {{- end }}
spec:
  {{- if .Values.ingress.tls }}
  tls:
  {{- range .Values.ingress.tls }}
    - hosts:
        {{- range .hosts }}
        - {{ . | quote }}
        {{- end }}
      secretName: {{ .secretName }}
  {{- end }}
  {{- end }}
  defaultBackend:
    service:
      name: web
      port:
        number: 80
  rules:
    {{- range .Values.ingress.hosts }}
    - host: {{ .host | quote }}
      http:
        paths:
          {{- range .http.paths }}
          - path: {{ .path }}
            pathType: Prefix
            backend:
              service:
                name: {{ .backend.service.name }}
                port:
                  number: {{ .backend.service.port.number }}
          {{- end }}
    {{- end }}
{{- end }}
```

模板中使用 range 遍历 ingress. hosts 列表，生成对应的 Ingress 规则。

相应地，在 values. yaml 文件中添加被引用的值，代码如下：

```
ingress:
  enabled: true
  annotations:
    nginx.ingress.kubernetes.io/rewrite-target: /
    kubernetes.io/ingress.class: "nginx"
    cert-manager.io/issuer: "letsencrypt-prod"
  hosts:
    - host: unit.ink
      http:
        paths:
          - path: /
            pathType: Prefix
            backend:
              service:
                name: web
                port:
                  number: 80
    - host: api.unit.ink
      http:
        paths:
          - path: /
            pathType: Prefix
            backend:
              service:
                name: api
                port:
                  number: 80
    - host: kibana.unit.ink
      http:
        paths:
          - path: /
            pathType: Prefix
            backend:
              service:
                name: kibana
                port:
                  number: 5601
  tls:
    - secretName: unit-tls
      hosts:
        - unit.ink
        - "*.unit.ink"
```

模板 templates/ingress. yaml 生成的 YAML 的文件内容如下：

```
# Source: accounts/templates/ingress.yaml
apiVersion: networking.k8s.io/v1
kind: Ingress
metadata:
  name: app-ingress
  annotations:
    cert-manager.io/issuer: letsencrypt-prod
    kubernetes.io/ingress.class: nginx
    nginx.ingress.kubernetes.io/rewrite-target: /
spec:
  tls:
    - hosts:
        - "unit.ink"
        - "*.unit.ink"
      secretName: unit-tls
  defaultBackend:
    service:
      name: web
      port:
        number: 80
  rules:
    - host: "unit.ink"
      http:
        paths:
          - path: /
            pathType: Prefix
            backend:
              service:
                name: web
                port:
                  number: 80
    - host: "api.unit.ink"
      http:
        paths:
          - path: /
            pathType: Prefix
            backend:
              service:
                name: api
                port:
                  number: 80
    - host: "kibana.unit.ink"
      http:
        paths:
```

```
        - path: /
          pathType: Prefix
          backend:
            service:
              name: kibana
              port:
                number: 5601
```

至此，用户认证项目本身的 Chart 定义就完成了，下面安装这个 Chart。

9.8.6 安装 Chart

在安装这个 Chart 之前可以先验证一下模板是否有问题，命令如下：

```
helm install  --generate-name --dry-run
```

在 Chart 的根目录下执行的安装命令如下：

```
helm install --generate-name .
```

运行命令后输出的信息如下：

```
NAME: chart-1627090994
LAST DEPLOYED: Sat Jul 24 09:43:18 2021
NAMESPACE: default
STATUS: deployed
REVISION: 1
TEST SUITE: None
```

由于 Chart 没有添加 NOTES.txt 文件，所以没有输出额外的帮助信息。

到这里应用的主体部分就安装完成了。

为了把应用暴露到外部，还需要使 Ingress 生效，这就需要安装一个 Ingress Controller。

接下来使用 Helm 安装 Nginx Ingress Controller，首先添加 Ingress Nginx 的 Chart 仓库，命令如下：

```
helm repo add ingress-nginx https://kubernetes.github.io/ingress-nginx
```

然后更新本地仓库，命令如下：

```
helm repo update
```

安装 Nginx Ingress Controller，命令如下：

```
helm install ingress-nginx ingress-nginx/ingress-nginx
```

安装成功后，会看到 app-ingress 会被分配到负载均衡地址，也就是访问应用的入口。

生产环境启用 HTTPS，可以使用 Helm 安装 cert-manager。

首先添加 cert-manager 官方的 Helm 仓库，命令如下：

```
helm repo add jetstack https://charts.jetstack.io
```

接着更新本地仓库缓存，命令如下：

```
helm repo update
```

安装 cert-manager，命令如下：

```
helm install \
  cert-manager jetstack/cert-manager \
  --namespace cert-manager \
  --create-namespace \
  --version v1.4.0 \
  --set installCRDs=true
```

命令中的--set installCRDs=true 表示安装 cert-manager 所需的一系列 Kubernetes CRD 资源。

运行命令后输出的信息如下：

```
cert-manager jetstack/cert-manager \
  --namespace cert-manager \
  --create-namespace \
  --version v1.4.0 \
  --set installCRDs=true
NAME: cert-manager
LAST DEPLOYED: Sat Jul 24 20:58:41 2021
NAMESPACE: cert-manager
STATUS: deployed
REVISION: 1
TEST SUITE: None
NOTES:
cert-manager has been deployed successfully!

In order to begin issuing certificates, you will need to set up a ClusterIssuer
or Issuer resource (for example, by creating a 'letsencrypt-staging' issuer).

More information on the different types of issuers and how to configure them
can be found in our documentation:

https://cert-manager.io/docs/configuration/
```

```
For information on how to configure cert-manager to automatically provision
Certificates for Ingress resources, take a look at the 'ingress-shim'
documentation:

https://cert-manager.io/docs/usage/ingress/
```

如果想在安装前查看安装的 YAML 文件，则可以使用 helm template 命令，命令如下：

```
helm template \
  cert-manager jetstack/cert-manager \
  --namespace cert-manager \
  --create-namespace \
  --version v1.4.0 \
  --set installCRDs=true
```

安装结束后，需要手动创建 Issuer，具体可以参考 8.8.2 节。新建文件 issuer-letsencrypt-prod.yaml，编辑内容如下：

```
apiVersion: cert-manager.io/v1
kind: Issuer
metadata:
 name: letsencrypt-prod
spec:
 acme:
   # The ACME server URL
   server: https://acme-v02.api.letsencrypt.org/directory
   # Email address used for ACME registration
   email: fishis@163.com
   # Name of a secret used to store the ACME account private key
   privateKeySecretRef:
     name: letsencrypt-prod
   # Enable the HTTP-01 challenge provider
   solvers:
   - http01:
       ingress:
         class: nginx
```

使用此文件创建 Issuer，命令如下：

```
kubectl apply -f issuer-letsencrypt-prod.yaml
```

等待一分钟左右，就可以完成证书签发的流程了。

第10章 服务网格

10.1 服务网格简介

分布式环境下服务间通信会因为网络等问题而失败，这就需要在应用中加入对服务间通信的控制。例如通信失败后自动重试、速率控制、负载均衡、故障转移等。除此之外，为了更好地监视和理解应用状态和对故障进行定位追踪及排除，需要对服务间通信进行监控。

服务网格(Service Mesh)就是解决服务间通信难题的最佳方案，通过代理服务间的流量以低侵入性的方式实现了流量管理。服务网格把服务通信控制与应用解耦，应用代码中不再需要处理与业务无关的通信控制。

开源的服务网格实现包括 Istio、Linkerd、Consul 等。

10.2 Linkerd

10.2.1 Linkerd 简介

Linkerd 是针对 Kubernetes 平台开发的轻量的开源服务网格，是 CNCF 孵化中的项目。使用 Linkerd 有助于故障排除，提高可用性和可观测性及安全性，并且 Linkerd 完全独立于应用代码，不需要对应用代码做出任何修改即可使用 Linkerd。最大限度地降低了引入服务网格带来的运维复杂性。

Linkerd 包含 3 个基本组件：UI、数据面和控制面。使用 Linkerd 前需要拥有一个正常运行的 Kubernetes 集群和 kubectl 命令行工具。

控制面提供数据面正常工作所需的命令和控制信号，以及用来配置、监控和运维 Linkerd 的 UI 和 API。

数据面即所有 Pod 中的流量代理容器构成的集合。数据面负责收集流量指标、监测流量并使用相应策略应对。代理容器与应用容器部署在相同的 Pod 中，这种模式称为边车模式。就像摩托车旁边挂了边车一样，边车代理在不修改应用容器的情况下，拦截了应用流量。

UI 即展示各项流量指标的用户界面。

Linkerd 的设计理念是简单、低成本、易用。下面分别介绍：

(1) 简单是 Linkerd 最重要的设计理念，简单并不意味着 Linkerd 不具有强大的功能也不意味着可以一键解决所有问题。简单体现在很少的运维操作和很低的认知负担。用户可以对 Linkerd 中的组件有清晰的认知，可以轻松地理解和预测 Linkerd 的行为。Linkerd 的行为的各方面都被明确划分，并被清晰地定义。例如它的控制面根据功能边界被清晰地划分为了 api 和 web 等组件。

(2) 低成本意味着 Linkerd 会尽最大可能降低它消耗的计算资源。尤其是 Linkerd 的数据面，这样才能在安装 Linkerd 后，应用的性能不会受到很大的影响。在流量波动的情况下，Linkerd 控制面的组件会谨慎地进行优雅的缩放。在数据面，Linkerd 代理不断追求提高性能和安全性，同时降低资源使用率。单个 linkerd-proxy 可以在 1s 内代理数千个请求并且使用不超过 10MB 的内存和四分之一的 CPU，同时 99%的请求的延迟控制在 1ms 内。

(3) 易用体现在 Linkerd 的使用无须复杂的配置，安装 Linkerd 不会对现有的应用造成破坏。为了实现零配置即可工作，Linkerd 做了大量的设计。例如 7 层协议自动检测、Pod 内 TCP 流量自动重路由等。当然如果要定制某些 Linkerd 的行为，则仍需进行配置。

下面使用 Linkerd 把 Kubernetes 集群中的应用服务网格化。

10.2.2　安装 Linkerd

安装 Linkerd 的过程比较简单，只需安装 Linkerd CLI，安装好 CLI 后，使用 CLI 将 Linkerd 控制面安装到 Kubernetes 集群。最后把 Linkerd 数据面添加到服务中就完成了服务网格化。

手动安装 CLI，命令如下：

```
curl -sL https://run.linkerd.io/install | sh
```

运行命令后输出的信息如下：

```
Download complete!

Validating checksum...
Checksum valid.

Linkerd stable-2.10.2 was successfully installed

Add the linkerd CLI to your path with:

  export PATH=$PATH:/Users/gsh/.linkerd2/bin

Now run:

  linkerd check --pre                     #validate that Linkerd can be installed
```

```
    linkerd install | kubectl apply -f -    # install the control plane into the 'linkerd' namespace
    linkerd check                            # validate everything worked!
    linkerd dashboard                        # launch the dashboard

Looking for more? Visit https://linkerd.io/2.10/next-steps
```

运行命令后输出信息中包含如何把CLI添加到PATH环境变量的指令。根据指令把Linkerd可执行文件添加到PATH中。

除了可以使用脚本手动安装Linkerd,还可以通过其他包管理器安装,例如使用Homebrew安装Linkerd,命令如下:

```
brew install linkerd
```

验证Linkerd是否可以正常工作,命令如下:

```
linkerd version
```

运行命令后输出的信息如下:

```
Client version: stable-2.10.2
Server version: unavailable
```

输出信息中包含了客户端版本号和服务器端版本号,Server version: unavailable的意思是无法获取服务器端版本号,这是因为到现在还没有在Kubernetes中安装Linkerd控制面。

安装Linkerd控制面之前检查Kubernetes集群,命令如下:

```
linkerd check --pre
```

运行命令后输出的信息如下:

```
kubernetes-api
--------------
√ can initialize the client
√ can query the Kubernetes API

kubernetes-version
------------------
√ is running the minimum Kubernetes API version
√ is running the minimum kubectl version

pre-kubernetes-setup
--------------------
√ control plane namespace does not already exist
```

```
√ can create non-namespaced resources
√ can create ServiceAccounts
√ can create Services
√ can create Deployments
√ can create CronJobs
√ can create ConfigMaps
√ can create Secrets
√ can read Secrets
√ can read extension-apiserver-authentication configmap
√ no clock skew detected

pre-kubernetes-capability
--------------------------
√ has NET_ADMIN capability
√ has NET_RAW capability

linkerd-version
----------------
√ can determine the latest version
√ cli is up-to-date

Status check results are √
```

输出信息显示所有的检查项都已经通过，如果有任意一项没有通过，则需要根据提供的相关链接及时修复，这样才能继续安装。

通过检查后，现在将 Linkerd 控制面安装到 Kubernetes 集群，命令如下：

```
linkerd install | kubectl apply -f -
```

命令中的 linkerd install 会输出安装控制面的配置文件，通过管道符输入 kubectl apply-f -命令，运行命令后输出的信息如下：

```
namespace/linkerd created
clusterrole.rbac.authorization.k8s.io/linkerd-linkerd-identity created
clusterrolebinding.rbac.authorization.k8s.io/linkerd-linkerd-identity created
serviceaccount/linkerd-identity created
clusterrole.rbac.authorization.k8s.io/linkerd-linkerd-controller created
clusterrolebinding.rbac.authorization.k8s.io/linkerd-linkerd-controller created
serviceaccount/linkerd-controller created
clusterrole.rbac.authorization.k8s.io/linkerd-linkerd-destination created
clusterrolebinding.rbac.authorization.k8s.io/linkerd-linkerd-destination created
serviceaccount/linkerd-destination created
role.rbac.authorization.k8s.io/linkerd-heartbeat created
rolebinding.rbac.authorization.k8s.io/linkerd-heartbeat created
clusterrole.rbac.authorization.k8s.io/linkerd-heartbeat created
```

```
clusterrolebinding.rbac.authorization.k8s.io/linkerd-heartbeat created
serviceaccount/linkerd-heartbeat created
customresourcedefinition.apiextensions.k8s.io/serviceprofiles.linkerd.io created
customresourcedefinition.apiextensions.k8s.io/trafficsplits.split.smi-spec.io created
clusterrole.rbac.authorization.k8s.io/linkerd-linkerd-proxy-injector created
clusterrolebinding.rbac.authorization.k8s.io/linkerd-linkerd-proxy-injector created
serviceaccount/linkerd-proxy-injector created
secret/linkerd-proxy-injector-k8s-tls created
mutatingwebhookconfiguration.admissionregistration.k8s.io/linkerd-proxy-injector-webhook-
config created
clusterrole.rbac.authorization.k8s.io/linkerd-linkerd-sp-validator created
clusterrolebinding.rbac.authorization.k8s.io/linkerd-linkerd-sp-validator created
serviceaccount/linkerd-sp-validator created
secret/linkerd-sp-validator-k8s-tls created
validatingwebhookconfiguration.admissionregistration.k8s.io/linkerd-sp-validator-webhook-
config created
podsecuritypolicy.policy/linkerd-linkerd-control-plane created
role.rbac.authorization.k8s.io/linkerd-psp created
rolebinding.rbac.authorization.k8s.io/linkerd-psp created
configmap/linkerd-config created
secret/linkerd-identity-issuer created
service/linkerd-identity created
service/linkerd-identity-headless created
deployment.apps/linkerd-identity created
service/linkerd-controller-api created
deployment.apps/linkerd-controller created
service/linkerd-dst created
service/linkerd-dst-headless created
deployment.apps/linkerd-destination created
cronjob.batch/linkerd-heartbeat created
deployment.apps/linkerd-proxy-injector created
service/linkerd-proxy-injector created
service/linkerd-sp-validator created
deployment.apps/linkerd-sp-validator created
secret/linkerd-config-overrides created
```

安装过程的用时取决于 Kubernetes 集群的网络速度，可能会花费一两分钟。

检查 linkerd 控制面是否安装成功，命令如下：

```
linkerd check
```

在安装成功的情况下，运行命令后输出的类似信息如下：

```
kubernetes-api
--------------
```

```
√ can initialize the client
√ can query the Kubernetes API

kubernetes-version
------------------
√ is running the minimum Kubernetes API version
√ is running the minimum kubectl version

linkerd-existence
-----------------
√ 'linkerd-config' config map exists
√ heartbeat ServiceAccount exist
√ control plane replica sets are ready
√ no unschedulable pods
√ controller pod is running

linkerd-config
--------------
√ control plane Namespace exists
√ control plane ClusterRoles exist
√ control plane ClusterRoleBindings exist
√ control plane ServiceAccounts exist
√ control plane CustomResourceDefinitions exist
√ control plane MutatingWebhookConfigurations exist
√ control plane ValidatingWebhookConfigurations exist
√ control plane PodSecurityPolicies exist

linkerd-identity
----------------
√ certificate config is valid
√ trust anchors are using supported crypto algorithm
√ trust anchors are within their validity period
√ trust anchors are valid for at least 60 days
√ issuer cert is using supported crypto algorithm
√ issuer cert is within its validity period
√ issuer cert is valid for at least 60 days
√ issuer cert is issued by the trust anchor

linkerd-webhooks-and-apisvc-tls
-------------------------------
√ proxy-injector webhook has valid cert
√ proxy-injector cert is valid for at least 60 days
√ sp-validator webhook has valid cert
√ sp-validator cert is valid for at least 60 days
```

```
linkerd-api
-----------
√ control plane pods are ready
√ can initialize the client
√ can query the control plane API

linkerd-version
---------------
√ can determine the latest version
√ cli is up-to-date

control-plane-version
---------------------
√ control plane is up-to-date
√ control plane and cli versions match

Status check results are √
```

接下来安装 Linkerd 的扩展 viz，viz 会将 Prometheus 和仪表盘及指标组件安装到 Kubernetes 集群，命令如下：

```
linkerd viz install | kubectl apply -f -
```

运行命令后输出的信息如下：

```
namespace/linkerd-viz created
clusterrole.rbac.authorization.k8s.io/linkerd-linkerd-viz-metrics-api created
clusterrolebinding.rbac.authorization.k8s.io/linkerd-linkerd-viz-metrics-api created
serviceaccount/metrics-api created
serviceaccount/grafana created
clusterrole.rbac.authorization.k8s.io/linkerd-linkerd-viz-prometheus created
clusterrolebinding.rbac.authorization.k8s.io/linkerd-linkerd-viz-prometheus created
serviceaccount/prometheus created
clusterrole.rbac.authorization.k8s.io/linkerd-linkerd-viz-tap created
clusterrole.rbac.authorization.k8s.io/linkerd-linkerd-viz-tap-admin created
clusterrolebinding.rbac.authorization.k8s.io/linkerd-linkerd-viz-tap created
clusterrolebinding.rbac.authorization.k8s.io/linkerd-linkerd-viz-tap-auth-delegator created
serviceaccount/tap created
rolebinding.rbac.authorization.k8s.io/linkerd-linkerd-viz-tap-auth-reader created
secret/tap-k8s-tls created
apiservice.apiregistration.k8s.io/v1alpha1.tap.linkerd.io created
role.rbac.authorization.k8s.io/web created
rolebinding.rbac.authorization.k8s.io/web created
clusterrole.rbac.authorization.k8s.io/linkerd-linkerd-viz-web-check created
clusterrolebinding.rbac.authorization.k8s.io/linkerd-linkerd-viz-web-check created
```

```
clusterrolebinding.rbac.authorization.k8s.io/linkerd-linkerd-viz-web-admin created
clusterrole.rbac.authorization.k8s.io/linkerd-linkerd-viz-web-api created
clusterrolebinding.rbac.authorization.k8s.io/linkerd-linkerd-viz-web-api created
serviceaccount/web created
rolebinding.rbac.authorization.k8s.io/viz-psp created
service/metrics-api created
deployment.apps/metrics-api created
configmap/grafana-config created
service/grafana created
deployment.apps/grafana created
configmap/prometheus-config created
service/prometheus created
deployment.apps/prometheus created
service/tap created
deployment.apps/tap created
clusterrole.rbac.authorization.k8s.io/linkerd-tap-injector created
clusterrolebinding.rbac.authorization.k8s.io/linkerd-tap-injector created
serviceaccount/tap-injector created
secret/tap-injector-k8s-tls created
mutatingwebhookconfiguration.admissionregistration.k8s.io/linkerd-tap-injector-webhook-
config created
service/tap-injector created
deployment.apps/tap-injector created
service/web created
deployment.apps/web created
```

完成 viz 的安装后，可以使用 linkerd check 再次验证是否一切正常。

如果一切正常，接下来就可以查看 Linkerd 的仪表盘了，命令如下：

```
linkerd viz dashboard &
```

这样会建立一个 linkerd-web Pod，然后到本机的端口转发，& 符号会让进程进入后台运行。

运行命令后输出的类似进程号如下：

```
[1] 26918
```

接着输出的类似信息如下：

```
Linkerd dashboard available at:
http://localhost:50750
Grafana dashboard available at:
http://localhost:50750/grafana
Opening Linkerd dashboard in the default browser
```

最后会自动打开浏览器页面并显示 Linkerd 仪表盘，如图 10-1 所示。

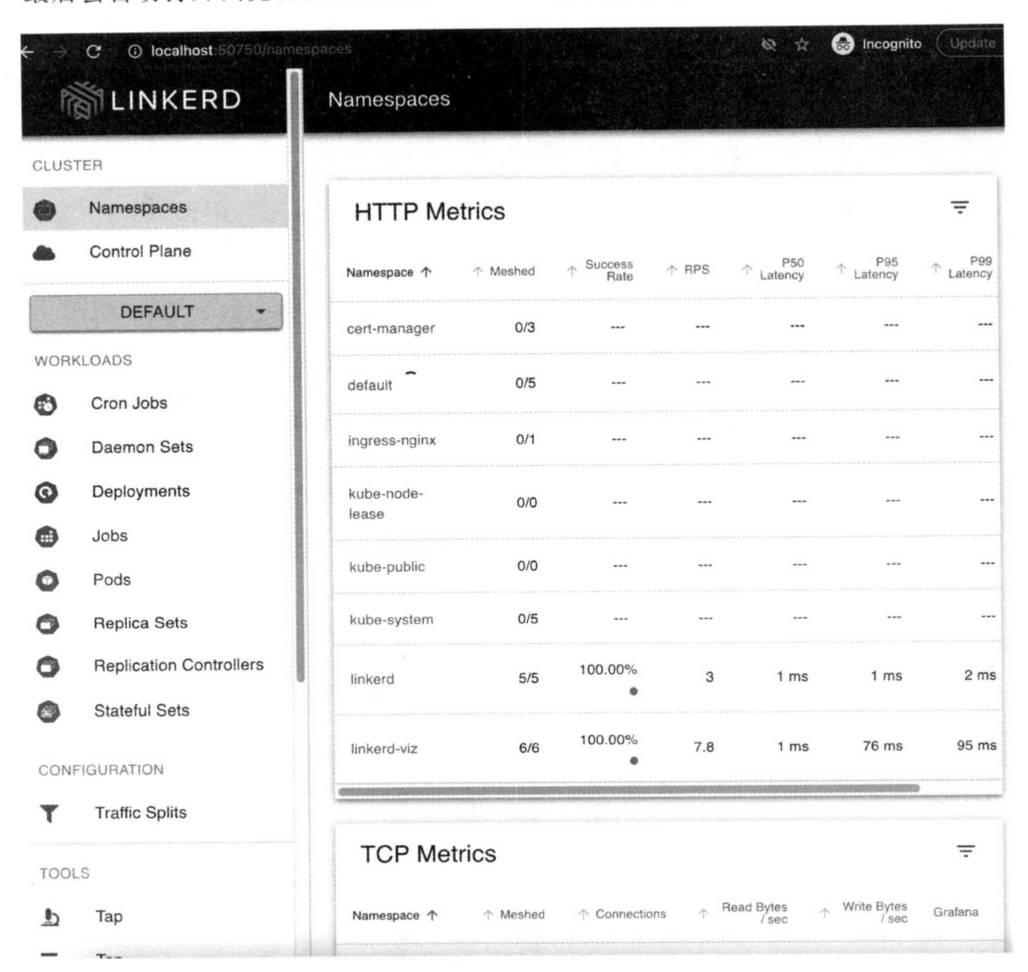

图 10-1　Linkerd 仪表盘首页

10.2.3　网格化

在应用中加入 Linker 数据平面以代理流量，命令如下：

```
kubectl get -n default deploy -o yaml \
  | linkerd inject - \
  | kubectl apply -f -
```

这个命令会查询 default 命名空间下的所有 Deployment，然后把 Deployment 以 YAML 格式输出到 linkerd inject 命令中进行处理，最后把处理结果再次应用到 Kubernetes 集群中。

linkerd inject 命令会修改在 Deployment 中的 Pod 模板，在其中添加一条注解，代码如下：

```
template:
  metadata:
    annotations:
      linkerd.io/inject: enabled
```

这个 linkerd.io/inject：enabled 注解会触发 Linkerd 在 Pod 中注入一个流量代理容器。

运行上述命令后输出的信息如下：

```
deployment "api" injected
deployment "elasticsearch" injected
deployment "kibana" injected
deployment "mysql" injected
deployment "web" injected

deployment.apps/api configured
deployment.apps/elasticsearch configured
deployment.apps/kibana configured
deployment.apps/mysql configured
deployment.apps/web configured
```

输出显示 Linkerd 数据面代理已经注入所有 default 命名空间下的 Deployment 中的 Pod 内。运行命令后刷新浏览器中的 Linkerd 仪表盘页面，页面中会显示 HTTP 指标，如图 10-2 所示。

HTTP Metrics

Namespace ↑	↑ Meshed	↑ Success Rate	↑ RPS	↑ P50 Latency	↑ P95 Latency	↑ P99 Latency	Grafana
cert-manager	0/3	---	---	---	---	---	
default	5/5	100.00% ●	3.28	1 ms	25 ms	4.61 s	

图 10-2　Linkerd 仪表盘中的 HTTP 指标

页面中显示 default 命名空间下的 HTTP 指标数据已经被 Linkerd 展示出来了。证明 Linkerd 已经被成功安装到应用中了。

10.2.4　代理自动注入

Linkerd 可自动将代理添加到具有 linkerd.io/inject：enabled 注解的 Pod 中，这个功能

称为代理注入。基于 Kubernetes admission webhook 实现。

查看 Linkerd 的 ValidatingWebhookConfiguration，命令如下：

```
kubectl get validatingwebhookconfigurations.admissionregistration.k8s.io \
linkerd-sp-validator-webhook-config -o yaml
```

运行命令后输出的信息如下：


```
apiVersion: admissionregistration.k8s.io/v1
kind: ValidatingWebhookConfiguration
metadata:
  annotations:
    kubectl.kubernetes.io/last-applied-configuration: |
      {"apiVersion":"admissionregistration.k8s.io/v1","kind":"ValidatingWebhookConfiguration",
"metadata":{"annotations":{},"labels":{"linkerd.io/control-plane-component":"sp-
validator","linkerd.io/control-plane-ns":"linkerd"},"name":"linkerd-sp-validator-
webhook-config"},"webhooks":[{"admissionReviewVersions":["v1","v1beta1"],"clientConfig":
{"caBundle":"...","service":{"name":"linkerd-sp-validator","namespace":"linkerd",
"path":"/"}},"failurePolicy":"Ignore","name":"linkerd-sp-validator.linkerd.io",
"namespaceSelector":{"matchExpressions":[{"key":"config.linkerd.io/admission-webhooks",
"operator":"NotIn","values":["disabled"]}]},"rules":[{"apiGroups":["linkerd.io"],
"apiVersions":["v1alpha1","v1alpha2"],"operations":["CREATE","UPDATE"],"resources":
["serviceprofiles"]}],"sideEffects":"None"}]}
  creationTimestamp: "2021-06-26T01:12:09Z"
  generation: 1
  labels:
    linkerd.io/control-plane-component: sp-validator
    linkerd.io/control-plane-ns: linkerd
  name: linkerd-sp-validator-webhook-config
  resourceVersion: "8847"
  uid: 962f59a3-ff00-4faa-abf2-fa0564ba2f3c
webhooks:
- admissionReviewVersions:
  - v1
  - v1beta1
  clientConfig:
    caBundle: ...
    service:
      name: linkerd-sp-validator
      namespace: linkerd
      path: /
      port: 443
  failurePolicy: Ignore
  matchPolicy: Equivalent
  name: linkerd-sp-validator.linkerd.io
```

```
  namespaceSelector:
    matchExpressions:
    - key: config.linkerd.io/admission-webhooks
      operator: NotIn
      values:
      - disabled
  objectSelector: {}
  rules:
  - apiGroups:
    - linkerd.io
    apiVersions:
    - v1alpha1
    - v1alpha2
    operations:
    - CREATE
    - UPDATE
    resources:
    - serviceprofiles
    scope: '*'
  sideEffects: None
  timeoutSeconds: 10
```

为了简洁起见，在上述输出中删去了 caBundle 字段的值。

Linkerd 会在 Pod 中注入 2 个容器，分别是 linkerd-init 和 linkerd-proxy，下面分别介绍。

linkerd-init 是一个 Kubernetes InitContainer，它会对 iptables 进行配置，以便使进出的流量都通过代理转发。

linkerd-proxy 容器是转发流量的代理。

如果想禁用自动注入，则可以在 Pod 中添加的注解如下：

```
linkerd.io/inject: disabled
```

在 10.2.3 节中完成了服务网格化，Linkerd 数据面代理已经自动注入 Pod 中，以 api Deployment 中的 Pod(api-54f476c57d-n75n4)为例，查看 Pod 中的容器，命令如下：

```
kubectl get pod api-54f476c57d-n75n4 \
 -o jsonpath='{.spec.containers[*].name}'
```

运行命令后输出的信息如下：

```
api linkerd-proxy
```

输出信息显示 Pod 中注入了代理容器 linkerd-proxy。

在注入代理的时候，Linkerd 使用 Init Container 完成对 iptables 的配置，实现截获 Pod 上的所有流量。查看 Pod(api-54f476c57d-n75n4)中的 Init Container，命令如下：

```
kubectl describe pod api-54f476c57d-n75n4
```

输出信息中的 Init Containers 字段如下：

```
Init Containers:
  linkerd-init:
    Container ID:   docker://0c36e29d1a8e0579d286c9136b718412bd04e4220495848dcc8e6aa1253e
1cb7
    Image:          cr.l5d.io/linkerd/proxy-init:v1.3.11
    Image ID:       docker-pullable://cr.l5d.io/linkerd/proxy-init@sha256:9a7d17d2bce72
74f5abfd611d34e24460a59c55e55ff93440c31ab6b60b542b2
    Port:           <none>
    Host Port:      <none>
    Args:
      --incoming-proxy-port
      4143
      --outgoing-proxy-port
      4140
      --proxy-uid
      2102
      --inbound-ports-to-ignore
      4190,4191
    State:          Terminated
      Reason:       Completed
      Exit Code:    0
      Started:      Sat, 26 Jun 2021 17:40:53 +0800
      Finished:     Sat, 26 Jun 2021 17:40:53 +0800
    Ready:          True
    Restart Count:  0
    Limits:
      cpu:     100m
      memory:  50Mi
    Requests:
      cpu:        10m
      memory:     10Mi
    Environment:  <none>
    Mounts:
      /run from linkerd-proxy-init-xtables-lock (rw)
      /var/run/secrets/kubernetes.io/serviceaccount from default-token-vfcmb (ro)
```

现在流量由 Linkerd 代理，所以 Linkerd 可以统计各项流量指标，查看流量指标，命令如下：

```
linkerd -n default viz stat deploy
```

运行命令后输出的信息如下：

```
NAME MESHED          SUCCESS    RPS      LATENCY_P50   LATENCY_P95   LATENCY_P99   TCP_CONN
api   1/1            100.00%    0.3rps   1ms           1ms           1ms           2
elasticsearch 1/1    100.00%    2.0rps   1ms           14ms          19ms          6
kibana 1/1           100.00%    0.3rps   1ms           1ms           1ms           1
mysql 1/1            100.00%    0.3rps   1ms           1ms           1ms           3
Web 1/1              100.00%    0.3rps   1ms           1ms           1ms           1
```

输出信息包括请求成功率、请求速率和延迟分布等流量指标。

Linkerd CLI 还提供了 tap 子命令，用于支持实时统计每个请求的数据，以 api Deployment 为例，进一步查看实时请求流的相关数据，命令如下：

```
linkerd -n default viz tap  deploy api
```

运行命令后输出的信息如下：

```
req id=0:0 proxy=in  src=192.168.59.231:41566 dst=192.168.48.240:80 tls=true :method
 =GET :authority=api :path=/
rsp id=0:0 proxy=in  src=192.168.59.231:41566 dst=192.168.48.240:80 tls=true :status
 =200 latency=1132µs
end id=0:0 proxy=in  src=192.168.59.231:41566 dst=192.168.48.240:80 tls=true duration
 =50µs response-length=18B
req id=0:1 proxy=in  src=192.168.59.231:42204 dst=192.168.48.240:80 tls=true :method
 =GET :authority=api :path=/
rsp id=0:1 proxy=in  src=192.168.59.231:42204 dst=192.168.48.240:80 tls=true :status
 =200 latency=785µs
...
```

输出信息显示了 api Deployment 中的每个请求的相关数据，包括源 IP、目标 IP、HTTP 状态码和延迟。

在浏览器中的仪表盘页面也可以查看应用流量指标数据。在首页单击 default 命名空间，会看到流量拓扑图和以 Deployment、Pod 等为单位的流量统计数据。

流量拓扑如图 10-3 所示。

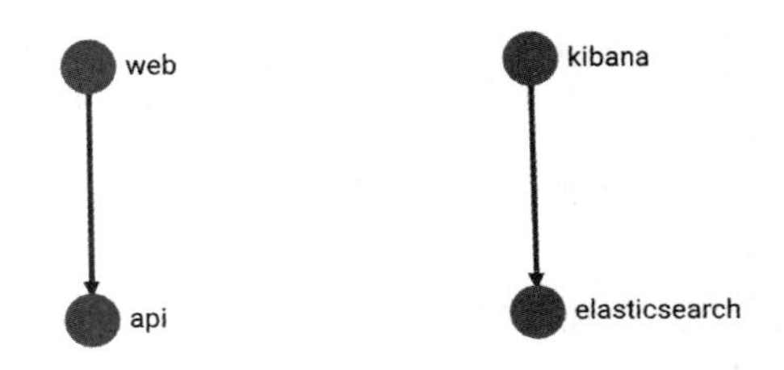

图 10-3　Linkerd 仪表盘中的应用流量拓扑

Deployment 流量指标数据如图 10-4 所示。

Deployment

Deployment ↑	↑ Meshed	↑ Success Rate	↑ RPS	↑ P50 Latency	↑ P95 Latency	↑ P99 Latency	Grafana
api	1/1	100.00% ●	0.33	1 ms	4.80 s	4.96 s	
elasticsearch	1/1	100.00% ●	1.98	1 ms	20 ms	28 ms	
kibana	1/1	100.00% ●	0.3	1 ms	1 ms	1 ms	
mysql	1/1	100.00% ●	0.3	1 ms	1 ms	1 ms	
web	1/1	100.00% ●	0.3	1 ms	1 ms	1 ms	

图 10-4　Linkerd 仪表盘中的 Deployment 流量指标

Pod 流量指标数据如图 10-5 所示。

Pod

Pod ↑	↑ Meshed	↑ Success Rate	↑ RPS	↑ P50 Latency	↑ P95 Latency	↑ P99 Latency	Grafana
api-54f476c57d-n75n4	1/1	100.00% ●	0.33	1 ms	9.00 s	9.80 s	
elasticsearch-687d57c6d4-rrqn4	1/1	100.00% ●	1.97	1 ms	20 ms	28 ms	
kibana-7b88f9ff6-lvqb6	1/1	100.00% ●	0.3	1 ms	1 ms	1 ms	
mysql-7bb8d88f66-2c5lw	1/1	100.00% ●	0.3	1 ms	1 ms	1 ms	
web-59cd78dc5b-gjx5z	1/1	100.00% ●	0.3	1 ms	1 ms	1 ms	

图 10-5　Linkerd 仪表盘中的 Pod 流量指标

单击表格中每一行末尾的图标可以进入 Grafana，进入页面后便可以可视化的方式查看更为详细的实时流量指标。

以 api Deployment 为例，单击 Grafana 图标后跳转到如图 10-6 所示的页面。

图 10-6　Grafana 流量指标相关图表

10.2.5　暴露仪表盘

为了避免每次使用 linkerd viz dashboard 命令才能查看仪表盘，可以使用 Ingress 暴露仪表盘。

下面演示通过 Nginx Ingress 暴露仪表盘的过程。

为了保护 Ingress，采用 basic auth 来认证请求，使用 Secret 保存用户名和密码。

假设用户名和密码都是 admin，使用 htpasswd 生成凭证文件，命令如下：

```
htpasswd -c auth admin
New password:
Re-type new password:
Adding password for user admin
```

使用生成的 auth 文件包含的数据创建 Secret，命令如下：

```
kubectl create secret generic web-ingress-auth --from-file=auth
```

查看生成的 Secret 信息，命令如下：

```
kubectl get secrets web-ingress-auth -o yaml
```

运行命令后输出的信息如下：

```
apiVersion: v1
data:
  auth: YWRtaW46JGFwcjEkdEdUN1FpRk8kcVRhMGk3cHRrRkF0WmJtdmx2UHlzLgo=
kind: Secret
metadata:
  creationTimestamp: "2021-07-24T07:14:20Z"
  name: web-ingress-auth
  namespace: default
  resourceVersion: "42021"
  selfLink: /api/v1/namespaces/default/secrets/web-ingress-auth
  uid: 7c6e2ba8-e00a-4ee6-b132-40c16807abda
type: Opaque
```

接下来就可以在 Ingress 中使用这个 Secret 了，从中可以获取 basic auth 认证凭据。新建文件 linkerd-ingress.yaml，编辑内容如下：

```
apiVersion: networking.k8s.io/v1
kind: Ingress
metadata:
  name: web-ingress
  namespace: linkerd-viz
  annotations:
    kubernetes.io/ingress.class: 'nginx'
    nginx.ingress.kubernetes.io/upstream-vhost: $service_name.$namespace.svc.cluster.
local:8084
    nginx.ingress.kubernetes.io/configuration-snippet: |
      proxy_set_header Origin "";
      proxy_hide_header l5d-remote-ip;
      proxy_hide_header l5d-server-id;
    nginx.ingress.kubernetes.io/auth-type: basic
    nginx.ingress.kubernetes.io/auth-secret: web-ingress-auth
    nginx.ingress.kubernetes.io/auth-realm: 'Authentication Required'
spec:
  rules:
    - http:
        paths:
          - path: /
            pathType: Prefix
            backend:
```

```
            service:
              name: web
              port:
                number: 8084
```

使用此文件创建 Ingress，命令如下：

```
kubectl apply -f linkerd-ingress.yaml
```

运行命令后输出的信息如下：

```
ingress.extensions/web-ingress created
```

查看 Ingress 信息，命令如下：

```
kubectl get ing -n linkerd-viz
```

运行命令后输出的信息如下：

```
NAMECLASSHOSTS ADDRESS                                    PORTS   AGE
web-ingress <none> *                  \
a9126288ed6724dcaa6d73fd829e754d-d1a2cc1418cd96f9.elb.ap-east-1.amazonaws.com   80
    7m2s
```

输出信息中 Address 列显示了 Ingress 分配到的负载均衡器地址。

浏览器访问 Ingress 的地址，会弹出认证窗口，在 Username 栏填入 admin，在 Password 栏也填入 admin，然后单击 Sign In 按钮即可登录，如图 10-7 所示。

图 10-7　basic auth 对话框

然后就可以正常访问仪表盘了，如 10-8 所示。

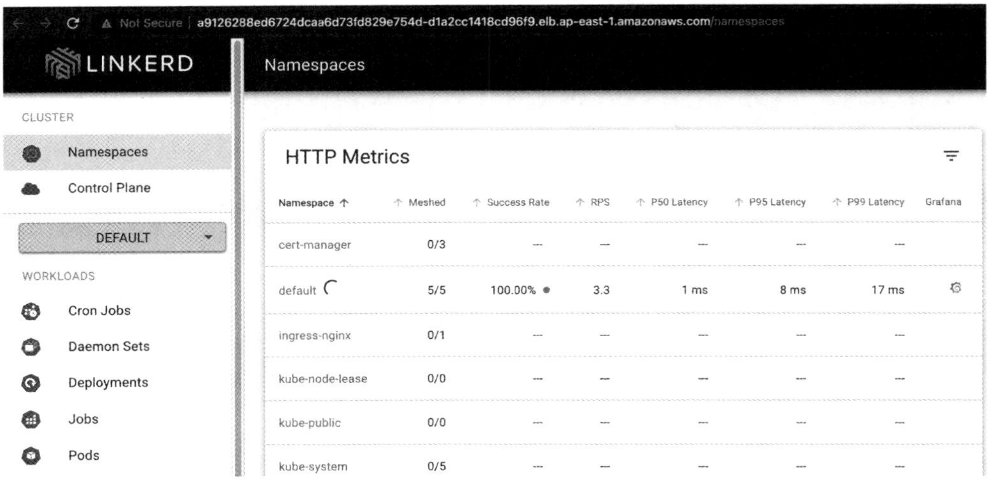

图 10-8 通过 Ingress 访问 Linkerd 仪表盘

为了方便，还可以使用为 Ingress 的规则指定 host，这样就可以通过 DNS 域名访问了。

第11章 云原生现状和展望

11.1 云原生在企业的落地情况

随着云原生理念和相关技术的蓬勃发展,以容器、Kubernetes 等为代表的云原生技术被越来越多的公司引入生产环境。云原生产业生态日趋成熟,进入快速发展期。云原生化需求从行业巨头不断下沉到中小企业,云原生已经成为常态化技术。

云原生使企业可以充分利用云计算带来的低成本且弹性的计算能力,最大化云计算的利用效率,更多聚焦业务发展,节省 IT 基础设施的运营成本,革命性地提升软件应用的可用性、稳定性、业务响应能力及 IT 运营的自动化程度,从而为软件企业带来巨大的竞争优势。

CNCF 会定期对用户进行调研。2020 年 CNCF 在中国进行的第 4 次云原生调研(参考 https://www.cncf.io/blog/2021/04/28/cncf-cloud-native-survey-china-2020/)得出了以下结论:

(1) 在 2020 年,68%的组织在生产过程中使用了容器,比 2 年前增长了 240%。

(2) 在生产环境中使用 Kubernetes 的比例从 72%提升到了 82%。

另外从互联网岗位招聘需求也可以看出来云原生技术的流行,在招聘网站中搜索 Kubernetes 或者云原生可以看到非常多的岗位需求,包括阿里巴巴、百度、小米、京东、美团等知名互联网企业及诸多中小型企业都有云原生相关岗位。这也从侧面印证了云原生带来的巨大价值。

以阿里巴巴为例,2020 年双 11,阿里巴巴实现了核心系统全面云原生化,这是一个重大的技术突破。带来资源效率、研发效率、交付效率的三大提升,万笔交易的资源成本 4 年间下降 80%,研发运维效率平均增效 10%以上,规模化应用交付效率提升了 100%。可以说,阿里巴巴在 2020 年双 11 完成了全球最大规模的云原生实践。

云原生技术不仅在阿里巴巴内部大规模普及,阿里云原生还支撑了中国邮政、申通快递、完美日记、世纪联华等客户,稳定高效应对双 11 大促的流量。

以物流行业为例,申通快递将核心系统搬到云上,采用阿里云容器服务,IT 成本降低了 30%。

以大型商超为例,世纪联华基于阿里云函数计算(FC)弹性扩容,业务峰值 QPS 超过 2019 年双 11 的 230%,研发效率交付提效超过 30%,弹性资源成本减少 40%以上。

在 2020 年疫情期间,基于阿里云容器解决方案,钉钉 2h 内扩容 1 万台云主机,支撑 2 亿上班族在线开工。申通快递将核心系统搬到阿里云上,并进行应用容器化和微服务改造,在日均处理订单 3000 万的情况下,IT 成本降低 50%;采用了阿里云原生 PaaS 平台的中国联通号卡应用,开卡业务效率提升了 10 倍,需求响应时间缩短了 50%,支撑访问量由 1000 万上升至 1.1 亿。

11.2 云厂商对云原生的支持

可以说云原生是为了云计算模型而设计的,云提供的弹性计算资源通过云原生技术可以更便捷地释放技术价值。

主流云厂商对作为云原生技术代表的容器和 Kubernetes 等做了很好的支持,提供了托管的容器和 Kubernetes 集群环境,进一步降低了用户使用云原生技术的门槛。

例如亚马逊提供 Kubernetes 集群服务 Elastic Kubernetes Service(EKS)和容器服务 Elastic Container Service(ECS)。

谷歌提供 Kubernetes 集群服务 Google Kubernetes Engine(GKE)。

阿里云提供 Kubernetes 集群服务容器服务 Kubernetes 版(ACK)和容器服务,即弹性容器实例 Elastic Container Instance(ECI)。

这类支持云原生的产品让用户可以按需使用计算资源,快速扩容缩容,提高业务响应速度,避免资源闲置,从而节省 IT 开支,并且进一步屏蔽了底层服务器等基础设施的管理复杂性,让用户可以更专注于业务。

云原生可以实现云厂商和用户双赢的局面,对用户而言,云原生使计算资源使用效率最大化,降低 IT 运营成本,这种按需付费的集约化的使用方式所节省的成本可以投入业务创新中,进一步提高生产力。对云厂商而言,云原生使更多用户采用云产品,进一步提高规模效益,所以云厂商乐于看到云原生技术的蓬勃发展,并积极参与云原生的生态建设,许多知名的云厂商为云原生开源社区做出了巨大的贡献,积极参与云原生技术的研发和推广。

2020 年 CNCF 在中国进行的第 4 次云原生调研结论表明 CNCF 在中国有 60 个会员,占到全部会员总数的 8%以上。继美国和德国之后,中国是 CNCF 项目第三大贡献者基地(贡献者和提交者)。在所有会员中,PingCAP 和华为分别以将近 85 000 项和超过 66 000 项贡献,位列第六和第八大贡献者。

CNCF 已有 11 个来自中国的 CNCF 项目:BFE(百度)、Chaos Mesh(PingCAP)、ChubaoFS(京东)、CNI-Genie(华为)、Dragonfly(阿里巴巴)、Harbor(VMware 中国)、KubeEdge(华为)、OpenKruise(阿里巴巴)、OpenYurt(阿里巴巴)、TiKV(PingCAP)和 Volcano(华为)。

11.3　云原生趋势展望

未来云计算将无处不在，会融入我们生活的方方面面，而云原生则让云计算变得标准、开放、简单、高效、触手可及。云原生成为云计算价值触达用户的最优路径。

云原生化是大势所趋，它与传统 IT 相比具有巨大优势，可以更好地帮助企业平滑、快速、渐进式地落地上云。可以预测在未来企业加快数字化转型的过程中，云原生一定会变成现代业务最重要的基础设施。

通过云原生企业可以高效使用云计算，聚焦于自身业务发展，另一方面开发者基于云原生的技术也可以提升效率，并将精力更多地聚焦于业务而不是底层设施。

云原生正在成为新基建落地的重要技术推手。只有拥抱云原生，才不会被时代淘汰。如何更好地拥抱云计算、拥抱云原生架构、用技术加速创新，将成为企业数字化转型及升级成功的关键。

任何一项技术变革本质上都是为了解放劳动力，提升生产效率。技术变革的初衷都是更高效地创造社会财富，帮助更多人享受美好的生活。这也是每一位技术人员的幸福感与成就感的重要来源。云原生正是在软件领域正在发生的深刻技术变革，它使云计算更高效地赋能各行各业，持续创造巨大的社会财富。

图书推荐

书　　名	作　　者
鸿蒙应用程序开发	董昱
鸿蒙操作系统开发入门经典	徐礼文
鸿蒙操作系统应用开发实践	陈美汝、郑森文、武延军、吴敬征
华为方舟编译器之美——基于开源代码的架构分析与实现	史宁宁
鲲鹏架构入门与实战	张磊
华为 HCIA 路由与交换技术实战	江礼教
Flutter 组件精讲与实战	赵龙
Flutter 实战指南	李楠
Dart 语言实战——基于 Flutter 框架的程序开发(第 2 版)	亢少军
Dart 语言实战——基于 Angular 框架的 Web 开发	刘仕文
IntelliJ IDEA 软件开发与应用	乔国辉
Vue+Spring Boot 前后端分离开发实战	贾志杰
Vue.js 企业开发实战	千锋教育高教产品研发部
Python 人工智能——原理、实践及应用	杨博雄主编，于营、肖衡、潘玉霞、高华玲、梁志勇副主编
Python 深度学习	王志立
Python 异步编程实战——基于 AIO 的全栈开发技术	陈少佳
Python 数据分析从 0 到 1	邓立文、俞心宇、牛瑶
物联网——嵌入式开发实战	连志安
智慧建造——物联网在建筑设计与管理中的实践	[美]周晨光(Timothy Chou)著；段晨东、柯吉译
TensorFlow 计算机视觉原理与实战	欧阳鹏程、任浩然
分布式机器学习实战	陈敬雷
计算机视觉——基于 OpenCV 与 TensorFlow 的深度学习方法	余海林、翟中华
深度学习——理论、方法与 PyTorch 实践	翟中华、孟翔宇
深度学习原理与 PyTorch 实战	张伟振
ARKit 原生开发入门精粹——RealityKit+Swift+SwiftUI	汪祥春
Altium Designer 20 PCB 设计实战(视频微课版)	白军杰
Cadence 高速 PCB 设计——基于手机高阶板的案例分析与实现	李卫国、张彬、林超文
Octave 程序设计	于红博
SolidWorks 2020 快速入门与深入实战	邵为龙
SolidWorks 2021 快速入门与深入实战	邵为龙
UG NX 1926 快速入门与深入实战	邵为龙
西门子 S7-200 SMART PLC 编程及应用(视频微课版)	徐宁、赵丽君
三菱 FX3U PLC 编程及应用(视频微课版)	吴文灵
全栈 UI 自动化测试实战	胡胜强、单镜石、李睿
pytest 框架与自动化测试应用	房荔枝、梁丽丽
软件测试与面试通识	于晶、张丹
深入理解微电子电路设计——电子元器件原理及应用(原书第 5 版)	[美]理查德·C. 耶格(Richard C. Jaeger)、[美]特拉维斯·N. 布莱洛克(Travis N. Blalock)著；宋廷强译
深入理解微电子电路设计——数字电子技术及应用(原书第 5 版)	[美]理查德·C. 耶格(Richard C. Jaeger)、[美]特拉维斯·N. 布莱洛克(Travis N. Blalock)著；宋廷强译
深入理解微电子电路设计——模拟电子技术及应用(原书第 5 版)	[美]理查德·C. 耶格(Richard C. Jaeger)、[美]特拉维斯·N. 布莱洛克(Travis N. Blalock)著；宋廷强译

图书资源支持

感谢您一直以来对清华大学出版社图书的支持和爱护。为了配合本书的使用，本书提供配套的资源，有需求的读者请扫描下方的“书圈”微信公众号二维码，在图书专区下载，也可以拨打电话或发送电子邮件咨询。

如果您在使用本书的过程中遇到了什么问题，或者有相关图书出版计划，也请您发邮件告诉我们，以便我们更好地为您服务。

我们的联系方式：

地　　址：北京市海淀区双清路学研大厦 A 座 714

邮　　编：100084

电　　话：010-83470236　010-83470237

资源下载：http://www.tup.com.cn

客服邮箱：tupjsj@vip.163.com

QQ：2301891038（请写明您的单位和姓名）

教学资源・教学样书・新书信息

人工智能科学与技术
人工智能|电子通信|自动控制

资料下载・样书申请

书圈

用微信扫一扫右边的二维码，即可关注清华大学出版社公众号。